Entomological Society of Canada

The Canadian Entomologist

Volume 29

Entomological Society of Canada

The Canadian Entomologist
Volume 29

ISBN/EAN: 9783337844936

Printed in Europe, USA, Canada, Australia, Japan

Cover: Foto ©berggeist007 / pixelio.de

More available books at **www.hansebooks.com**

The

Canadian Entomologist

VOLUME XXIX.

LIBELLULA TRIMACULATA.

EDITED BY THE

Rev. C. J. S. Bethune, M.A., D.C.L.,

HEAD MASTER OF TRINITY COLLEGE SCHOOL,

PORT HOPE, ONTARIO.

ASSISTED BY

Dr. James Fletcher and W. H. Harrington, Ottawa: H. H. Lyman, Montreal: and Rev. T. W. Fyles, South Quebec.

London, Ont.:
The London Printing and Lithographing Company, Limited.
1897.

LIST OF CONTRIBUTORS TO VOLUME XXIX.

ASHMEAD, W. H.........................WASHINGTON, D. C.
BAKER, CARL F... ....AUBURN, ALABAMA.
BANKS, NATHAN.........................WASHINGTON, D. C.
BARNES, DR. WILLIAM................ ......DECATUR, ILLINOIS.
BETHUNE, REV. C. J. S. (The Editor), F.R.S.C...PORT HOPE, ONT.
BRAINERD, DWIGHT..................... ... MONTREAL.
COCKERELL, T. D. A.........................MESILLA, NEW MEXICO.
COOLEY, R. A............. ....AMHERST, MASS.
COQUILLETT, D. W.............. .. ..WASHINGTON, D. C.
DUNNING, S. N.........................HARTFORD, CONN.
DYAR, DR. HARRISON G.........................WASHINGTON, D. C.
FALL, H. C............ .. ..PASADENA, CAL.
FISKE, W. F......... ....MAST YARD, N. H.
FLETCHER, DR. JAMES, F.R.S.C.... ....OTTAWA.
FRENCH, PROF. G. H.... .. ..CARBONDALE, ILL.
FYLES, REV. T. W., F.L.S.... ....SOUTH QUEBEC.
GILLETTE, C. P............. ....FORT COLLINS, COL.
GODING, DR. F. W....... ....RUTLAND, ILL.
GROTE, A. RADCLIFFE... ....HILDESHEIM, GERMANY.
HANHAM, A. W.... ....WINNIPEG.
HARRINGTON, W. H., F.R.S.C...............OTTAWA.
HUNTER, W. D...... ....LINCOLN, NEBRASKA.
KEEN, REV. J. H.........................MASSETT, Q. C. ISLANDS, B.C.
KING, GEORGE B.. ....LAWRENCE, MASS.
KIRKLAND, A. H......................MALDEN, MASS.
LYMAN, HENRY H., M.A ....MONTREAL.
MOFFAT, J. ALSTON.... ....LONDON, ONT.
MURTFELDT, MISS MARY E.................KIRKWOOD, MISSOURI.
NEEDHAM, JAMES G.......... ....ITHACA, N. Y.
OSBORN, PROF. HERBERT . ...AMES, IOWA.
PATTON, W. HAMPTON.................HARTFORD, CONN.
ROBERTSON, CHARLES............... ....CARLINVILLE, ILL.
SCUDDER, DR. S. H.........................CAMBRIDGE, MASS.
SKINNER, DR. HENRY.........................PHILADELPHIA.
SLINGERLAND, M. V.........................ITHACA, N. Y.
SMITH, PROF. JOHN B........ ....NEW BRUNSWICK, N. J.
SNYDER, ARTHUR J.................. .. ....BELVIDERE, ILL.
STEVENSON, CHARLES.......................MONTREAL.
TOWNSEND, C. H. TYLER........ ....FRONTERA, MEXICO.
VAN DUZEE, E. P.........................BUFFALO, N. Y.
WALKER, E. M.... ....TORONTO.
WICKHAM, PROF. H. F......... ....IOWA CITY, IOWA.
WEBSTER, PROF. F. W..................WOOSTER, OHIO.
WINN, A. P....... .. ..MONTREAL.

THE CRINKLED FLANNEL MOTH, MEGALOPYGE
CRISPATA. PACKARD.

The Canadian Entomologist.

Vol. XXIX.   LONDON, JANUARY, 1897.   No. 1.

THE CRINKLED FLANNEL MOTH (MEGALOPYGE CRISPATA, PACK.).

BY M. V. SLINGERLAND, CORNELL UNIVERSITY, ITHACA, N. Y.

September 3rd, 1895, I received several nearly full-grown specimens of the curious, sluglike caterpillars of this beautiful moth, so aptly named by Professor Comstock, "the crinkled flannel moth." The cunning brown caterpillars were placed in a cage here at the insectary, where they fed freely on apple leaves, although they were feeding on quince when found at Worcester, Mass. Since Dr. Packard described the insect in its different stages in 1864, its life-history has been worked out in detail by Dr. Lintner (Ent. Contrib., II., p. 138, 1870), and recently by Dr. Packard (Proc. Am. Phil. Soc. for 1894, p. 275). In this last paper Dr. Packard has described and figured in detail the extra two pairs of abdominal legs (seven pairs in all) possessed by the caterpillars, and some curious lateral glandular processes.

It is now our practice here at the insectary to photograph, so far as possible, every stage, phase, and habit of any insect that we may study. It is not often, however, that we have as good a subject as the crinkled flannel moth proved to be. The main object of this note is to introduce some of the lifelike pictures we were able to secure of this interesting and beautiful insect.

As shown at *d* on the plate, three of the cunning little caterpillars posed for their photograph, which represents their natural size and brings out their characteristic appearance much better than any other figures we have seen. They spun their tough brown cocoons (represented natural size at *a* on the plate), with the tightly fitting and ingenious door at one end, on September 5th. Upon prying open the door of one cocoon, the male pupa (shown natural size at *b* on the plate) was revealed. As the cage was kept in our warm office, the development of the insect was doubtless abnormally accelerated, for on December 21st and 24th the pupæ pushed open the little doors, worked their way nearly out of the cocoon, and the moths emerged. We aimed our "Premo" at one of the male moths as it was resting quietly and naturally on the muslin cover of the cage, with the result as shown at *c* on the plate. We were somewhat loath to kill such a pretty, daintily bedecked creature, but —— well, he now fills an honoured place in our collection here at the University. Figure *e* on the plate well represents this pretty creature (twice natural size) as he now looks in the collection. Imagine the lighter portions of the figure to be of a delicate straw-yellow colour and the darker waves and crinkles of a rich brown shade, and you have a faint conception of this crinkled flannel moth.

I do not know that the insect has ever done enough damage to make it of economic importance. It certainly has a wide range of food plants, as shown by Mr. Beutenmüller (Ent. Americana, III., 180), who lists twenty-five different plants, and the cranberry has since been added in Massachusetts. Briefly stated, its life-history seems to be as follows: The eggs are laid about July 1, and hatch in a week or ten days; the caterpillars feed during July and August, pupating in September; some of the moths may emerge in the fall, but doubtless most of them hibernate as pupæ, the moths appearing in June and some laying their eggs.

TORONTO BRANCH OF THE ENTOMOLOGICAL SOCIETY OF ONTARIO.

It is with much gratification that we announce the formation of a branch of our Society in Toronto. In the month of February last a number of entomologists in Toronto, feeling their isolation and need of co-operation, met together and decided to form an organization for the promotion of the study of entomology. They accordingly established "The Toronto Entomological Society," with Mr. E. V. Rippon as President, and Mr. Arthur Gibson, Secretary. Regular meetings have been held on the first and third Fridays of each month, and recently a room has been engaged at 451 Parliament Street, where the books and collections are kept and the meetings held, and which is open at all times for the use of the members. For the last ten months the Society has been very successful and its members full of enthusiasm; much satisfactory work has been accomplished, and great pleasure has been derived by the members from meeting with kindred spirits, comparing specimens, discussing questions that arise from time to time, and giving and receiving much assistance in many ways.

Recently the desirability of affiliating with the old-established Entomological Society of Ontario was brought before the members, and after full deliberation it was decided to become incorporated with it as a "Branch," in accordance with the terms of our Constitution. It will therefore be known, from the beginning of the New Year, as " The Toronto Branch of the Entomological Society of Ontario." It is hoped that every one interested in entomology, living in Toronto or the neighbourhood, will join the " Branch," and thus become members of our Society. The next meeting will be held on Friday evening, January 8th, at 8 o'clock, when visitors will be heartily welcomed.

The Montreal Branch has been in active operation for over twenty-three years, and held its 200th meeting a few months ago. We hope that in time to come the Toronto Branch may be able to boast of a similar record, and that each year as it goes by may find it growing and prospering, and doing good work for the furtherance of the science of entomology in the Dominion of Canada.

BREPHOS MIDDENDORFI, MEN.

On April 25th, 1896, I made a very lucky capture of a perfect specimen of this rare and beautiful moth. The afternoon being sunshiny and warm — one of our first spring days — I had gone out to look for beetles in a piece of wood along the Red River, a few miles from the city. This locality had proved rich in Carabidæ in 1894, about the same date. Greatly to my disgust, I found the place transformed, all logs and "brush" having been cleared away the previous season, and hardly a beetle of any kind was to be found.

The moth in question was first seen to alight on the bank of a cutting leading down to the river ; when disturbed from there by my investigations as to its identity, it flew up and down the roadway for a little while, and then hovered about some patches of mud, occasionally resting on the mud in the sunshine, very much after the manner of some of our butterflies. By this time I had got near enough to it to discover that it was something quite new to me, and my desire to capture it was therefore increased ten-fold. I had no net with me ; in fact, I was only provided with a rather narrow-necked cyanide bottle for Coleoptera (the neck of my bottle was not an inch in diameter). That I was able, after several futile attempts, to get the mouth of the bottle down over it as it sat in the road, without damaging it in any way, was a matter of surprise at the time and congratulation whenever I have thought of it since. I certainly never made a more lucky capture. To Prof. John B. Smith I am indebted both for the identification and for his generosity in returning the specimen to me.   A. W. HANHAM, Winnipeg, Man.

ON THE MEXICAN BEES OF THE GENUS AUGOCHLORA.

BY T. D. A. COCKERELL, MESILLA, N. M.

The Mexican species of this beautiful genus may be readily separated by the following table :—

A. Hind spur of hind tibia minutely ciliate or simple. = AUGOCHLORA, s. str.
   1. Entirely copper colour, with tints of carmine.... ...*flammea*, Sm.
   2. Head and thorax dark indigo blue, abdomen black with some green
      reflections...............................*nigrocyanea*, Ckll.
   3. Head and thorax green...........................4.
   4. Abdomen black, size small...................*seminigra*, Ckll.
      Abdomen crimson.................................*ignita*, Sm.
      Abdomen green, without hair-bands........................5.
   5. Hind margins of abdominal segments broadly black ; large blue-
      green species, with fuscous nervures........*Binghami*, n. sp. ♂.
      Hind margins of abdominal segments narrowly or not black; smaller,
      more yellowish-green species................................6.
   6. Small, wings dusky, nervures fuscous............*aurifera*, n. sp.
      Medium size, nervures dull testaceous...................7.
   7. Face broad, emargination of eyes deep ....*labrosa*, Say.
      Face narrow, emargination of eyes shallow...........*pura*, Say.
B. Hind spur of hind tibia pectinate. = AUGOCHLOROPSIS,
      subg. nov.............................. ......(type, *subignita*).
   1. Head and thorax black, abdomen ferruginous........*aspasia*, Sm.
      Head and thorax green...............................2.
   2. Abdomen crimson...........................*subignita*, Ckll.
      Abdomen brassy, with dense short fulvous pubescence beyond
      basal segment................................*aurora*, Sm.
      Abdomen green, of the same colour as head and thorax, with two
      narrow bands of yellow pubescence............*splendida*, Sm.
C. Hind spur of hind tibia not yet described.
   1. Bright green, agreeing only with *splendida* in having abdominal
      hair-bands, but these are white................ .*viridana*, Sm.
   2. Small piceous species ; margin of mesothorax, postscutellum, most
      of enclosure of metathorax, and bases of second and third abdomi-
      nal segments shining green................*tisiphone*, Gribodo.

    A. labrosa is cited from Mexico by its describer, but I have not seen it from that country.   Mr. Robertson sends it to me from Illinois.   There are two species found in Texas, which may be expected also across the

Mexican border. One of them is what passes for *A. sumptuosa*, Sm., in this country, and indeed agrees with Smith's description ; but Col. Bingham finds that a co-type in the British Museum belongs to Section A above (spur minutely ciliate), while our insect belongs to Sect. B. It is just possible that the B. M. co-type is not identical with the true type of *sumptuosa* ; if this is not so, our *sumptuosa* will have to be renamed. The other Texan species referred to was recorded by Cresson as *A. lucidula*, Sm., but it differs from that, and is referable to *A. humeralis*, Patton, of which it may perhaps constitute a geographical race. I have several specimens collected by Prof. C. H. T. Townsend at Beeville, Texas, Aug. 29, 1896, on a species of Compositæ. Col. Bingham's studies at the British Museum show that *A. humeralis*, which belongs to Sect. B, cannot be identical with *A. fervida*, Sm., as Robertson has supposed, since that belongs to Sect. A. Also, Patton was wrong in referring *lucidula*, Sm., which belongs to Sect. B, to *viridula*, Sm., which is of Sect. A. I will now describe the two new species indicated above :—

Augochlora Binghami, n. sp. (subg. *Augochlora*, s. str.)—♂. Length about 12 mm., brilliant bluish-green, the face a yellower green. Face narrowing below, eyes deeply emarginate ; sides of face with conspicuous, partly appressed, silky white pubescence ; cheeks with long white hairs. Clypeus, supraclypeal area and middle of vertex with sparse, inconspicuous black hairs. Clypeus rather prominent, subcancellate with very large close punctures, its anterior margin and the upper half of the labrum whitish, mandibles wholly dark. Vertex finely and very closely punctured. Antennæ reaching to base of wings, piceous, flagellum obscurely rufescent beneath, last joint conspicuously hooked. Mesothorax shining, with very distinct rather small close punctures, much densest at the sides, where a minute cancellation results. Parapsidal grooves distinct. Prothoracic keel fairly strong. Enclosure of metathorax fairly well defined, irregularly wrinkled, its hind margin gently curved, not angled. Posterior truncation roughened, bounded below at sides by an acute ridge, which ascending rapidly fails. Pubescence of thorax sparse, grayish-white, black and inconspicuous on dorsum. Tegulæ shining piceous, anteriorly whitish, basally green and punctured. Wings smoky-hyaline, apical margin darker, stigma dull testaceous, nervures fuscous, marginal cell minutely appendiculate. Legs green with black tarsi, pubescence short and pale. Abdomen shining, closely punctured, hind margins of segments broadly purplish-black. No hair-bands, but a very fine glittering pile all over, longer pale hairs at base of first segment, sparse black hairs on dorsum of hindmost segments and at tip. Punctuation of second segment conspicuously closer than that of first. Venter piceous, first three segments with blue reflections. End of third segment with a large dark brown brush of hair, shaped like the tail of a fish ; i. e. deeply emarginate, the sides diverging and ending in a point.

Hab.—San Rafael, Vera Cruz, March 13, on flowers of plant No. 4, which is papilionaceous (C. H. T. Townsend).

This beautiful species is named after Lt.-Col. Bingham, without whose notes on the British Museum types I should not have attempted this paper.

Augochlora aurifera, n. sp. (subg. *Augochlora*, s. str.)— ♀ . Length about 7½ mm , green ; head and thorax dullish, rather a bluish-green ; abdomen shining, a yellower green, with the hind margins of the segments very narrowly coppery. Face fairly broad, emargination of eyes deep. Pubescence of head and thorax sparse and inconspicuous, dirty whitish, some black hairs on thoracic dorsum ; lower part of face in certain lights canescent. Clypeus with close punctures of unequal size, supracypeal area more finely punctured, vertex coarsely granular. Labrum and margin of clypeus black. Mandibles notched within, stout, rufescent medially. Glossa very long and narrow, coming to a fine point. Antennæ black, flagellum slightly rufescent beneath. Mesothorax very closely, finely, and uniformly punctured. Enclosure of metathorax conspicuously longitudinally, or rather radiately, sulcatulate. Truncation shining, finely malleate, with a median groove. Tegulæ shining piceous, the margin subhyaline. Wings smoky, stigma dull testaceous, nervures fuscous, marginal cell appendiculate. Legs piceous-black, with brownish pubescence ; only the anterior femora show any green. Abdomen shining, with minute, not very close, punctures ; pubescence very sparse, no hair-bands. It requires a strong lens to see the abdominal punctures.

Hab.—San Rafael, Vera Cruz, March 9, on flowers of plant No. 6, referred by Dr. Rose to the genus *Melopodium*. The hind legs, base of thorax and abdomen, and ventral surface of abdomen, carry considerable quantities of the orange pollen. Another specimen differs by being much bluer, the punctuation a little coarser, the stigma fuscous; but it is evidently the same species. It is from San Rafael, March 14, on flowers of plant No. 5, a *Vernonia*. Both were collected by Prof. C. H. T. Townsend.

THE COLEOPTERA OF CANADA.

BY H. F. WICKHAM, IOWA CITY, IOWA.

XIX. THE CHRYSOMELIDÆ OF ONTARIO AND QUEBEC — (*Continued*).
TRIBE IX. —GALERUCINI.

This tribes includes a number of species which are, as a rule, easily distinguished by the peculiar appearance given by their soft integuments and usually somewhat elongate form. A number of them are pubescent, while others, on account of the peculiar sculpture of the surface, are quite opaque, the effect on the eye being, at first glance, the same in each case. The elytra in our species are longer than the abdomen, the prothorax is margined, the antennæ approximate, inserted on the front, the hind legs with rather slender thighs, not fitted for leaping. It will be remembered in this connection that I consider the Halticini as a distinct tribe.

Many of the Galerucini are extremely injurious, the striped cucumber beetle being well known and dreaded by gardeners ; its congener, *Diabrotica longicornis*, which has lately been found by Mr. Harrington in the Eastern Provinces, is a notorious pest to corn in the United States. In the Northeastern States the imported elm-leaf beetle, *Galerucella xanthomelæna*, Schr., is doing much mischief, but I cannot find that it is reported from Canada. If found, it may be distinguished from all our other species of *Galerucella* by the colour of the antennæ, which are piceous above and pale beneath, while the elytra are comparatively finely and equally punctate. It is yellowish above, the head with one dark spot, the thorax with three, the elytra with a short inner stripe (sometimes wanting), and a long one from the humerus ; legs pale, each femur with a small dark spot.

The tribe has recently been worked up in an excellent paper by Dr. Horn, and this has been closely followed and freely used in the preparation of the following pages. In order to avoid the constant repetition of quotation marks and statements of acknowledgments, it is well to say that the differential characters brought out are in almost every case those used by the Doctor, and that while I have not scrupled to change the arrangement of his tables where it seemed to me more likely to serve the purpose of the present article, I have, on the other hand, found it impossible to improve on most of his expressions, and have therefore used them entire. With this acknowledgment of the source of whatever is good in the paper, we may proceed to separate the genera occurring in Canada, thus :—

A. Anterior coxal cavities open behind.
    b. Claws simple or bifid.
        c. Tibiæ without terminal spurs ; epipleura of elytra extend-
           ing nearly to apices.
            d. Antennæ longer than one half the body ; claws deeply bifid.
               Third antennal joint shorter than fourth ; large
               species ....*Trirhabda.*
               Third joint longer than fourth ; small
               species........................... ......*Galerucella.*
            dd. Antennæ less than half as long as body ; claws simple or
                narrowly bifid ....*Monoxia.*
        cc. Tibiæ (middle and posterior) with terminal spurs, outer edge
            more or less carinate ....*Diabrotica.*
    bb. Claws appendiculate (*i. e.* with broad dilatation at base).
        Epipleura not distinct, tibiæ without spurs .....*Phyllobrotica.*
        Epipleura distinct, all the tibiæ with spurs.......*Luperodes.*
AA. Anterior coxal cavities closed behind.
        Large species, tarsal claws bifid, tibiæ without spurs...*Galeruca.*
        Smaller species, claws appendiculate, tibiæ with spurs..*Cerotoma.*

I have omitted *Scelolyperus* from the above table, although the
Southern Californian *S. maculicollis,* Lec., is in the Society list. The
genus belongs in the group with open anterior coxal cavities, appendiculate claws and well-defined epipleura. In the scheme it would
precede *Luperodes,* from which it differs in having no tibial spurs. The
species above mentioned is about one-fourth of an inch in length, head
and under surface black, thorax either yellow with three dark spots or
entirely black, elytra bluish or greenish. Antennæ two-thirds as long as
the body, piceous, with three basal joints pale beneath.

TRIRHABDA, Lec.

Large insects, of rather elongate-oblong form, usually of somewhat
opaque surface, the thorax in most cases spotted, the elytra bluish,
greenish, or brownish, with yellowish stripes. They are to be taken
during the summer months by sweeping rank herbage in lanes and
meadows, and may often be taken in numbers on the golden-rod. Dr.
Horn has thus separated our species :

A. Surface of body without any trace of metallic lustre in the markings,
   these being opaque or brownish.

b. Elytral punctures so dense as to be indistinct as such.
    Yellow vittæ of elytra attenuate to apex. .30–.40
        in.............................. *tomentosa*, Linn.
    Yellow vittæ broad, parallel and entire. .28–.38
        in ....*canadensis*, Kby.
bb. Elytral punctures dense, but distinctly separate. Elytra normally
    vittate as in *canadensis*. .26–.36 in..........*virgata*, Lec.
AA. Surface of body with metallic lustre ; if not in the markings of the
    elytra, at least on those of the head and thorax. Punctuation
    of elytra comparatively rough.
        Elytra entirely blue, except border. .20–.32
            in...............................*flavolimbata*, Mann.
        Elytra with outer border and discal vitta yellow.
            .20–.28 in ....*convergens*, Lec.

GALERUCELLA, Crotch.

This genus, as now understood, contains species formerly distributed partially in *Adimonia* and partially in *Galeruca*. Many
of them are quite common, and are to be found in the sweepings
of meadows, on water lilies, *Sagittaria*, *Eupatorium*, or occasionally on
the leaves of deciduous trees, as in the case of *G. cavicollis*, which I have
taken abundantly on wild cherry. All but three of the North American
species have been recorded from Canada, and Dr. Horn's table is here
reproduced almost in full, though some portions are transposed, and the
remainder made to include the non-vittate specimens of *G. americana*, so
as to render identification a trifle more easy when reference cannot be
had to detailed descriptions. The limit of variation in some of the
vittate forms is very wide, and has resulted in the multiplication of
nominal species. It is believed that the table will now cover any cases
likely to be met with in the Provinces of Ontario and Quebec. In case
of the occurrence there of the elm-leaf beetle, a reference to the first page
of this article will result in its proper identification.
A. Colour red.
        Elytra more coarsely punctured, intervals between punctures dis-
        tinct, surface shining. .18–.22 in............*cavicollis*, Lec.
        Elytra finely and densely punctured, surface rather opaque.
        .18–.22 in ....*rufosanguinea*, Say.
AA. Colour yellowish, brownish or piceous, elytra vittate or not.
    b. Elytra normally vittate.

c. Elytra scarcely explanate at sides, middle coxæ separated.

Elytra convex, coarsely punctate ; thorax more or less shining, spotted indistinctly if at all. .14–.26

in ....*americana*, Fabr.

Elytra less convex, more closely and less coarsely punctate,
thorax opaque with three spots. .20–.24

in ....*sexvittata*, Lec.

cc. Elytra distinctly explanate, middle coxæ contiguous.

Sutural vitta joined by next at or behind the
middle. .14–.20 in.................*notulata*, Fabr.

Vitta next to the sutural very short, basal. .14–.20

in...............................*notata*, Fabr.

bb. Elytra not vittate, often with lighter side margin.

d. Form convex, elytra coarsely punctate... ...*americana*, var.

dd. Form not notably convex.

e. Middle coxæ separated, thorax angulate at middle, sub-
sinuate behind, hind angles obtuse. .18–.24

in.. ....*nymphææ*, Linn.

ee. Middle coxæ contiguous, hind angles of thorax distinct.

Thorax coarsely, not very closely, punctate..*notulata*, var.

Thorax densely punctured and opaque. .18–.22

in.............................*decora*, Say.

MONOXIA, Lec.

M. consputa, Lec. *(guttulata)*, has been recorded on the Society's
list. It is a small insect, .14–.18 in. long, of a somewhat oblong form,
resembling some *Galerucella*, but with shorter antennæ ; yellowish or
reddish-yellow in colour, elytra often with numerous very small black
spots. It is common on the plains to the westward, but I have seen no
specimens from Ontario or Quebec, and it is just possible that an
immaculate specimen of *Galerucella notulata* has been mistaken for it.

DIABROTICA, Chevr.

Here belongs the striped cucumber beetle *(D. vittata*, Fabr., Fig.
1), so common on and often injurious to cucumber and
squash vines. It is a little less than one-fourth of an inch in
length, yellow above ; head, scutellum, and three elytral
stripes (one common sutural, one discal on each wing-cover)
black. Basal joints of antennæ partially yellowish, legs with
dark tarsi and knees, front tibiæ and tips of middle and hind tibiæ also

FIG. 1.

dark. The twelve-spotted *Diabrotica*, *D. 12-punctata*, Fabr. (Fig. 2), in life is pale greenish above, turning to yellowish in old cabinet specimens ; antennæ dark, with three basal joints pale, head black, scutellum dark, each elytron with six black spots. Legs dark, basal half of femora pale. Size a little greater than the preceding. Mr. Harrington has recently found *D. longicornis*, Say, in the Eastern Provinces. It may easily be distinguished by its smaller size and immaculate green, fading to yellowish, elytra.

FIG. 2.

PHYLLOBROTICA, Chevr.

These are very pretty insects, marked with yellow and black. Two have been recorded from Canada, but as there is a chance of error in determination I herewith include *limbata* as well, since its other recorded distribution seems to indicate a more northern range than is found in *discoidea*. All three have yellow head and thorax. Dr. Horn thus defines them :

Elytra yellow, with two oval piceous spots on each (Fig. 3), .22-.28 in............................ ....*decorata*, Say.

Elytra piceous, sides and suture yellow.

Thorax with moderately deep fovea each side. .14-.26 in...... ....*discoidea*, Fabr.

Thorax with transverse depression. .14-.26 in...*limbata*, Fabr.

LUPERODES, Motsch.

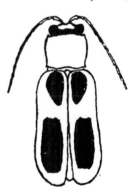

Contains one Canadian species, *L. meraca*, Say, an elongate insect, .20 in. long, dark blue or blue-black above, piceous beneath, thorax nearly equal in length and breadth, hind angles acute and prominent, disk convex, smooth, elytra sparsely punctate. Legs yellow, basal half of femora piceous. It has been reported by Mr. Chittenden as feeding on the witch-hazel, while on another occasion he found it in great numbers on the flowers of the wild rose, the petals of which served as food.

FIG. 3.

GALERUCA, Geoff.

G. externa, Say, represents the genus in North America, and while commoner to the eastward, has been reported from Canada. It is a robust insect, easily known from our other Galerucini by the large size 27-.44 in.) and broadly oval form. The colour is blackish, outer margin of elytra yellowish, upper surface coarsely and closely punctate. Elytra with four more or less well-marked costæ. The food-plant of this beetle is still unknown to me, for, while I have collected a great many specimens, they were always found under logs or stones.

CEROTOMA, Chevr.

Represented by *C. trifurcata*, Forst. (*caminea*, Fabr.), resembling somewhat the common *Diabrotica 12-punctata* in form, but shorter. Head and under side of body black, upper surface of thorax and eiytra yellowish or occasionally red. Elytra marked with black, as shown in fig. 4, this pattern being often reduced or added to by the greater or less extension of

FIG. 4.

the yellow. Length, .14-.20 inch. Mr. Chittenden records the bush-clover, *Lespedeza*, as a food plant, and remarks that legumes form the chief food of the species. My own captures have been, for the most part, made by overturning boards and chips in patches of meadow land during the middle of spring.

A GENERIC REVISION OF THE HYPOGYMNIDÆ (LIPARIDÆ).

BY HARRISON G. DYAR, NEW YORK.

Before the generic names of our moths can become permanent, it is necessary that all the described genera should be compared, but specially the older genera of Europe. To make a beginning in this matter, I have drawn up the following synoptic table of the Hypogymnidæ, based on the characters used in Hampson's Moths of India, adding thereto the species found in Europe and in North America. The types of the genera are recognized as determined by Kirby.

Probably but few, if any, changes will be necessary from this list, as the African and South American species for the most part belong to other genera, or else have later dates than the generic names here defined.

I exclude two genera given by Hampson, viz. : Retarda and Thiacidas. The latter seems to me to be a Noctuid, perhaps one of the Apatelidæ, while the former has the venation of the Tineides and is without frenulum ; it probably represents a new family type.

In the Tentamen, Hübner gives the three plural terms, Hypogymnæ, Leucomæ, and Dasychiræ, all referring to this family. As these appear to be the first plural terms, one of them must stand for the family. The term Liparidæ as used by Herrich-Schaffer, Kirby, etc., and the Lymantriidæ of Hampson cannot stand. Grote at first used Dasychiræ, as in his list of 1882. Later he selected Leucomidæ (Syst. Lep. Hild., 1895), and finally Hypogymnidæ (Syst. der Nord. Schmett., 1896). Dasychiridæ is unavailable as the generic term becomes synonymous, and the first of Hübner's terms may best be retained.

Two new generic terms seem necessary. The two European species of Ocneria are not congeneric, as one has two pairs of spurs on the hind tibiæ and the other but one. The latter may be separated under the term Parocneria, type *detrita*, Esp. The same is the case with our species of Notolophus. All the European species which I have seen, and our *antiqua* and *vetusta*, have one pair of spurs, as stated by Hampson. The larvæ have black heads. Two other species, *leucostigma* and *definita*, have two pairs of spurs, and may be called Hemerocampa. The larvæ have pale heads.

I add to the synopsis a partial list of species. Kirby and Hampson may be consulted for details, and for the genera not specifically mentioned.

1. Primaries with vein 10 from the accessory cell....................2.
  Primaries with vein 10 from beyond the accessory cell.... *Mardara*.
  Primaries without accessory cell, or rarely with one with vein 10 before the accessory cell or joined to vein 11................12.
2. Palpi porrect.................... ....3.
  Palpi upturned...11.
3. Hind tibiæ with no spurs............. .... *Varmina*.
  Hind tibiæ with one pair of spurs...........................4.
  Hind tibiæ with two pairs of spurs.........................6.
4. Female with well-developed wings.........................5.
  Female with the wings useless, largely aborted..... *Hypogymna* (2).
  Female with aborted wings................... *Notolophus* (3).
5. Robust, the palpi not or but slightly exceeding the front.. *Gynaephora* (1).
  Fragile with small body, the palpi considerably exceeding the front...*Pantana*.
6. Primaries short and broad.................................7.
  Primaries more produced...................................9.
7. Female with well developed wings.........................8.
  Female with aborted wings.... ....*Hemerocampa* (7).

8. Fore tarsi with lateral tufts of hair on the joints..........*Cifuna.*
   Fore tarsi without these tufts...........................*Aroa.*
9. Fore tarsi smooth haired ; palpi long...............*Laelia* (4).
   Fore tarsi more roughly haired or tufted.................. 10.
10. Palpi not reaching beyond the front.................*Orgyia* (5).
    Palpi reaching beyond the front......................*Olene* (6).
11. Palpi slight, closely approximated to the front and not
    reaching the vertex............................*Daplasa.*
    Palpi reaching the vertex.........................*Numenes.*
    Palpi reaching above the vertex..................... .....*Pida.*
12. Primaries with veins 7 to 10 stalked......................13.
    Primaries with veins 8 to 10 stalked......................23.
    Primaries with vein 10 from the cell, or rarely stalked with 11....24.
13. Palpi upturned...14.
    Palpi porrect.......... ....17.
14. Primaries with the apex rounded..... ....15.
    Primaries with the apex acute..................... *Topomesa.*
15. Primaries with vein 10 given off near the apex..........*Heracula.*
    Primaries with vein 10 given off nearer the cell than vein 7.....16.
16. Female with well developed wings.................*Lymantria* (8).
    Female with aborted wings.........................*Enome* (9).
17. Posterior tibiæ with two pairs of spurs....... ....18.
    Posterior tibiæ with one pair of spurs.......................21.
18. Palpi short...19.
    Palpi long ....22.
19. Vein 5 of secondaries absent....................*Leucoma* (12).
    Vein 5 of secondaries present...........................20.
20. Primaries with vein 10 given off near the apex......*Euproctis* (13).
    Primaries with vein 10 nearer the cell, or with vein 7........*Cispia.*
21. Palpi very minute............................. ....*Perina.*
    Palpi rather long.............................*Parocneria* (10).
22. Antennæ of female with long pectinations................*Imaus.*
    Antennæ of the female with short pectinations........*Ocneria* (11).
23. Vein 5 of secondaries near lower angle of cell ; palpi
    very long...............................*Dactylorhyncha.*
    Vein 5 near upper angle of cell ; veins 3 and 4 united....*Gazelina.*
24. Palpi porrect ....25.
    Palpi upturned.... ....*Arctornis* (14).

25. Palpi long ; hind tibiæ with two pairs of spurs............*Himala.*
    Palpi short ; hind tibiæ with one pair of spurs...............26.
26. Secondaries with veinlets between vein 1 and margin..*Dendrophelps.*
    Secondaries without supplementary veinlets........*Stilpnotia* (15).

1. Genus GYNÆPHORA, Hübner.

    Type *selenitica*, Esp. Also, *ladacensis*, Moore (Hampson I., 435, as Lachana) ; *rossii*, Curt., and probably *groenlandica*, Hom., which I have not seen.

2. Genus HYPOGYMNA, Hübner.
    Type *morio*, Linn.

3. Genus NOTOLOPHUS, Germ.

    Type *antiqua*, Linn. Also, *gonostigma*, Linn.; *ericiæ*, Germ.; *postica*, Walk.; *viridescens*,Walk.;*turbata*, Butl.; *vetusta*, Boisd.; *cana*, Hy. Edw.; *gulosa*, Hy. Edw.

4. Genus LAELIA, Stephens.
    Type *coenosa*, Hübn. Also 12 Indian species.

5. Genus ORGYIA, Ochs. (= *Dasychira*, Hübn.)
    Type *fascellina*, L. Also *pudibunda*, L.

6. Genus OLENE,Hübn(= ‖ *Dasychira*, Hampson = *Parorgyia*,Packard).
    Type *mendosa*, Hübn. Also *abietis*, Den. & Sch. ; *cinnamomea*, G. R.; *achatina*, A. S.; *leucophæa*, A. S.; *plagiata*, Walk. ; and 18 Indian species.

7. Genus HEMEROCAMPA, Dyar.
    Type *leucostigma*, A. S. Also *definita*, Pack.

8. Genus LYMANTRIA, Hübn.
    Type *monacha*, L. Also *dispar*, L., and 14 species from India.

9. Genus ENOME, Walk.

    Type *ampla*, Walk. Also ten other Indian species. Hampson makes this a section of Lymantria, but I regard it as a higher group.

10. Genus PAROCNERIA, Dyar.
    Type *detrita*, Esp.

11. Genus OCNERIA, Hübn.
    Type *rubea*, Fab.

12. Genus LEUCOMA, Hubn., Tent. (= *Porthesia*, Steph.)
    Type *similis*, Fuessl. Also two Indian species.

13. Genus EUPROCTIS, Hübn. (= *Artaxa*, Wlk.)

Type *chrysorrhœa*, L.   Also fifty-three Indian species.   See Hampson for the generic synonymy.

14. Genus ARCTORNIS, Germ. (= ‖ *Leucoma*, Steph. = ‖ *Laria*, Schr.) Type *L–nigrum*, Müll.   Also eight Indian species.

15. Genus STILPNOTIA, Westw. & Hump. (= *Leucosia*, Ramb. = *Charala*, Moore = *Caragola* Moore = *Nymphyxis*, Grote.)

Type *salicis*, Linn.   Also six Indian species listed under Caviria, Walk., which, however, is a South American genus, and not strictly congeneric with the Indian forms.

CATALOGUE OF THE PHYTOPHAGOUS AND PARASITIC HYMENOPTERA OF VANCOUVER ISLAND.

BY W. HAGUE HARRINGTON, F. R. S. C., OTTAWA.

The following list is based upon a very interesting collection made, chiefly at Cedar Hill, near Victoria, by the Rev. G. W. Taylor, F.R.S.C., but includes such other species as I have found described, or recorded from Vancouver Island.   Even with such additions it is a short list in comparison with those that could be compiled from much less extensive areas in Ontario.   British Columbia has, as yet, had but few resident entomologists, and its rich fauna is, in consequence, but poorly known. Butterflies and beetles have been fairly well collected, but in other directions there are almost unexplored fields for investigation.

I have found but little literature relating to the Hymenoptera of Vancouver Island, and but scanty records of species captured there. Lord, in his interesting narrative of a Naturalist in British Columbia, has an appendix enumerating the insects secured by him, with descriptions of a few new species.   Cresson, in a paper entitled Descriptions of Ichneumonidæ, chiefly from the Pacific Slope of the United States and British North America (Proc. Acad. Nat. Sci., Phil.; Nov., 1878), described about twenty-five species from the Island, contained in the collections of the late distinguished entomologists, Mr. H. Edwards and Mr. Crotch. The late Abbé Provancher described a few species in the CANADIAN ENTOMOLOGIST (Vol. XVII., p. 114), and in the Additions to his Petite Faune Entomologique du Canada credits the Island with some thirty-five species, mostly new forms contributed by Mr. Taylor and Mr. Fletcher. The types of some of those species are now in my collection, through Mr. Fletcher's kindness, and have been found very useful for comparison.

Kirby, in his List of the Hymenoptera in the British Museum, records several species of Tenthredinidæ and Uroceridæ. To Mr. Taylor, however, is due a large proportion of our knowledge of the Hymenopterous fauna. In Vol. XVI. and XVII. he published a list of eighty-one species, from the vicinity of Victoria, and he continued to collect there and sent specimens to Mr. Fletcher and myself until he came to reside in Ottawa a few years ago. He then brought his collection with him to this city, and on his return to the Pacific Coast he placed all the remaining Hymenoptera in my hands, on the condition that I should prepare a list of them for publication, in revision and enlargement of his own earlier list, in which there are some errors in determination.

The collection has proved to be a most interesting one, and to contain quite a number of new insects. It is, as might be expected, deficient in the smaller forms, such as Cynipidæ, Braconidæ, Chalcididæ, and Proctotrypidæ. As time has permitted, I have proceeded with the determination of these insects, and have published descriptions (CAN. ENT., Vol. XXVI.) of some new species. The Aculeata require further study, especially such genera as Andrena, Halictus, Osmia, etc., before a satisfactory list can be made of them. Mr. Taylor is now resident at Nanaimo, and it is to be hoped that his duties will afford him opportunity to collect in that district. The publication of a list (even though imperfect) of the recorded species may perhaps stimulate others to join with him in a more systematic collection of the Hymenoptera of Vancouver Island, which offers so rich a field for study. The fauna is evidently a very extensive one, containing many species occurring in the Pacific States, while in the northern portion of the Island and on the mountains there should be a large intermingling of species inhabiting Alaska and the Rocky Mountains. It would not require much effort to increase many-fold the number of species at present known. The order Hymenoptera is so rich in species, and the conditions of the occurrence of the species are so varied, that it will long be possible to discover forms new to science, even in Ontario, where the fauna is so much better known. In the vast and diversified regions of the Pacific Slope, such new and undescribed species must be almost unlimited.

TENTHREDINIDÆ.

Trichiosoma Taylori, *Prov.*—Common on the Island and throughout B. C. I took it at New Westminster, and have examples from Tacoma (Wickham) and the Rocky Mountains (Bean). Probably only a

Western form of *T. triangulum*, under which name Taylor records it. Cocoons very frequently parasitized.

Trichiosoma vittellina, *Linn.*—Kirby *(List Hym. Brit. Musm., Vol. I., p. 10)* records a ♂ of this European species from the Island (Dr. Lyall) and a ♀ from the Rocky Mountains. Perhaps all our forms belong to one boreal species. They certainly do not vary so much as the insects included in *Cimbex americana*.

Abia Kennicotti, *Nort.*—One ♀ received by Mr. Fletcher, dated 4th June.

Hylotoma McLeayi, *Leach.*—One ♀ received by Mr. Fletcher, dated 2nd June.

Euura sp.—Two specimens in condition not favorable for determination.

Cladius pectinicornis, *Fourc*; *Cladius isomera*, Harris.—One ♀ from Mr. Wickham.

Pontania nevadensis, *Cress. (Nematus).*—Marlatt ; Rev. N. A. Nematinæ, p. 30.

Pteronus mendicus, *Walsh (Nematus).*—Two ♀ received by Mr. Fletcher ; also one ♀ from Mr. Wickham.

Pteronus vancouverensis, *Marlatt.*—Rev. N. A. Nematinæ, p. 70.

Pachynematus coloradensis, *Marlatt.*—One ♀ received by Mr. Fletcher.

Pachynematus palliventris, *Cress. (Nematus).*—One ♀ received by Mr. Fletcher apparently belongs to this species.

Dolerus collaris, *Say.*—One ♀.

Dolerus sericeus, *Say.*—Eight ♀, seven ♂ ; a very common species, generally more robust and pubescent than Ottawa examples.

Monophadnus atratus, *Hargtn.*—Type ♂ in my coll.

Phymatocera nigra, *Hargtn.*—One ♂.   April.

Hoplocampa halcyon, *Nort.*—Taylor ; Can. Ent., Vol. XVI., p. 92.

Labidia opimus, *Cress.*

   Allantus opimus, Cr.; *Labidia columbiana*, Prov.—Originally described from V. I. collection of Crotch ; redescribed from Taylor's collection. Appears to be common. Four ♀, four ♂. The *A. originalis* of Taylor's list, and probably identical with that species.

Allantus elegantulus, *Cress.*—Five ♀, one ♂ ; June.   Also to Fletcher, four ♀, two ♂ ; labelled May and June.

Taxonus parens, *Prov.*—Type ♂ in my coll.   Probably the ♂ of *Strongylogaster rubripes*, Cress., from Col.

Strongylogaster distans, *Nort.*—Common in April and May. I have eight ♀ and six ♂ specimens, and Mr. Fletcher has six ♀ s. The abdomen of the male is entirely red, except base of first segment and basal plates, but the female has the remaining segments more or less marked with basal black spots.

Strongylogaster (?) marginata, *Prov.*
Selandria marginata, Prov.—Type ♀ in my coll. Mr. Fletcher has also six ♀ and four ♂ from Cedar Hill. May and June.

Tenthredo erythromera, *Prov.*—Type ♀ in my coll.

Tenthredo nigrisoma, *Hargtn.*—Types ♀ in my coll. One taken by Taylor, 5th June, 1888 ; the other, also at Victoria, by Wickham.

Tenthredo nigricosta, *Prov.*—Type ♀ in my coll.

Tenthredo rubricus, *Prov.*
Allantus rubricus, Prov. — Type ♀ and another in my coll.; one also examined for Mr. Fletcher. The antennæ are not those of an Allantus, and the insect is apparently a variety of *T. mellina*, with antennæ slightly shorter and pale markings less conspicuous.

Tenthredo ruficoxa, *Prov.*—Type ♀ in my coll.

Tenthredo rufopedibus, *Nort.*—Recorded by Taylor as common in spring, but not in his collection ; probably the species I have determined as *T. variata.*

Tenthredo terminalis, *Prov.*—Type ♀ in my coll.

Tenthredo variata, *Nort.*—Three ♂ specimens. May and June. Mr. Fletcher has also one ♂.

Tenthredo varipicta, *Cress.* — Prov.; *Add. Faune Hym.*, p. 14. Two females taken 28th May and 4th June, received by Mr. Fletcher.

Tenthredopsis Evansii, *Hargtn.*—Mr. Fletcher has one ♂ taken in May.

Synairema pacifica, *Prov.*—Type ♀ in my coll. Apparently a species of Macrophya ; the coxæ are shorter than usual, but the femora reach to tip of abdomen. Head coarsely punctured ; in shape and sculpture resembling Macrophya ; antennæ wanting. Thorax coarsely but more sparsely punctured, and scutellum polished, with a few shallow punctures. Appears to be closely related to *M. bicolor*, Cress., but has first segment black.

Pamphilius pacificus, *Nort.* — Kirby ; *List Hym. Brit. Musm., Vol. I., p. 348.*

Macroxyela, sp. nov.? One ♀ labelled as captured on oak. May 12th, 1896.

UROCERIDÆ.

Urocerus abdominalis, *Harris.*—Two specimens ; probably males of *albicornis* or *flavicornis.*

Urocerus albicornis, *Fabr.*—One ♀.

Urocerus apicalis, *Kirby.*—*List Hym. Brit. Musm., Vol. I., p. 377*, ♂ ; probably the male of *cæruleus.*

Urocerus cæruleus, *Cress.*— ♀ described from V. I. coll., H. Edw.   Mr. Fletcher has taken it at New Westminster, B. C.

Urocerus caudatus, *Cress.*—One ♀ and one ♂.

Urocerus cyaneus, *Fabr.*—One ♀.

Urocerus flavicornis, *Fabr.*—One ♀. Recorded by Taylor as "common in autumn."

Urocerus flavipennis, *Kirby.* — Five ♀. A large, handsome insect, but probably a form of *albicornis.*

Urocerus Morrisoni, *Cress.*—One ♀. This is doubtless a *var.* of *caudatus.*

Urocerus varipes, *Smith.*—One ♀. Very close to *cyaneus.*

ORYSSIDÆ.

Oryssus Sayi, *Westw.*—One ♀. Also a ♂ of *var. occidentalis*, Cress.

CYNIPIDÆ.

Ibalia ensiger, *Nort.*—One ♀ received by Mr. Fletcher.

Onchyia Provancheri, *Ashm.*—One ♀ ; 4th June.

EVANIIDÆ.

Aulacus pacificus, *Cress.*— ♀ described from V. I. coll., Crotch.

ICHNEUMONIDÆ.

Ichneumon atrox, *Cress.*—One ♀ ; 6th June. Also one ♀ to Mr. Fletcher.

Ichneumon cæruleus, *Cress.*—Taylor ; CAN. ENT., Vol. XVI., p. 91. One ♀ to Mr. Fletcher.

Ichneumon cestus, *Cress.*—Three ♀. Species was described from V. I. coll., H. Edw.   A common species, easily recognized by single black band on abdomen.   Mr. Fletcher has numerous examples from Mr. Danby.

Ichneumon compar, *Cress.*— ♀ described from V. I. coll., H. Edw.

Ichneumon creperus, *Cress.*—Three ♂.

Ichneumon difficilis, *Cress.*—This insect was described from Cal., but a *var.?* is noted from V. I. coll., H. Edw.

Ichneumon inconstans, *Cress.?*—One ♂.

Ichneumon infucatus, *Cress.*—Cat. Hym. N. Am., p. 185. One ♂ received by Mr. Fletcher.

Ichneumon insolens, *Cress.*—Taylor, *loc. cit.:* "One specimen bred from chrysalis of *Vanessa antiopa.*"

Ichneumon lividulus, *Prov.*—One ♀ received by Mr. Fletcher, labelled *Ich. grandis*, determined by Mr. Brodie. Seems, from the partially rufous legs, etc., to belong rather to this species.

Ichneumon longulus, *Cress.*—Taylor, *loc. cit.* A specimen so labelled, received by Mr. Fletcher, is, however, only the ♂ of *cestus*, varying a little from typical coloration.

Ichneumon nuncius, *Cress.*— Three ♂ s; also four received by Mr. Fletcher.

Ichneumon occidentalis, *Hargtn.*—Type ♀ in my collection.

Ichneumon otiosus, *Say.*—Taylor, *loc. cit.:* "My only specimen was unfortunately destroyed during the process of examination."

Ichneumon rufiventris, *Brullé.*—One ♀ labelled *insolens* apparently belongs to this species.

Ichneumon russatus, *Cress.*—Two ♀ s. Type was from V. I. coll., H. Edw.

Ichneumon sagus, *Cress.*—One ♀ received by Mr. Fletcher.

Ichneumon salvus, *Cress.*—The ♂ was described from V. I. coll., H. Edw.

Ichneumon scibilis, *Cress.*—One ♂.

Ichneumon seminiger, *Cress.*—Taylor, *loc. cit.* Not seen.

Ichneumon sequax, *Cress.*—Type ♀ was from V. I. coll., H. Edw. Taylor *(loc. cit.)* says: "Very common; one specimen was bred from the chrysalis of a Lycæna.".

Ichneumon Taylori, *Hargtn.*—Type ♀ in my collection.

Ichneumon vancouverensis, *Prov.*—Type ♂ was from coll. Taylor, who says *(loc. cit.)*, "This fine insect is abundant, and I have bred it in some numbers from the pupa of a Bombyx." Not seen, but answers to description of *neutralis*, Cr., from Cal.

Ichneumon variegatus, *Cress.*—One ♂ to Mr. Fletcher.

Hoplismenus pacificus, *Cress.*— ♀ ♂ described from V. I. coll., H. Edw.

Amblyteles hudsonicus, *Cress.*—Two ♀ s. One of these is a *var.* with the head and thorax above rufous. Mr. Fletcher also has one ♀.

Amblyteles nubivagus, *Cress.?*—One ♂ *var.?*

Amblyteles perluctuosus, *Prov.*—One ♀.

(TO BE CONTINUED.)

BOOK NOTICES.

Rules for regulating nomenclature with a view to secure a strict application of the law of priority in entomological work; compiled by Lord Walsingham and John Hartley Durrant (Merton rules). Longmans, Green & Co., London., New York, and Bombay; 2nd Nov., 1896; 18 pages. Price sixpence.

The rules are for the most part a good statement of current practice, with the suggestion of a considerable number of signs to facilitate brevity of reference without loss of accuracy. These may advantageously be adopted.

Rules 7, 20, 21, 24, 25, 29 and 30 imply a much more rigidly classical attitude in regard to names than is prevalent in America. The authors would have all names according to the rules of Latin orthography, and would change those that are not, even so radically as *gypsodactylus* for *cretidactylus*. Names with similar sound are rejected ; *e. g.*, Ucetia invalidates Eusesia ; also those which involve a false proposition, or are offensive politically, morally, or by irreverence.

Rule 12 defines publication as including the possibility of purchase. If the rule be not extended, it would invalidate all species published in Government or private papers which are distributed without charge.

The definition of a genus by designation of type without description is not referred to, and apparently is condemned by implication.

The case of restriction of a heterotypical genus to one type by the successive removal of species to other genera by subsequent authors is not explicitly stated, and might well be added to rule 42.

A few rules about the formation of family names might have been added, for example :

1. Family names shall be formed by adding —idæ to the stem of some genus included in the family.

2. The generic name so used must be a valid one.

3. The first generic name used in a plural form shall be the one so used for the family type unless it be invalid, in which case the next generic name included in the family, which has been used in a plural sense, shall be substituted according to the rule of priority.

HARRISON G. DYAR.

MONOGRAPH OF THE BOMBYCINE MOTHS: *I. Notodontidæ;* by Alpheus
S. Packard, M. D., National Academy of Sciences, Vol. VII.

This magnificent work is, without doubt, an immense credit to the
author, and will take a permanent place among the triumphs of American
Lepidopterology. It is not my intention to discuss matters of general
classification or nomenclature here. My reasons for differing on certain
points as to the latter have all been given elsewhere, and the merits of
the Comstock–Dyar classification have been insisted upon by Dr. Dyar.
Dr. Packard's work, as a whole, with its superb technical execution, has a
value which could have been only enhanced by his attention to points of
nomenclature, which I believe cannot be properly contradicted, and by
his adhesion to a scheme of general classification, which I believe cannot be adequately gainsaid. I can here, out of my present limited
knowledge, merely mention a few points, which may be of general or
only of particular interest. There are a few errors in authorities. I do
not know why my *Notodonta stragula* and *Schizura leptinoides* and *S.
eximia* are given to Grote and Robinson (plates). Nor do I know why
my name is placed in brackets after *Heterocampa Belfragei.* I described
the latter as a *Heterocampa,* and have no responsibility for its having
been placed under *Litodonta,* a reference which never occurred to me.
I differ from Dr. Packard as to the validity of *Litodonta.* The costa is
straighter, the primary fuller outwardly over internal angle, apex sharper,
while the antennal structure is decisive, as compared with *Heterocampa
subrotata;* the orange spots are peculiar. *H. subrotata* is a miniature
obliqua, and is placed next in my list. *H. celtiphaga* is founded on
obscurely marked and small specimens, probably not different specifically.
Litodonta may be a more specialized form, from the character of the
female antennæ; the discovery of the larva will be attended with interest.
The unhappy influence which Mr. Walker has exercised is very apparent,
and the synonymy of *Schizura ipomeæ* exhibits this at its worst. I do
not insist upon the validity of *S. telifer* as a species; the black streaks
are very distinct in both sexes and our nomenclature was invented to
designate such forms, if not as species then as varieties. With regard to
Hyparpax, and in connection with Dr. Packard's remarks upon *H.
perophoroides,* I again draw attention to my previous statements as to
Abbot and Smith's plate, that the figure of the female *aurora* at least
approaches that form. The late Mr. Hy. Edwards sent me at one time
a damaged specimen (I think without head or feet) of a well-sized pink

and yellow moth from Colorado, resembling this genus or *Anisota rubicunda* in colours.   I would not describe it, but returned it as a probably new Noctuid.   The figure of *Euhyparpax* distantly recalls the specimen, which must be in coll. Central Park Museum.   The figure (Plate VI., 14) certainly does not look like a Ptilodont, rather like an Agrotid, but, especially an uncoloured figure, may be deceptive.

A short classification of the *Melalophidæ* may be found in ' Entomologist's Record,' VIII., 107, but I find since that *Phalera*, Hübn. Verz., 147, 1816, is preoccupied by *Phaleria*, Latreille, 1804.   Another name must be used for the genus of *bucephala* and the subfamily of which I made it the type.   As to *Datana,* I rather missed an allusion to the fact that Grote and Robinson first drew attention that there were many closely allied species, and to the characters of the uneven margin, differences in the lines and general tinting which serve to distinguish the moths.   One paper in Vol. VI. of the Proceedings Ent. Soc., Phil., was an answer to the criticism passed by the late Mr. Walsh upon our previously described *Datana perspicua.*   There is still a memorandum in my note-book of a reference in this genus which I do not seem to have published and which I do not find in either Packard or Dyar.

A. RADCLIFFE GROTE, A. M.

PRELIMINARY NOTES ON THE ORTHOPTERA OF NOVA SCOTIA; by Harry Piers.   Transactions of the N. S. Institute of Science, Vol. IX., 1896.

So little attention is paid to Entomology in the Maritime Provinces that we gladly welcome this contribution to the subject and are much pleased that Mr. Piers intends to devote some years to the study of the order Orthoptera.   The paper before us gives some very interesting notes on the habits and range of fourteen common species of cockroaches, crickets, and locusts, and describes more at length the ravages committed by *Melanoplus atlanis* on Sable Island, a hundred miles off the coast of Nova Scotia in the Atlantic Ocean.   C. J. S. B.

REV. THOMAS W. FYLES, F. L. S.

The Canadian Entomologist.

Vol. XXIX.   LONDON, FEBRUARY, 1897.   No. 2.

THE REV. THOMAS W. FYLES, F. L. S.

We have much pleasure in presenting to our readers the excellent portrait of our colleague, the Rev. Thomas W. Fyles, who has been for many years an active member of the Entomological Society of Ontario. Though living at South Quebec, he has regularly attended the annual meetings at London, travelling many hundreds of miles in order to do so, and has invariably delighted those present with his excellent papers. He was a member of the Council from 1882 to 1888, when the change in the Act of Incorporation required the directors to be resident within certain districts of the Province of Ontario. Three times he has represented the Society as their delegate to the Royal Society of Canada at Ottawa, and he has been a member of the editing committee of the CANADIAN ENTO-MOLOGIST since 1889. While filling the arduous position of chaplain to the immigrants landing in Canada, under the auspices of the Society for Promoting Christian Knowledge, he devotes any spare moments that he can get to the study of entomology. He has succeeded, with an energy and enthusiasm worthy of admiration, in forming an extensive collection of insects, and acquiring a knowledge of the science beyond what is ordinarily met with. That he may long continue to carry on his excellent work, both in his official position and in his scientific pursuits, is the hearty wish of all his friends.

A PARASITE OF HEMIPTEROUS EGGS.

BY T. D. A. COCKERELL, MESILLA, N. M.

The following description is offered of an insect to which I shall have occasion to allude in a forthcoming Bulletin, wherein such descriptive matter would be inappropriate.

Hadronotus mesillæ, n. sp.— ♂ . Length slightly over 1 mm.; black; coxæ black, legs otherwise rufous. Antennæ dark rufous, arising just above mouth, delicately pubescent ; pedicel oval, shining, punctured, conspicuously shorter than the long first flagellar joint ; second flagellar joint shorter than the first, but fully twice as long as broad ; third to fifth joints oval, shorter than the second, the third slightly longer than the following, all longer than broad. Head short, broadly transverse, slightly broader than thorax ; lateral ocelli separated from the eyes by a space about equal to their own diameter ; a depression in front of middle ocellus. Frons and face minutely reticulated by grooves, reminding one of crocodile hide. Thorax subglobular, somewhat broader than long, with very sparse short pubescence ; anterior part of mesothorax very indistinctly subreticulately sculptured, its anterior margin with a distinct row of pits. Hind portion distinctly but very delicately and minutely reticulated with raised lines. Scutellum smooth, with a few hairs ; hind margins of scutellum and postscutellum with a row of pits. Abdomen short and broad, carinated at sides, smooth, rather shiny. Wings hyaline, quite hairy, fringe short, nervures rufofulvous ; marginal vein short, not half length of stigmal.

Habitat.—Las Cruces, New Mexico ; bred from eggs of some Hemipteron, apparently Pentatomid. The eggs are barrel-shaped, pale gray with a white base and a white ring at top, the lid with a white central ringlet, and its suture white. Only one specimen was bred, and the tips of its antennæ are broken off, but the species differs at once, by its reticulate sculpture and other characters, from all those described by Mr. Ashmead in his Monog. Proctotrypidæ or in his work on the Hymenoptera of St. Vincent. Another parasite of Pentatomid eggs occurs in the Mesilla Valley, namely, *Trissolcus euschisti*, Ashm. (a Mesilla example det. Ashm.). With us, I believe it is a parasite on the eggs of *Brochymena obscura*, H. S., which abounds in orchards.

NOTES ON VANESSA INTERROGATIONIS.

BY W. F. FISKE, MAST YARD, N. H.

I remember about ten years ago to have taken several large specimens of a Grapta, probably *G. interrogationis*, but they were lost without being identified. I saw no more of the species until August, 1895, when I took a fine example of the form *Fabricii*. It proved to be the forerunner of a " wave " of the species, and from that date until frost a number were seen, perhaps in all twenty or more, but all but two of them were of the form *Fabricii*. This spring I watched the hibernating butterflies closely, hoping to obtain a fertile female and rear a brood of larvæ, but although there were many *G. comma* and *j-album*, and a few *progue* and *faunus*, on the wing throughout April, I did not observe one *interrogationis* amongst them. By the middle of May the other species of Grapta had disappeared, or were represented by a few specimens worn almost beyond recognition. I had about given up meeting with *interrogationis* that spring, when on the 16th of May I captured a large but badly worn *umbrosa* fluttering over lilac blossoms. I was surprised that it should be of this form instead of the more common *Fabricii*, but what was my astonishment to see four or five more of the same form the same day. During the rest of May and first part of June the species was common, but not one *Fabricii* was seen. A large female was captured while ovipositing on elm, and netted over a branch of that tree. She deposited a large number of eggs indiscriminately on leaves, branch, and net, in most cases singly, but in a few instances in "chains" of three or four. In order not to disturb the eggs, I let the net remain as it was until the larvæ should hatch, and then, thinking that the larvæ would do better in the open air, left it until they had passed the second moult, when on removing it I found only eight remaining. These pupated without further accident, and on the 13th of July and the few days following five imagoes emerged —three *Fabricii* and two *umbrosa*. This was after the larger part of the brood of *j-album* had emerged and several weeks after the first brood of *comma*, and as the former species is probably but single brooded here, I was not expecting a second brood of *interrogationis*. It was with some surprise, therefore, that a large colony of young larvæ were discovered in the latter part of August feeding on the heads of hops. Later several other colonies were found on hop and elm, and a number of larvæ were transferred to my breeding-boxes and carried successfully to pupation, but as many of the pupæ rotted, only about thirty imagoes, all *Fabricii*, were obtained. The last specimen, delayed by a long continued "spell" of severe weather, did not emerge until November 6th, after being in the pupa state nearly six weeks and freezing at least once. It was smaller and darker than the average, but not otherwise remarkable.

Now, the question which I wish answered is, Where did the large number of *umbrosa* come from that appeared here so suddenly in May? They certainly did not breed here, because every specimen seen was badly worn, and they could not have flown in any such numbers either the same spring or the fall before, and besides, the fall before it was *Fabricii* that was in the majority. The only explanation which I can offer is that they migrated thither from some other locality, probably in the South. *Pyrameis atalanta* appeared about the same time in very large numbers, but as the species has always been more or less common,

I did not think it so remarkable. The first brood of larvæ of this species are usually so scattering that it is difficult to find them. This summer they were so numerous as to completely strip large clumps of nettle, so that numbers of larvæ must have perished for want of food. Some large and healthy bunches of nettle were so weakened by the larvæ of this species and of *Vanessa Milberti* repeatedly stripping them of every green leaf that they have probably died.

LARVAL STAGES OF AMPHION NESSUS (Cr.).
BY WILLIAM BEUTENMÜLLER, NEW YORK.

Egg. Pale green, almost globular; very similar to that of *Everyx myron*, but smaller. Young larvæ collected at Greenwood Lake, New Jersey, June 25th. Length, 1 mm.

Stage I.—Pale apple-green, with numerous minute white dots and a narrow white subdorsal stripe along each side, beginning at the anterior part of the first segment and running to the base of the caudal horn, which is black, and brown at the base. Length, 9 mm. Moulted June 28th.

Stage II.—Very much like the preceding stage, but the white dots and the subdorsal stripe are much heavier and more distinct. Caudal horn jet black, reddish-brown basally. Head with a narrow white stripe on each side. Length, 13 mm. Moulted July 1st.

Stage III.— Much like the last stage, but the stripes on the head are continuous with the ones on the subdorsum; the third and fourth segments are now considerably swollen and thicker than the remaining segments. Caudal horn black, reddish-brown at the base. Spiracles black. Length, 17 mm. Moulted July 4th.

Stage IV.—Same as the last stage. Length, 22 mm. Moulted July 7th.

Stage V.—The general colour is now dirty orange-brown, speckled with small smoky-black dots. On the junction of the segment along the dorsum is a smoky-black spot, and along the sides is a series of oblique smoky-black bands, the last one running to the base of the caudal horn, which is black. From the head to the end of the third segment are three black stripes, one on the dorsum and one on each side on the subdorsum. Head dirty purplish-brown, with a whitish stripe on each side. Under side darker than above. Length, 45 mm. Full-grown July 18. When fully fed the larva spins a rude cocoon between a few leaves on the ground.

Food-plants: Grape and Virginia creeper.

THE COLEOPTERA OF CANADA.

BY H. F. WICKHAM, IOWA CITY, IOWA.

XX. THE CHRYSOMELIDÆ OF ONTARIO AND QUEBEC — *(Continued)*.

TRIBE IX. —GALERUCINI (Sub-tribe HALTICINI).

The "jumping beetles," or "flea beetles," constitute the above sub-tribe, and are separated from the genuine Galerucini by the fact that the hind thighs are greatly enlarged and thickened for leaping. Most of the species are quite small, though a few are of moderate size for this family, and a considerable number of them are prettily coloured. They are of great importance from an economic standpoint, a number of them being quite injurious. The identification of some of the members of this group is attended with considerable difficulty, yet most of the genera have a peculiar facies, which, once grasped, renders the proper location of additional specimens tolerably certain.

The sub-tribe has recently been worked up in detail, as far as the North American species are concerned, by Dr. Horn, from whose paper on the "Halticini of Boreal America" most of the tables and specific diagnoses have been condensed. His paper has rendered possible an intelligent survey of the group — something heretofore lacking in the American literature on the subject. The diagrams representing elytral markings are reproduced from the figures given in his article.

A. Last joint of hind tarsi globosely inflated ; elytra with confused punctuation, surface glabrous. Size, large or moderate. *Œdionychis.*

AA. Last joint of hind tarsi not globose, usually slender, sometimes thickened when viewed laterally.

    b. Anterior coxal cavities open behind. Mesosternum visible.

        c. Prothorax without antebasal transverse impression, hind tibiæ faintly or not grooved.

            d. Moderate or large sized species, first joint of hind tarsi short, as compared with tibiæ, and rather broad. *Disonycha.*

        dd. Small species, first joint of hind tarsi long and slender.

            Hind tibiæ grooved on outer edge, first joint of hind tarsi as long as one-half the tibia .... *Longitarsus.*

            Hind tibiæ not grooved, slightly excavated near tip; first joint of hind tarsi about one-third as long as tibia .... *Phyllotreta.*

cc. Prothorax with antebasal impression, which is transverse, usually feeble and not distinctly limited at each extremity .... *Haltica.*

bb. Anterior coxal cavities closed behind.

e. Antennæ 11-jointed, approximate at base.

f. Posterior tibiæ sinuate near the apex, the sinuation limited above by a distinct tooth ; first two ventral segments connate, but with distinct suture ; thorax without antebasal impression............... ....*Chætocnema.*

ff. Posterior tibiæ without either sinuation or tooth.

g. Thorax with distinct antebasal transverse impression, usually well limited at its ends. Elytra punctato-striate.

h. Elytra glabrous.

Form more or less ovate ; antennæ moderate...... . ...*Crepidodera.*

Form elongate, parallel ; antennæ as long or longer than body.....................*Orthaltica.*

hh. Elytra with rows of setæ on interstices, giving a pubescent appearance. Form short, ovate; antennæ not elongate.......................*Epitrix.*

gg. Thorax without *transverse* antebasal impression.

i. Spur of hind tibia small and slender.

Thorax with short, deep *longitudinally* impressed line each side ; elytra punctato-striate, paler at tip.............................*Mantura.*

Thorax without impression, elytral punctuation confused. .... *Systena.*

ii. Spur of hind tibiæ broad, emarginate at tip....*Dibolia.*

ee. Antennæ 10-jointed, hind tibiæ prolonged beyond the insertion of the tarsus, which is placed rather on the outer side, above the apex.....................*Psylliodes.*

ŒDIONYCHIS, Latr.

The species of this genus are of large or moderate size (for Halticini) and are readily recognizable on account of the inflated or globose claw-joint of the hind tarsi. Some of them are of bright colours and handsomely marked. The Canadian forms are thus separated by Dr. Horn :

A. Antennæ stouter, scarcely one-half the length of the body; species larger and more convex, front of head oblique, elytra never explanate at sides.

b. Elytra entirely blue, green, violaceous, blackish or testaceous.
  c. Body never entirely black beneath.
    Elytra bright blue or green, thorax smooth ; body beneath entirely pale. .18–.28 in............*gibbitarsa*, Say.
    Elytra violaceous or greenish-black, thorax more or less distinctly punctate, body beneath in great part dark, thorax yellowish with a large piceous space or M-like mark blackish. .16–.28 in........ ......*vians*, Ill.
  cc. Body entirely black beneath, upper surface dull black, impunctate. .18–.22 in....................*lugens*, Lec.
 bb. Elytra with pale margin, disk violaceous or bluish.
    Thorax and elytra coarsely and closely punctate. .20–.26 in...........................*thoracica*, Fabr.
    Thorax and elytra indistinctly punctured ; elytra brilliant violaceous. .20–.24 in................*flavocyanea*, Cr.
AA. Antennæ slender, equal to or greater than one-half the length of the body ; front of head vertical ; elytra with explanate margin.
    d. Elytra broadly oval, sides much arcuate, coarsely punctate ; may be yellowish with indistinct vittæ, or black with only the margin pale. .14–.20 in..*limbalis*, Mels.
   dd. Elytra with sides feebly arcuate or nearly parallel ; yellowish, with indistinct brown spots and bands or with the disk entirely piceous.
     e. Thorax very coarsely punctured; elytra with a more or less evident costa extending from humeri to apex, yellowish with blackish spots which sometimes coalesce to form an X, behind which is an irregular transverse band. .14–.16 in. (fig. 5)....*sexmaculata*, Ill.

FIG. 5.

    ee. Thorax finely punctured or smooth.
      Head coarsely punctate, punctures closely placed ; yellowish ; elytra with base, suture, and often two spots on each, brown. .14–.16 in....... ....*suturalis*, Fabr.
      Head sparsely punctate or nearly smooth ; thorax often entirely yellow, or may be piceous with the margin pale ; elytra piceous with yellow margin,

rarely with two large yellowish spots on each. .14–.15 in.................... *quercata*, Fabr.

While *flavocyanea* is included in the above table, on account of its being recorded in the Society's list, it has probably been identified in error, since it is a Southern species.

DISONYCHA, Chevr.

Also contains large or moderate sized species, some of them even exceeding *Œdionychis*, which they often resemble in markings, but they may easily be separated therefrom by the claw-joint of the hind tarsi not being swollen. They separate thus :

A. Elytra not striped.

> Thorax yellow with three black spots arranged in the form of a triangle ; under surface of body and the legs black .20–.25 in..................................*triangularis*, Say.

> Thorax yellow, not spotted; abdomen yellow, femora usually yellow at basal half. .21–.23 in... ....*xanthomelæna*, Dalm.

AA. Elytra striped.

> b. Form very elongate; elytra vaguely grooved ; thorax somewhat uneven.

> > Body beneath black, except sides of thorax, which are margined with yellow.   Black spot on disk of thorax very large.......................var. *limbicollis*, Lec.

> > Body beneath partly black, abdomen paler at sides and apex, thorax with under surface entirely yellow, discal spot on upper surface smaller. .26–.30 in......*pennsylvanica*, Ill.

> bb. Form not very elongate ; elytra and thorax even, the former with discal and submarginal vittæ.

> > c. Abdomen densely punctured, conspicuously pubescent.

> > > d. Head coarsely punctured from side to side. .22–.36 in..........................*quinquevittata*, Say.

> > dd. Head smooth at middle.

> > > Elytral vittæ rather broad, head and body beneath more or less clouded with darker, labrum piceous. .22–.26 in............. ....*crenicollis*, Say.

> > > Elytral vittæ narrow, head and body beneath always pale yellow, labrum pale. .20–.26 in. ...*caroliniana*, Fabr.

cc. Abdomen very sparsely punctured, pubescence scarcely visible. Thorax smooth, head rough, epipleura black. .20–.22 in.................... ....*glabrata*, Fabr.

It is quite likely that *glabrata* may have been recorded in error ; the species called *5-vittata* is the one everywhere identified as *alternata*, and so recorded in the Canadian lists ; while *crenicollis* and *caroliniana* are inserted in the table, with the characters assigned them by Dr. Horn, since it is, to my mind, likely that one of these is the species which was mistaken for *glabrata* by the Canadian recorder.

HALTICA, Geoffr.

The species belonging here are of moderate size, none of them with markings of any sort on the upper surface of the body, which is blue, green or bronzed, and usually shining. The thorax is marked near the base with a transverse more or less distinctly impressed line, which has been used as a means of differentiating species. The following table is a tolerably close copy of that of Dr. Horn, and will serve to distinguish the recorded Canadian forms with some degree of accuracy.

A. Elytra with a prominent lateral plica along the lateral submargin, giving the appearance of a double margin. .20–.24 in.... ....*bimarginata*, Say.

AA. Elytra not plicate.

Thorax with deep antebasal groove extending *completely* across.

Larger (.16–.20 in.) usually blue, form robust, thorax distinctly wide at base. ., ...*chalybea*, Ill.

Smaller (.12–.16 in.) metallic, brassy, blue, green or bronze. Elytra distinctly sparsely punctate at base, more faintly toward apex...................... ....*ignita*, Ill.

Thorax with transverse antebasal groove, which is not entire.

Transverse impression, ending in a fovea on each side. .18 in. ....*evicta*, Lec.

Transverse impression gradually evanescent at either end.

Impression deep, humeri of elytra well marked, thorax relatively coarsely punctate. Elytral punctuation coarser than usual. Colour more or less coppery, sometimes nearly blue. .14–.18 in....................*carinata*, Germ.

Impression feeble, almost obliterated: humeri rounded, thorax sparsely punctulate, elytra scarcely visibly punctate, colour bright green to dark blue. .14–.18 in.......*foliacea*, Lec.

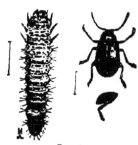

It should be remarked that *evicta* is a Pacific Coast species (found in Oregon), of which I have seen no Canadian examples ; while *foliacea* is Southern, occurring in Texas, Colorado, New Mexico, and Arizona. *H. inærata*, Lec., is synonymous with *ignita*.

(Fig. 6 represents the larva and beetle of *H. chalybea*, and a leg of the latter, showing the greatly thickened thigh.)

Fig. 6.

CREPIDODERA, Chevr.

The best known species of this genus is *Crepidodera helxines*, a bright metallic blue or green flea beetle, very commonly found on willows. All of the members belonging here are quite small, and do not resemble each other at all closely, so that reference should be had to the generic characters (as laid down in the table of genera) before trying to place any presumed *Crepidoderas* by the following specific analysis, which is that of Dr. Horn :

Form oblong-oval; elytra uniform in colour with the head and thorax, surface metallic, blue or green ; thoracic punctuation abundant, intermixed. .09-.13 in....... ....*helxines*, Linn.

Form oval, narrowed in front ; colour piceous, with slight aeneous lustre, apical third of elytra indeterminately testaceous. .08-.10 in................................... *modeeri*, Linn.

Form broadly oval and convex; colour rufotestaceous, without metallic lustre ; abdomen piceous, prothorax not distinctly punctured. .06-.07 in...... ....*atriventris*, Mels.

EPITRIX, Foudras.

Contains one Canadian species, *E. cucumeris*, Harr., the "cucumber flea beetle" (fig. 7), which is often found very abundant on potato vines. It is a small (.06 to .08 in.), ovate, slightly oblong beetle, nearly black in colour, the legs reddish or brownish, femora often darker. It may easily be told from any of the *Crepidoderas* or other genera which might otherwise resemble it, in our fauna, by the fact that the upper surface is pubescent. The thoracic punctures are well separated from each other ; the elytral striæ, especially near the suture, very feeble.

Fig. 7.

Orthaltica, Crotch.

O. copalina, Fabr., is an elongate-parallel insect, of shining surface, brownish or blackish in colour. .08–.10 in. long. The antennæ are more elongate than usual in the Halticini, equaling about two-thirds of the length of the body in the male, somewhat shorter in the female. The antennæ and legs are rufotestaceous, the thorax is broader than long, sides arcuate, margin finely serrate, punctures coarse and deep, but not densely placed. Elytra with nine striæ of closely-placed coarse punctures, intervals narrower than the striæ. I have found this species in abundance on the flowers of sumach.

Systena, Clark.

The species of this genus are rather elongate, somewhat depressed or only moderately convex in form. The antennæ are about one half the length of the body. Some of them are injurious to cruciferous plants. Two of the Canadian species are dark, the other two pale or vittate. They may be separated thus :

Black, head reddish. .14–.20 in...................*frontalis*, Fabr.

Black, head not red ; joints 3, 4, 5 of antennæ testaceous. .18 in.........................*hudsonias*, Forst.

Elytra pale or striped.

Surface shining, punctuation fine ; may be entirely pale, or the elytra may be vittate. Under side of body and sides of thorax often piceous. .12–.18 in.........*tæniata*, var. *blanda*, Mels.

Surface subopaque, punctuation coarse, close and deep. .14–.16 in.........................*marginalis*, Ill.

Longitarsus, Latr.

Three species have been reported from the region under discussion. They all belong to the division of the genus in which the fourth antennal joint is not longer than the second, and are distinguished by the use of the following characters in the table of Dr. Horn :

Surface entirely shining, form robust, elytral humeri well marked, punctuation rather coarse. Colour blackish. .07 in...*erro*, Horn.

Surface more or less alutaceous, thorax always so, form more elongate, humeri not prominent.

Elytra not shining, punctuation very indistinct ; colour yellowish-testaceous. .07–.08 in...................*testaceus*, Mels.

Elytra shining, punctuation coarse ; colour dark rufotestaceous to nearly piceous. .08 in...................*melanurus*, Mels.

GLYPTINA, Lec.

Species of this genus will almost certainly be found in Canada. They have the elytral punctuation disposed in rather regular striæ, while in *Longitarsus* the punctures are confused. Otherwise there is considerable similarity between the two genera, as far as aspect is concerned.

PHYLLOTRETA, Foudras.

Contains a few species only, the Canadian ones all being of a piceous colour, more or less aeneous or greenish, shining, the elytra marked with yellow stripes or spots. (*P. vittata*, fig. 8.) Often injurious by their great abundance; they are to be seen on the leaves of horse-radish, wild mustard, and various allied plants, wild or cultivated. It should be noted that the record for *lepidula* ought to be carefully verified, since the species is Californian. *P. sinuata* has been included in the table, though not actually known to occur in Canada.

A. Fifth joint of antennæ much enlarged (♂) or longer than the sixth
  (♀). Elytra usually vittate, rarely spotted.
  b. Elytral vitta simple, narrow, nearly straight, but incurved
    at the apex. .08-.10 in . . . .*lepidula*, Lec.
  bb. Elytral vitta sinuous, more or less dilated or appendiculate at ends.
    Vitta incurved at base, approaching the scutellum ; intermediate portion sometimes wanting, leaving the apical parts in the form of spots (fig. 9a.). .08 in. . .*vittata*, Fabr.
    Vitta parallel with suture at its basal half. .10 in. (fig. 9b). . . . .*sinuata*, Steph.
AA. Fifth joint of antennæ not modified ; fifth joint not longer than sixth in either sex. Piceous, not metallic. Each elytron with two oval yellow spots, one humeral, the other near the apex. .08-.10 in . . . .*bipustulata*, Fabr.

FIG. 8.

a   b

FIG. 9.

MANTURA, Steph.

Represented by *M. floridana*, Cr., an oval, somewhat elongate, moderately convex beetle, of a brownish colour, faintly bronzed above ; thorax without transverse antebasal impression, longitudinal basal impressions deep and triangular. Elytra indefinitely paler at apical third. Legs reddish, hind femora darker, each of the tibiæ with a terminal spur. In colour this species somewhat resembles *Crepidodera modeeri*, Linn., but that insect has a moderate transverse antebasal impression on the prothorax. Length, .08 in.

CHÆTOCNEMA, Steph.

This is a large genus, well represented in the United States. The Canadian list contains only three species, one of which (*alutacea*, Cr., known from Georgia and Florida) may be erroneously cited, leaving only *denticulata* and *parcepunctata* as undoubted natives. Several are known from the Lake Superior region, and some of them must undoubtedly occur in Ontario. Following Dr. Horn's arrangement, these recorded forms may thus be known ; all of them belonging to the group in which the sides of the thorax are not obliquely truncate at the front angles.

Head distinctly punctate ; upper surface of body bright bronze or brassy; elytral striæ of coarse deep punctures, the scutellar series usually irregular, the remainder not confused. Form oval, not elongate, clypeo-frontal region subopaque. .08-.10 in....*denticulata*, Ill.
Head impunctate.

Thorax with entire basal marginal line, which is not defined by punctures ; legs entirely piceous, surface subopaque. .06-.08 in.......................................*alutacea*, Cr.
Thorax finely and sparsely punctate, with basal marginal row of distinct punctures, surface shining. Femora piceous, tibiæ and tarsi brownish or rufotestaceous. .06 in....*parcepunctata*, Cr.

DIBOLIA, Latr.

The form of the spur of the hind tibiæ (broad with a distinct emargination at tip) will in itself define the genus. *D. borealis*, Chevr. (= *ærea*, Melsh.), is recorded from Canada and is about .12 in. long, oval, convex, robust, the surface bronzed, elytral striæ of coarse punctures ; anterior and middle legs and hind tibiæ reddish.

PSYLLIODES, Latr.

Antennæ ten-jointed, inserted against the inner border of the eye, hind tarsi inserted before the end of the tibiæ and slightly to the outer side, first joint more than half the length of the tibia. The Canadian species is *P. punctulata*, Mels., a bronzed beetle .08-.10 in. long, of elongate-oval, rather convex form, thorax at base not narrower than the elytra, which are punctato-striate, the punctures coarse and deep, closely placed. The male has the last ventral distinctly impressed.

ON LEDRA PERDITA, A. & S.

BY CARL F. BAKER, AUBURN, ALABAMA.

On page 577 of their great work on the Hemiptera, Amyot and Serville describe two species of *Ledra*. One, *L. aurita*, the well-known species of Europe, was characterized from specimens collected near Paris. I have specimens of it now before me. Its size, the broad membranous prolongation of the head, the ear-shaped horns on the thorax, together with other details of structure, separate it widely from any other homopterous insect. The other species described, *L. perdita*, though equally unique in form, was characterized under circumstances which, for such eminent scientists as Amyot and Serville, seem extraordinary. After a three-line description, they remark : " L'exemplaire unique d'après lequel cette espèce a été figurée, ayant été détruit, nous la décrivons d'après la figure." Unfortunately, the figure, number five on plate II., is very poor. The species is credited to " Amérique septentrionale."

Since that time the species has never again been recognized, although often noticed in hemipterological literature. Mr. Van Duzee, in his "Catalogue of the Jassoidea," lists it as an unquestionable *Ledra*, and gives its habitat as Pennsylvania, on the authority of Amyot and Serville.

It is perfectly evident from the figure that the species is not a *Ledra*. It lacks utterly the characteristic head structure of *Ledra aurita*. It is equally evident that the figure is that of a Membracid belonging in the Centrolinæ, near *Microcentrus caryæ*, Fh. Indeed, Dr. Goding tells me Fitch himself noticed this resemblance.

During several years past I have been receiving quantities of material in Homoptera from many localities in Pennsylvania and throughout the East. This material is the result of careful work by good collectors, and contains immense series of the native Membracids and Jassids. In the examination of this material I have been constantly on the watch for *Ledra perdita*. Lately it has occurred to me in several specimens from Pennsylvania, New Jersey, and Indiana, collected by Messrs. Dietz, Liebeck, and Weith. There is nothing else among all the American material I have examined that is at all like this species, with the single exception of *Microcentrus caryæ*, and that lacks the long ear-shaped horns on the thorax. So peculiar in form is it that there is not a possibility of confusing it with anything else in our fauna.

And not until another species from the same region shall have been dis-covered, having closer affinities with it than has *Microcentrus caryæ*, will there be any reasonable grounds for doubting that this, which I so refer, was the form which Messrs. Amyot and Serville described under the name *perdita*.

I forwarded specimens of the species to Dr. Goding, and was much surprised to learn that it was identical with his *Centruchus Liebeckii*, also from Pennsylvania, described on page 471 of the List of N. A. Membracidæ. In a letter he cites the genus as " *Centruchoides*," which I suppose to be a manuscript name founded on this species. I, however, believe this species (which in future must be known as *perdita*, A. & S.) to be congeneric with the *caryæ* of Fitch. I have specimens of *caryæ* with rudiments of thoracic horns. Outside of this character the species are very closely related.

I have yet to see a true *Ledra* from either North or South America.

SOME NEW SPECIES AND VARIETIES OF LEPIDOPTERA FROM THE WESTERN U. S.

BY WM. BARNES, M. D., DECATUR, ILL.

Argynnis Charlottii, n. sp.

♂.—Upper surface very much like *Cybele;* differs from *Leto* in the lighter shade of the ground colour and the much darker and more exten-sive basal area. This area is sharply limited at the outer edge and extends to the median row of markings, which on the hind wings are quite obscured by it. The apical region is not so clear as in *Leto*, the row of round spots in the outer belt continuing of large size up to the costa, and the dark blotch lying just within the upper three spots is very prominent, as in *Cybele*.

Under surface clearer, brighter, and markings less heavy than in *Leto*. The marginal brown shading very faint, and the submarginal row of crescents, which on the secondaries are very narrow but well silvered, have but a very fine edging of the same shade. The dark basal area stops sharply at the median row of silvered spots, as in *Cybele*, and is not present on their outer side, as in *Leto*.

♀.—Upper surface closely resembles *Leto*, the ground colour and basal area being the same. The markings are, however, not so heavy and the submarginal row of lunules do not so completely enclose the row of spots of the ground colour. On the under surface the markings are not so heavy as in *Leto*; the apical region is clearer, the three or four brown spots so conspicuous in *Leto* being here wanting or but faintly indicated. The outer belt on the secondaries presents the same clean-cut character as in the male, owing to the absence of the brown shadings to its inner and outer sides.

Types.—1 ♂ and 2 ♀s in my collection, from Glenwood Springs, Colo.

This species stands intermediate between *Leto* and *Cybele*. The locality has been thoroughly worked for several years and no typical *Leto* taken there. I have *Leto* from Utah, California, Nevada, Oregon, Idaho, Montana, and British Columbia, and they are uniform in their points of difference from the form here described.

Melitæa Gillettii, n. sp.

♂ expands 1½ inches; head and thorax black; abdomen black above, beneath yellowish-white; palpi and legs dark red; antennæ fuscous; club yellow; wings, ground colour black, markings dull red and white, veins black. Primaries above show a wide margin of the ground colour, in which are two rows of spots; the margin red, very faint, scarcely discernible except towards apex; the second row is white, small and not very prominent; the third row is red, the spots are large, quadrate and completely fill the intercellular spaces, thus giving the appearance of a broad red band cut by the black veins; the fourth row is rather irregular, white and joined opposite the cell by a demi-row from costa; two red and two white spots in cell; two white spots and one red in subcellular space.; basal area rather obscured with black.

Secondaries above have the four outer rows as on primaries, the marginal red row even fainter, two red and one white spot in cell and a white subcellular spot. The under surface shows but little of the black ground colour, it being reduced to the veins and lines between the rows of spots, which are all rather quadrate in shape, filling the intercellular spaces, thus giving a well-marked, banded appearance. The marginal band is red and is followed by the white, red, and white bands as on upper surface. The cellular and subcellular spots on primaries same as above, only larger and more distinct. On basal area of secondaries there are four white spots, separated by an irregular shaped red area, the result of a fusion of the red spots.

Described from seven ♂s taken in Yellowstone Park, Wyoming, July 18.

This species is very closely allied to *M. Iduna*, Dalm., of Lapland, but in that species the antennæ are black and the red band not half so wide. That a species so distinct from any other thus far described from N. A. should be turned up at this late day is remarkable, and shows the possibilities of many other interesting discoveries when the Park region is thoroughly explored.

Melitæa nubigena, var. *capella*.

In the Henry Edwards collection are specimens of a Melitæa separated under the above name ; but in so far as I know, no description was ever published. The variations of *nubigena* are without number, yet they all come into one of three general classes. In Western Colorado and Utah the tendency is towards a gradual increase of the white at the expense of the red and black, producing forms allied to *Wheeleri*, Hy. Edw. Farther north in the Yellowstone region the tendency is to darker forms, the black replacing the red to such an extent that the spots are small and round, set in a black ground. Around Manitou and Denver forms occur which are of a solid brick red, the white being entirely gone and the black reduced to the veins and fine cross lines, the latter even being wanting in portions of the wings. On the primaries the spots at the costal end of the third row are the last to lose the white colour, and in most of the specimens there are traces of it remaining there. In some few males there is none whatever. The fourth row on the secondaries preserves the whitish colour the longest, but not so tenaciously as is the case on the primaries. In some specimens which have entirely lost the white, the black ground colour still remains well marked, while in others there is considerable fusion of the red spots, while considerable of the white is retained. It is to those dark red forms that Hy. Edwards applied the name *capella*, and I take pleasure in retaining the name proposed by him.

Described from eight pairs in my collection and others among my duplicates.

Colias pelidne, var. *Skinneri*.

Male, expanse 1½ to 1¾ inches ; upper surface of a greenish-yellow shade somewhat darker than *Scudderi*, lightly dusted with dark scales over costal two-thirds of primaries ; marginal bands not so broad and cut less deeply by the yellow nervules than is the case in *Scudderi*. The inner margin of the border varies, being almost entire in some specimens, dentate in others, and in a few erose. The discal spot on primaries black, much more distinct than in either *Scudderi* or *pelidne*. In some few specimens the spot is centered with a few yellowish scales, the spot on secondaries about the same as in *Scudderi*. Under side of primaries yellow, paler along the inner margin, thickly dusted with dark scales over costal two-thirds from base to just within the line where the inner margin of the black border of the upper surface shows through; discal mark faint—scarcely discernible in many specimens. Secondaries thickly dusted with dark scales over the basal three-quarters, discal spot prominent, dark brown ring, centre silvered or white, more or less covered with roseate scales; costa and fringes, except at inner angle of primaries, roseate. Antennæ roseate; club roseate below, brown above; collar, head, legs, and a spot at base of secondaries, roseate; palpi roseate above, yellow beneath; thorax and abdomen dark above, covered with yellow hairs, yellow beneath.

Female, expanse 1⅝ to 1⅞ inches; greenish-yellow or white, about evenly divided. Border well marked, varying greatly in extent. In some specimens, on the primaries it is broad, and entirely encloses a row of spots of the ground colour; in others, while equally broad, it is uniformly dark; from these there are all gradations down to one in which the black is restricted to the apical region, and to pear-shaped spots at the ends of the veins. On the secondaries the border is usually well marked, and extends in some almost to anal angle; in some examples, however, it is confined to the outer angle, as three or four blotches. The upper surface is less dusted with dark scales than in the male, the under surface about the same, the discal spots, fringes and other characters as in the male.

Described from 15 males and 7 females — three of which are yellow, three white, and one intermediate — taken in Yellowstone National Park, and at Arangie, Idaho, in July.

Mr. Bean, in CANADIAN ENTOMOLOGIST, Vol. XXII., p. 127, mentions specimens of a Colias intermediate between *Scudderi* and *pelidne*, and it is probable that this is the same, but as I have none of his material, and he gives no description of it, I am not certain.

Thymelicus Edwardsii, n. sp.

Upper surface bright golden-yellow, fringe dark brown within, lighter outwardly. Beneath primaries yellowish, except inner margin, which is shaded with black; hind wings yellow over the anal margin for one-third the width of the wing, rest grayish-yellow.

Type.—One male, taken near Denver, Colorado.

CATALOGUE OF THE PHYTOPHAGOUS AND PARASITIC HYMENOPTERA OF VANCOUVER ISLAND.

BY W. HAGUE HARRINGTON, F. R. S. C., OTTAWA.

(Continued from page 21.)

Amblyteles subrufus, *Cress.*—One ♀ labelled *Ich. sequax*, received by Mr. Fletcher, appears to belong to this species. It is certainly not *sequax*.

Amblyteles suturalis, *Say.*

> *Amblyteles superbus,* Prov.— Thirteen ♀ s, including Provancher's type, which Mr. Davis has found to equal *suturalis* (CAN. ENT., Vol. XXVII., p. 287). They are yellower than Ottawa specimens, with sutural bands of abdomen weaker and sometimes wanting.

Amblyteles subfuscus, *Cress.*—Two ♀ s.

Trogus buccatus, *Cress.*— ♀ described from V. I. coll., H. Edw.

Trogus Edwardsii, *Cress.*— ♂ described from same coll.

Trogus Fletcheri, *Hargtn.*—Type ♀ in my coll.

Platylabus pacificus, *Hargtn.*—Type ♀ in my coll.

Hemichneumon vancouverensis, *Hargtn.*

> *Hypocryptus vancouverensis,* Hargtn.— Type ♂ in my coll. Mr. Davis informs me that this species belongs to Hemichneumon.

Phæogenes discus, *Cress.*—One ♀ .

Phæogenes fungor, *Nort.*—Two ♀ s.

Phæogenes sectus, *Prov.*— One ♂ . The species was described from coll., Taylor.

Centeterus canadensis, *Hargtn.*—Three types ♀ in my coll.

? Herpestomus attenuatus, *Prov.*

> *Phygadeuon attenuatus,* Prov.—Taylor, *loc. cit.* Not seen.

Herpestomus orbus, *Prov.*

> *Phæogenes orbus,* Prov.—The ♀ of this species was described by Provancher from a specimen sent to him by Mr. Fletcher. Not seen.

Phygadeuon crassipes, *Prov.*—One ♀ . Differs from description in colour of ovipositor.

? Phygadeuon seminiger, *Hargtn.*

> *Semiodes seminiger,* Hargtn.—Type ♂ in my coll. Mr. Davis thinks this belongs to Phygadeuonini.

Phygadeuon nitidulus, *Prov.*—One ♀ . The ♀ of this species was described from coll., Fletcher.

Phygadeuon subspinosus, *Prov.*—Taylor ; *loc. cit.*   Not seen.

Cryptus extrematis, *Cress.*—Ten ♂ s sent to Mr. Fletcher are labelled as bred from Trichiosoma.

Cryptus flavipes, *Hargtn.*—Type ♀ in my coll.

Cryptus Fletcheri, *Prov.*— ♀ described from coll., Taylor.

Cryptus pentagonalis, *Prov.*—One ♂ .

Cryptus punicus, *Cress.*—Proc. Acad. Nat. Sci., Phil., 1878, p. 364.

Cryptus persimilis, *Cress.*—One ♂ .

Cryptus proximus, *Cress.*—Three ♀ s.

Cryptus resolutus, *Cress.*—One ♂ .

Cryptus robustus, *Cress.*— Taylor (*loc. cit.*), " Not uncommon." Not seen, and probably *proximus*.

Cryptus rufoannulatus, *Prov.*—Taylor, *loc. cit.*   One ♀ received by Mr. Fletcher.

Cryptus ultimus, *Cress.*—Two ♂ s.   Labelled as bred from Trichiosoma in April.

Cryptus, n. sp.?—One ♀ near *vancouverensis*.

Cryptus vancouverensis, *Hargtn.*—Three types ♀ in my coll.

Cryptus victoriaensis, *Hargtn.*—Two types ♀ in my coll.   One ♀ also received by Mr. Fletcher.

Chæretymma Ashmeadii, *Hargtn.*—Type ♀ in my coll.   One ♀ also received by Mr. Fletcher.   This has annulate antennæ; the antennæ of type were missing.

Orthopelma occidentale, *Ashm.*—One ♀ .

Hemiteles crassus, *Prov.*—Taylor, *loc. cit.*   Not seen.

Hemiteles militææ, *Ashm.*—One ♀ .

Hemiteles occidentalis, *Hargtn.*—Type ♀ in my coll.

Hemiteles piceiventris, *Hargtn.*—Type ♀ in my coll.

Hemiteles scolyti, *Ashm.*—One ♀ .

Ophion bilineatum, *Say.*—Eighteen specimens.   These vary in size and colour, but apparently all belong to one species.

Ophion nigrovarium, *Prov.* (?)—Taylor (*loc. cit.*) notes that the single insect so determined for him was destroyed.

Anomalon Edwardsii, *Cress.*— ♀ described from V. I. coll., H. Edw.

Anomalon nigrum, *Prov.*—Taylor (*loc. cit.*): " Several bred from pupæ of Noctuæ." Not seen.

Campoplex laticinctus, *Cress.*—One ♀ .

Campoplex major, *Cress.*— ♀ described from V. I. coll., H. Edw.

Limneria argentifrons, *Cress.?*—One specimen, without abdomen, labelled *flaviricta*, but cannot be that species.

Limneria compacta, *Prov.*— ♀ described from coll., Taylor.

Limneria dubitata, *Cress.*—One ♀.

Limneria fugitiva, *Say.*—One ♀.

Limneria major, *Cress.*—One ♀. This is labelled *L. genuina*, Say, but there does not seem to any species described under that name, although Provancher also quotes it in his work.

Limneria valida, *Cress.*—One ♀.

Angitia americana, *Hargtn.*—Type ♀ in my coll.

Pyracmon vancouverensis, *Hargtn.*—Type ♀ in my coll.

Banchus superbus, *Cress.*

Banchus polychromus, Prov.—Two ♀ s. Provancher's type not seen, but it seems undoubtedly, from description, to be a somewhat immature example (in which the black is not fully developed) of this well-marked yellow and black species.

Mesoleptus fasciatus, *Prov.*— ? described from coll., Taylor.

Phobetes canadensis, *Hargtn.*—Type ♀ in my coll.

Mesoleius lætus, *Cress.*— ♂ described from V. I. coll., H. Edw.

Mesoleius truncatus, *Prov.*

Mesochorus truncatus, Prov.— ♀ described from coll., Taylor.

Tryphon communis, *Cress.*—Two ♂ s.

Syrphoctonus agilis, *Cress. (Bassus).*—Three ♀ s.

Syrphoctonus pacificus, *Cress. (Bassus).*— ♂ described from V. I. coll., H. Edw.

Coleocentrus occidentalis, *Cress.*— ♀ described from same coll.

Rhyssa persuasoria, *Linn.*—One ♂.

Ephialtes pacificus, *Hargtn.* —Three types ♀ and one ♂ in my coll. The male is a very small specimen.

Ephialtes thoracicus, *Cress.*— ♀ described from V. I. coll., H. Edw.

Ephialtes tuberculatus, *Fourc.*—Two ♀ s.

Ephialtes vancouverensis, *Hargtn.*—Type ♀ in my coll.

Theronia fulvescens, *Cress.* —Fourteen ♀ and four ♂ specimens. A common insect, infesting Clisiocampa, Orgyia, Menapia, etc.

Pimpla atrocoxalis, *Cress.*—One ♀.   From Clisiocampa.

Pimpla conquisitor, *Say.*—Two ♀s.

Pimpla ellopiæ, *Hargtn.*—Types ♂ ♀ in my coll.   Bred by Fletcher
from pupæ of *Ellopia somniaria*, a moth of which the larvæ are
most destructive to the foliage of oaks.

Pimpla inquisitor, *Say.*—Four ♀s.   Apparently the *P. indigatrix* of
list published by Taylor.

Pimpla pedalis, *Cress.*—One ♀.

Pimpla sanguinipes, *Cress.*—Four ♀s.

Pimpla tenuicornis, *Cress.*—One ♀.

Polysphincta texana, *Cress.*—Two ♀s.

Glypta erratica, *Cress.*—One ♀.

Arenetra pallipes, *Hargtn.*—Five types ♂ in my coll.   Common at
Victoria in March, April and May.   Four ♀s received by Mr.
Fletcher.

Cylloceria occidentalis, *Cress.*—Two ♂s.

Lampronota Edwardsii, *Cress.*—One ♀.   This was labelled *Coleocentrus
rufus*, Prov., and was entered under that name in Taylor's list.   The
species was described from ♀ in V. I. coll., H. Edw.

Lampronota pleuralis, *Cress.*—One ♀.

Lampronota segnis, *Cress.*—♀ described from V. I. coll., H. Edw.

Lampronota vivida, *Cress.*—♂ described from same coll.

Xorides occidentalis, *Cress.*—♀ described from same coll.

Euxorides vancouverensis, *Prov.*—The type ♀ was from Taylor's collection.   Not seen.

Xylonomus insularis, *Cress.*—♀ described from V. I. coll., H. Edw.

Aplomerus tibialis, *Prov.*

Platysoma tibialis, Prov.—One ♀ labelled as found under loose
bark.   The type ♀ was also collected by Taylor.

Ecthrus abdominalis, *Cress.*—One ♀.   Specimen also in coll. Geological
Survey.

Ecthrus (?) maurus, *Cress.*—♀ described from V. I. coll., H. Edw.

Bracon atripectus, *Ashm.*—Three ♀ and one ♂ specimens. The latter was labelled as type of *Bracon bisignatus*, Prov., but no description appears to have been published.

Bracon sanguineus, *Ashm.*—Two, ♀ ♂.

Doryctes pacificus, *Prov.*
   Phylax pacificus, Prov., CAN. ENT., Vol. XVII., p. 117, ♀; *Phylax niger*, Prov., *ibid.*, ♂.—Five ♀ and one ♂ specimen, which are considered by Ashmead to belong to the same species.

Microdus sanctus, *Say.*—One ♀.

Helcon frigidus, *Cress.*—One ♀.

Macrocentrus mellipes, *Prov.*—One ♀.

CHALCIDIDÆ.

Diomorus (?) Zabriskii, *Cress.*—One ♀.

Meraporus sp.—Six specimens.

PROCTOTRYPIDÆ.

Mesitius vancouverensis, *Ashm.*—♀ described from coll., Taylor.

Anteon puncticeps, *Ashm.*—♂ described from V. I. coll., Wickham.

Polymecus vancouverensis, *Ashm.*—♀ described from coll., Taylor.

TRIGONALIDÆ.

Trigonalys canadensis, *Hargtn.*—Type ♂ in my coll.

A NEW SPECIES OF PROTANDRENA, CKLL.

BY S. N. DUNNING, HARTFORD, CONN.

Protandrena Cockerelli, n. sp.— ♀. Length, 12 mm.; shining black. Upper half of clypeus, lower portion of supraclypeal area, and part of side pieces, bright yellow, all forming a band across the face one-half broader than high, and of equal breadth throughout; knees yellow spotted. Head rounded, broader than high, and covered with a short growth of gray hair, longer on cheeks and thickest at base of antennæ; lower half of clypeus and two small dots near lower edge of band black, not deeply or closely punctured; antennæ black at base, becoming brown towards the tip ; first joint of flagellum not quite as long as the second and third combined ; mandibles black ; vertex deeply but not very closely punctate. Thorax covered with gray hair, quite thick below and anteriorly ; mesothorax before deeply and a little more thickly punctured than vertex anteriorly, and the scutellum more largely but less closely punctate ; postscutellum similar to anterior mesothorax, while the metathorax is quite finely and closely punctate ; below the wings the thorax is closely and roughly punctured ; tegulæ and nervures rufous, the stigma with a light spot before ; wings hyaline, much clouded at tip, marginal cell truncate and strongly appendiculate. Abdomen with white basal hair bands ; first segment deeply but not thickly punctured ; second, third, and fourth not as deeply and more closely punctate ; fifth more deeply and quite roughly punctate, and with a heavy rufous hair band posteriorly ; abdomen below with long and not distinctly separated hair bands, more finely punctate than above. Legs hairy, all except first joint of anterior, and the last joint of the middle tarsi rufous ; hind tarsi black ; anterior spur one-half as long as first joint tarsi, middle spur two-thirds as long as first joint of middle tarsi, and lateral spurs shortest of all, rufous ; claws cleft with several teeth inside.

Described from one ♀ taken at Topeka, Kansas, in 1864, by Mr. J. E. Taylor, and numbered 1,043 in my collection. Prof. T. D. A. Cockerell (after whom I have named this species, as a slight token of respect and of my gratitude for his many favours) pronounces this to be a valid new species. I would adopt his table (as published on p. 92 of the Annals and Mag. Nat. Hist., July, 1896) as follows :

A. Large species.
    (1) Stigma ferruginous.
        (a) Hairy, tegulæ rufous, knees yellow......*Cockerelli*, Dun.
        (b) Not so hairy, tegulæ yellow spotted, 4 anterior knees
            yellow. .... *mexicanorum*, Ckll.
    (2) Stigma dark............................*asclepiadis*, Ckll.
B. Small species.
    (1) Tarsi piceous in ♀ .
        (a) Postscutellum and metathorax brownish...*maurula*, Ckll.
        (b) Postscutellum and metathorax black .....*trifoliata*, Ckll.
    (2) Tarsi rufous in ♀ , yellowish-white in ♂ ....*heteromorpha*, Ckll.

Mailed February 1st, 1897.

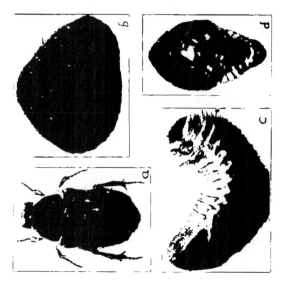

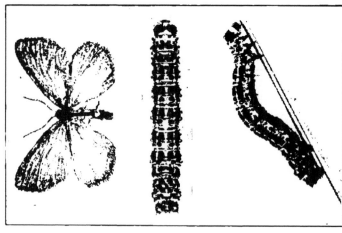

THE BILBERRY SPAN-WORM (D. _istaria_ _integraria_) and THE BUMBLE
FLOWER-BEETLE (_Euphoria_ _inda_).

The Canadian Entomologist.

Vol. XXIX.   LONDON, MARCH, 1897.   No. 3.

THE BLUEBERRY SPAN-WORM (DIASTICTIS INCEPTARIA, WALK.) AND THE BUMBLE FLOWER-BEETLE (EUPHORIA INDA, LINN.).

BY M. V. SLINGERLAND, CORNELL UNIVERSITY, ITHACA, N. Y.

On May 20th, 1896, I received the following letter from a correspondent in Mount Vernon, N. H.: "I enclose you worms that are making sad havoc with the blueberry crop in this section. They seem to be great feeders, completely stripping the bushes of leaves and blossoms, but do not touch the green berries after they begin to form. The berry fields look as though a fire had passed over them, and the worms have nearly ruined the blueberry crop in this vicinity.

"This blueberry (*Vaccinium pennsylvanicum*) needs no cultivation, only to burn over the old bushes every few years, when the new bushes will shoot up and bear the following year. There are hundreds of acres of land producing these berries in this and neighbouring towns, and so far as I can learn, about three-fifths of the crop has been destroyed by the worms."

Accompanying the letter were four nearly full-grown span-worms and one pupa. The larvæ were new to me, and their ravages described above also made them interesting from an economic standpoint. One was therefore photographed, about three times natural size ; both dorsal and lateral views of it are shown on the plate. When full-grown the larvæ measure about five-eighths of an inch in length and are peculiarly marked, as the figures show. The general colour of the body is light yellowish-purple. The dark portions are of a dead black colour. The sutures of the head are broadly margined with white, and a broad white band crosses the sides of the head. The mesal stripe on the dorsum is light yellow, as is also the narrow stripe extending along the subdorsal region through the large black areas. The broad stigmatal stripe is light orange-yellow, whitish below each large black area. Spiracles black. The large black subdorsal areas are in a broad light purplish stripe. The body is sparsely clothed with black hairs. The true legs are black, with yellow bands at the extremities of the joints. Venter yellowish, tinged with purple.

On May 22nd, one of the larvæ changed to a pupa on the soil in my cage. The worms would not eat the currant leaves placed in the cage. The pupa is of a very dark, shining brown colour, with the abdomen a little lighter and sparsely punctate.

As the other pupa and the larvæ had all died, the pupa just described was watched with much interest daily. At last, on the twelfth day (June 4), a dainty, modest little Quaker-gray moth emerged. It is shown, twice natural size, on the plate. About the only noticeable markings on the wings are one or two blackish spots on the costa of each front wing. The antennæ are quite stongly pectinated. The moth was at once sent to Mr. Hulst, who determined it as *Diastictis inceptaria*, Walk. In an illustrated communication to the "Rural New-Yorker" for July 25, 1896, I proposed that the insect be popularly known as the "Blueberry Span-worm," in recognition of its destructive work on that plant.

The moth was first described in 1862 (Cat. Brit. Mus., XXVI., 1667), from a Canadian specimen in the D'Urban collection. Dr. Packard again described it as *argillacearia* in 1874; this name was found to be synonymous with Walker's earlier name, *inceptaria*, by Mr. Moffat, as recorded by Mr. Hulst (Ent. News, VI., p. 11, 1895). Dr. Packard records the moth from Maine, Massachusetts, Pennsylvania, and Canada (Mon. of Geom., p. 258). He states that "it is very abundant in pine woods in Maine on a dry soil, rising and fluttering with rather a feeble flight, and soon settling again. In July, 1874, I captured thirty males before securing a female; the latter are apparently less ready to fly."

Heretofore nothing seems to have been known of the early stages of this Geometrid. Whether there is more than one brood of the caterpillars is not known. Doubtless the practice of burning over the blueberry fields every few years greatly checks the pest. The larvæ will probably quickly succumb to a Paris green spray, and a little united effort among those interested would soon control this blueberry span-worm.

THE BUMBLE FLOWER-BEETLE (*Euphoria inda*, Linn.).

This yellowish-brown beetle, with its wing-covers sprinkled all over with small, irregular black spots (shown at *a* on the plate, twice natural size), is our most common flower-beetle in the North. "It is one of the first insects to appear in the spring. It flies near the surface of the ground, with a loud, humming sound, like that of a bumble-bee, for which it is often mistaken. During the summer months it is not seen, but a new brood appears about the middle of September. The beetle is a general feeder, occurring upon flowers, eating the pollen ; upon cornstalks and green corn in the milk, sucking the juices ; and upon peaches, grapes, and apples. Occasionally the ravages are very serious." (Comstock's Manual for the Study of Insects, p. 565.)

Although this beetle is so common, and has been known for more than a hundred years, nothing was recorded of its earlier stages (beyond the fact that it occurred in its various stages in the nests of ants) until December, 1894. Then Mr. Chittenden (Insect Life, VII., 272) recorded the rearing of the beetle from larvæ found in manure on Long Island. When found, July 9th, the larvæ were encased in cocoons, and the last week in August these cocoons contained living adults.

On June 19th and July 8th, 1896, I received a large number of grubs from Mt. Kisco, N. Y. They were found in a manure pile that had not been disturbed since the preceding August, and from the soil beneath another pile made in October and moved in the following April. One of these grubs is represented, twice natural size, at c on the plate. When compared with a white grub (Lachnosterna, sp.), it was found to be considerably shorter and thicker-set ; its legs were not more than one-half as long, and its head was also much smaller than that of the white grub. The dull leaden hue of the body, due to the contents of the food-canal, indicated that its food consisted of dead vegetable matter rather than living roots, as in the case of the white grub. When they were placed on their feet or venter, they would crawl an inch or so and then roll over and crawl with considerable rapidity, with a wave-like motion, on their backs. I also found several similar grubs in a pile of rotting sod and manure which had not been disturbed for a long time. I have seen no evidence of their eating the roots of living plants.

The grubs were placed in cages containing rotting sod and manure, in which they quickly buried themselves. Twenty days later, July 28th, the grubs had changed to pupæ in earthen cocoons of the somewhat peculiar and definite shape shown, twice natural size, at b on the plate. Evidently the grub forms an earthen cell in the soil by rolling and twisting about, and then cements together the particles of earth composing the walls of the cell so as to form an earthen cocoon, which retains its form when removed from the soil. Each cocoon has a curious roughened or more granular spot on one side (the upper side in the figure).

The white pupa is shown, twice natural size, at *d* on the plate. In pupating, the larval skin is shed off the anal end in the same manner as caterpillars do. In the case of the Spotted Pelidnota *(Pelidnota punctata)*, however, the larval skin splits down the whole length of the back, retains the larval shape, and forms a covering for the pupa, which remains inside.

On August 13th, or sixteen days after pupæ were found in the cages, several beetles emerged. They continued to appear daily until September 10th; more (33) emerged on August 22nd than on any other day. They proved to be *Euphoria inda*, Linn.

This bumble flower-beetle evidently feeds only on decaying vegetable matter, as rotting sod or manure, and is thus destructive only in the beetle state. The beetles seem to do most of their injury soon after they emerge in the early fall. One correspondent wrote me that he collected forty-five of the beetles in one day on a single ripe peach. Doubtless the beetles hibernate, but whether egg-laying takes place in the fall or spring is not known. The fact that manure piled in August and October contained many nearly full-grown grubs the next June indicates that the eggs are laid and hatched in the fall, otherwise the grubs must develop very rapidly after hatching from eggs laid in the spring. There seems to be one brood of the insect in the course of a year. Hand-picking of the beetles is apparently the most practicable method of combating it when it is found working on ripe fruits or on green corn.

Since the above was written, some further notes on this insect (read by Dr. Lintner at the Buffalo meeting of A E. C. last August) have been published. Larvæ were sent to Dr. Lintner in chip manure in the latter part of June. On August 8th two beetles had emerged in his cage, and an examination of the earthen cells revealed other beetles and several pupæ. An instance is given which seems to indicate that there is a possibility that the grubs may have attacked growing corn, but the evidence is not conclusive.

BUTTERFLIES OF NORTH AMERICA.—Mr. Edwards is about to publish the last Part, the seventeenth, of the third volume of this magnificent work. It will contain three plates, illustrating Chionobas Iduna, Californica, Oeno, Varuna and Alberta, with their early stages, and the imago of C. Peartiæ. There will also be accounts of Papilio Brucei and Ajax, Neophasia Menapia, and Colias Eriphyle; and supplementary notes on a large number of other species, with title page and index.

DESCRIPTIONS OF SOME NEW GENERA AND SPECIES OF CANADIAN PROCTOTRYPIDÆ.

BY WILLIAM H. ASHMEAD, WASHINGTON, D. C.

The following new genera and species of Proctotrypidæ were all collected in Canada by Mr. W. Hague Harrington, of Ottawa.

SCORPIOTELEIA, gen. nov.

Abdomen with five visible segments; the last three segments long, slender, cylindrical, together as long as the second, and resembling the terminal segments of a scorpion; the third segment is about as long as the fourth and fifth segments united, the fifth pointed. Front wings with the marginal vein shorter than the marginal cell, and scarcely twice as long as the first abscissa of radius, which is slightly oblique. Antennæ 15-jointed, filiform, the first joint of flagellum the longest, about half the length of the scape, the following joints to the last very gradually shortening, the penultimate joint being about twice as long as thick, the last joint oblong-oval, one-half longer than the preceding.

(1) *Scorpioteleia mirabilis*, sp. n.

♀.—Length, 4 mm. Smooth, shining, pubescent; head and thorax black, collar and prosternum brownish; petiole and the large second abdominal segment brownish-piceous, the three terminal segments yellowish; mandibles, legs and basal four joints of antennæ ferruginous, the flagellum blackish towards apex; palpi yellowish.

The mesonotal furrows are deep, distinct; the scutellum has a large, deep fovea across the base; while the metanotum is smooth, tricarinate, with the posterior angles subdentate. Wings hyaline, pubescent, the tegulæ yellowish, the veins broad. Abdominal petiole longer than the metathorax, a little thicker towards base than at apex, striated, about three times as long as thick, rest of abdomen smooth, polished.

Hab.—Kettle Island, in Ottawa River, August 18, 1894.

STYLIDOLON, gen. nov.

Abdomen with six visible segments, the body of same being long and very slender, twice as long as the petiole, and gradually acuminate toward apex, which has a gentle upward curve; the second segment is scarcely longer than the petiole, the dorsum of same triangularly emarginated at apex; the third segment dorsally, on account of the emargination in the second, a little longer than the fourth and fifth, but ventrally it is not longer than these two segments united; the fifth is shorter than the fourth; the sixth is conically pointed, a little longer than the third. Front wings with the marginal vein as long as the marginal cell, or about 2½ times as long as the oblique first abscissa of radius. Antennæ 15-jointed, filiform, the first joint of flagellum about two-thirds the length of the scape, the following joints to the sixth gradually shortening; joints 7 to 11 much shorter, subequal, about twice as long as thick; the 12th very little longer than thick, the last joint thicker, ovate, nearly as long as the two preceding united.

(2) *Stylidolon politum*, sp. n.

♀.—Length, 3.5 mm. Polished black, shining, pubescent; tegulæ, scape and pedicel ferruginous, the flagellum black or brown-black. Wings hyaline, the veins dark brown. Legs rufous, the articulations paler or yellowish, the hind coxæ black or piceous black.

Hab.—Ottawa, May 13, 1896.

MIOTA, Förster.

(3) *Miota rufopleuralis*, sp. n.

♀.—Length, 2 mm. Polished, shining, pubescent; head black; dorsum of thorax and body of abdomen, except the tip, brown-black or piceous; mandibles, collar, sides of thorax and beneath, rufous; palpi, scape, pedicel, legs and petiole of abdomen, yellowish.

The antennæ are shorter than the body, the flagellum being brown-black; scape as long as flagellar joints 1 to 4 united, the first flagellar joint the longest, not more than thrice as long as thick, the joints beyond very gradually shortening, the three or four penultimate joints only a little longer than thick, the terminal joint conical, only a little longer than the preceding joint. Wings hyaline, the tegulæ yellowish, the veins brownish, the marginal vein very short, only a little longer than the first branch of the radius, or scarcely one-third the length of the radial cell.

Hab.—Hull, P. Q., August 14, 1894.

(4) *Miota Canadensis*, sp. n.

♀.—Length, 2.5 mm. Polished black; first three joints of antennæ, the tegulæ and legs brownish-yellow; palpi white.

The antennæ are not quite as long as the body; scape as long as flagellar joints 1 to 3 united, the first flagellar joint the longest, more than four times as long as thick; flagellar joints 7 to 12 hardly longer than thick. Wings hyaline, the veins brownish-yellow, the marginal vein about three times as long as the first abscissa of radius, or as long as the marginal cell.

Hab.—King's Mountain, Chelsea, P. Q., August 12, 1894.

ZELOTYPA, Förster.

(5) *Zelotypa fuscicornis*, sp. n.

♂.—Length, 2.5 mm. Polished black, pubescent; antennæ longer than the body, fuscous, the scape hardly as long as the pedicel and first joint of flagellum united, the latter excised at basal one-half. The flagellar joints 2 to 11 subequal, about four times as long as thick; legs brownish-yellow, the hind coxæ black. Wings hyaline, the veins brown, the marginal vein hardly two-thirds .the length of the marginal cell, or about one and a half times as long as the first abscissa of the radius. Petiole of abdomen rather stout, about two and a half times as long as thick, coarsely fluted.

Hab.—Hull, P. Q., July 23.

PANTOCLIS, Förster.

(6) *Pantoclis Canadensis*, sp. n.

♀.—Length, 2 mm. Polished black, pubescent, the body of abdomen more or less brownish piceous; antennæ, except the 7 or 8 terminal joints, and legs, brownish-yellow.

The scape is about as long as the first six joints of the flagellum united, the first joint of flagellum being a little longer and more slender than the pedicel, or about twice the length of the second joint; all joints of the flagellum, except the last, are submoniliform and gradually become thicker and broader, the six penultimate joints being a little wider than long, subpedunculate; the last joint is conical, a little longer than the preceding. Wings subhyaline, the veins dark brown, the radial cell rather small, triangular, a little longer than the oblique first abscissa of radius. Petiole of abdomen scarcely twice as long as thick, opaque, coarsely fluted.

Hab.—Ottawa, August 13, 1894.

(7) *Pantoclis similis*, sp. n.

♂.—Length, 2.6 mm. Polished black, pubescent; two basal joints of antennæ, the palpi, the tegulæ and the legs, including all coxæ, brownish-yellow.

The antennæ are shorter than the body, the scape being about as long as the pedicel and first joint of flagellum united; flagellum brown-black, the first joint the longest, not quite five times as long as thick, with the basal one-third strongly excised, the following subequal, but very gradually shortening, so that the three terminal joints are scarcely two and a half times as long as thick. Wings hyaline, the veins brownish,

the marginal vein about two-thirds the length of the marginal cell, or one-
half longer than the oblique first abscissa of radius. Petiole of abdomen
stout, two and a half times as long as thick, fluted.

Hab.—Russell's Grove, Hull, P. Q., August 5, 1894.

A NEW WATER-BUG FROM CANADA.

BY WILLIAM H. ASHMEAD, WASHINGTON, D. C.

The interesting new species of water-bug described below was received
some time ago from Abbé P. A. Bégin, of Sherbrooke, Canada. It was
captured swimming on a fresh-water stream some little distance above
Sherbrooke, and is of more than ordinary interest, from the fact that it
belongs to the genus *Halobatopsis*, Bianchi*, a genus not yet recognized in
the North American fauna, and only recently characterized, being based
upon the South American *Halobates platensis*, Berg., also a fresh-water
species.

Halobatopsis Béginii, n. sp.

♀. Length, 2.3 to 2.5 mm. Oval, velvety black ; a yellow dot or
spot on middle of pronotum anteriorly, a larger, somewhat triangular, yellow
spot, but more or less variable in shape and size, on the upper basal hind
angle of the mesopleura close to the base of the metapleura, while
beneath, the mesosternum anteriorly and posteriorly and along the median
furrow or suture is more or less broadly margined with yellow. Antennæ
scarcely two-thirds the length of body; the first joint subclavate, slightly
curved, shorter than the three following joints united, but distinctly
longer than joints 2 and 3 combined ; joints 2 and 4 subequal, longer
than the third, the latter being about three-fourths the length of the
second ; the fourth or last joint is fusiform. The legs in all my specimens
are broken, but are similar to those found in *Trepobates*, Uhler (= *Steph-
ania*, White), the middle legs being much the longest pair. The anterior
legs are very short, shorter than the body ; the femora, with their trochanters, being about as long as the tibiæ and tarsi combined ; the tarsi,
consisting of only a single joint, being a little longer than half the length
of tibiæ ; middle legs very long, their femora alone being as long or even
longer than the body, the tibiæ being fully one and a half times as long
as the femora, the tarsi about half the length of tibiæ. The hind legs in
all my specimens are broken, but the femora, which alone remain, are
much slenderer and considerably longer than those of the middle pair.

Hab.—Sherbrooke, P. Q., Canada. Dedicated to Abbé P. A. Bégin,
the discoverer of the species and a most valued correspondent.

*Ann. Musée Zool. l'Acad. Imp. des Sci. de St. Petersburg. 1896, p. 70.

MAMESTRA CIRCUMCINCTA, SMITH.

BY JOHN B. SMITH, SC. D., NEW BRUNSWICK, N. J.

The above species was described by me in the Proceedings of the U. S. National Museum, Vol. XIV., page 253, in my revision of the genus *Mamestra*. Recently Mr. Grote has questioned the distinctness of this species from *olivacea*. I could hardly credit this as being serious, and barely referred to the matter in the September, 1896, number of the CANADIAN ENTOMOLOGIST, page 240. In the December number, page 301, Mr. Grote returns to this subject, and again suggests that *circumcincta* may be either *olivacea* or *comis*. He refers to the fact that the description resembles that of both the species cited by him, and brings in Mr. Beutenmüller to testify to the fact that my species closely resembles *olivacea*. Mr. Beutenmüller is not a specialist in the Noctuidæ, and not entitled to an opinion that would carry decisive weight. Furthermore, it was not fair to Mr. Beutenmüller to ask him to make the comparison without first referring him to my description. Mr. Grote speaks as if the statement that *circumcincta*, or its description rather — for he has never seen the species — resembles *olivacea* was an important one and a discovery of his own. He does not refer to the fact that in my description I say: "the male resembling *olivacea* so strongly that I compared it closely at first, expecting a variety of this protean form." It seems to me it would be impossible to state more definitely the fact that I recognized the very close resemblance, superficially, between the species newly described by me and the very variable one long ago characterized. Mr. Grote also omits entirely the fact that the last sentence in the description and my comment on it reads: "The sexual characters, however, refer the species to the *renigera* group." On plate X., accompanying my paper, I delineate the sexual structures of *circumcincta* at figure 52, and of *olivacea* at figure 53. The two are so utterly different that it is simply impossible that one type should be a modification of the other. My species is, therefore, based upon a structural character primarily, and after that upon colour and markings. Now, if Mr. Grote will claim that these structural characters are not of specific value, then the question of whether my species may be *olivacea* is open for discussion. Until he takes this stand, these two species cannot be compared for a moment whatever their superficial similarity may be. I have asserted time and again that differences in sexual structure invariably indicate differences in species. Many other Entomologists have taken the same stand. Mr.

Grote has not, so far as I know, taken any stand in the matter, except so far as to deny the value of these characters for generic separation. If he is willing to assert that these structures have no specific value, then the question is an open one ; but I submit that to bring the matter before the readers of the CANADIAN ENTOMOLOGIST, as if there was a mere matter of colour and marking to be considered, is neither scientific nor honest. Before suggesting the identity of the two species he should have referred to the fact that I recognized their superficial resemblance, and separated them upon a distinct structural character.

One other point in Mr. Grote's paper is worth noting. In the matter of *Agrotis crassa*, Mr. Grote excuses his failure to recognize the true character of the frontal structure by stating that neither he nor the Museum with which he is connected possesses a microscope. He does not distinctly say so, but it would seem as if neither did they possess an ordinary hand lens of from $\frac{1}{2}$ to $\frac{3}{4}$ inch focal length, which is all that is necessary to recognize external structures of Noctuid moths serving for the division of genera. If not even the simplest and most necessary appliances for study are at hand, is any man justified in making assertions on points concerning which he cannot have any possible certainty? But even without the optical assistance to which I have referred, surely either Mr. Grote or the Institution at Hildesheim has in its possession a little camel's-hair brush, and with this, or even the frayed end of an ordinary wooden toothpick, the scales from the front can be sufficiently removed to enable one to recognize the frontal structure with the unassisted eye. One who makes assertions as to structure, should at least take every means within his power to make certain that they are accurate. Mr. Grote evidently has not done this, and in every assertion that he has made, concerning the identity of genera in this *Feltia* matter, I have proved him wrong. To escape from the necessity of considering his genus *Carneades* a synonym of *Agronoma*, he seems now to be willing to recognize the distinctness of the division that I have called *Porosagrotis*, basing it, however, upon the fact that the antennæ in the typical species are pectinated. This he considers a good generic character, differing in that point from all the authors who have written on this genus. Unfortunately, the genus *Carneades* contains species with antennæ pectinated and antennæ serrated, and so also does the genus that I have called *Porosagrotis*. There is no line of distinct demarcation between these two types of antennal structure, so that I could not utilize them even for divisions within the genus. The ordinary type of antenna in *Caruades* is what Mr. Grote has called brush-like, and consists of joints with more or less marked lateral projections, bearing on all sides stiff, bristly hair. It is the form that is called "bristle-tufted" by other authors. The lateral projections vary in size in the species, and when they become evident to the naked eye the antenna is called pectinated. The pectinations may be long or short, and the distinction between a shortly-pectinated antenna and one that is merely "brush-like" is entirely a matter for the individual judgment of the author who uses the term, as the two forms grade into one another imperceptibly. Mr. Grote cannot escape either admitting that the sexual character that I have made use of to separate *Porosagrotis* is a good one for the generic purposes or admitting that *Agronoma* must supercede *Caruades*. It does not make any difference to me which he chooses, because it does not distress me, as Mr. Grote says it does him, to have any name proposed by me relegated into the synonym, whenever there is scientific cause for it set forth by one whose methods of work and accuracy of research entitle him to the confidence of those for whom he writes.

MONODONTOMERUS IN APPALACHIA.

BY W. H. PATTON, HARTFORD, CONN.

MONODONTOMERUS STIGMA (Fabr.).
M. viridæneus, Prov., Canada.

Common in New England. In the District of Columbia I have reared it from the cell of *Melitoma euglossoides*, var. *taurea*, Say.

The genus *Oligosthenus* cannot remain separated, the fine dentitions of hind femora being more or less indistinct.

A frequent variety has no cloud about stigma. The abdomen varies in the amount of purple.

A *male* taken by me at Hartford, Conn., Aug., 1895, differs decidedly from the male of *M. montivagus*, Ashm., described by Mr. Cockerell in the CAN. ENT., XXVIII, 127, May, 1896. My male measures 3 mm. in length. It has no cloud about stigma; the abdomen is purple, except apex and most of the first segment. The scape is slender, as in the female; the flagellum is as in the female. Hind coxæ and femora much more swollen than in the female, tooth longer, no denticulations. The abdomen is short, broad; dorsum flat, shining. The descriptions of the females do not differ specifically.

THE COLEOPTERA OF CANADA.

BY H. F. WICKHAM, IOWA CITY, IOWA.

XXI. THE CHRYSOMELIDÆ OF ONTARIO AND QUEBEC — (*Concluded*).

Tribe X.—HISPINI.

The form alone of these little beetles is amply sufficient for their separation from the other tribes of Chrysomelidæ. They are more or less wedge-shaped, the elytra often broadly and squarely truncate behind and with rows of deep punctures, sometimes costate as well. Only two of the North America genera have been recorded from our territory, *Microrhopala*, with 8-jointed antennæ (owing to the fact that the last four joints are closely connate), and *Odontota*, in which the antennæ are 11-jointed. The middle tibiæ are straight in both of these genera.

MICRORHOPALA, Chevr.

A. Elytra with only eight series of punctures.

    b. Head usually red, thorax red, elytra blue-black with side margin and discal vitta red. .21–.25 in...... ......*vittata*, Fabr.

    bb. Head, thorax and elytra unicolorous (bluish, greenish or bronzed). Punctures of the outer rows of elytra larger than inner. .20 in ....*excavata*, Newm.

    Punctures of outer rows like those of the inner. .22–.25 in.................................*cyanea*, Say.

AA. Elytra with more than eight series of punctures on a part of their length, the fourth interval bearing four rows near the apex. Form more elongate. .12 in.................*porcata*, Mels.

ODONTOTA, Chevr.

A. Elytral punctures in ten rows; more or less distinctly costate.

    Elytra reddish or yellowish, with black sutural stripe. .24–.26 in (fig. 10)...... .....*dorsalis*, Thunb.

    Elytra blackish, humeri sometimes reddish.

        Body beneath black, thorax in part and humeri of elytra red. .22–.28 in.......*scapularis*, Oliv.

        Body beneath and thorax red, elytra black. .24 in....................... .. *bicolor*, Oliv.

FIG. 10.

    Elytra rosy or reddish yellowish, much broader at apex, and with serrate, explanate margin, the disc indistinctly marked with dark spaces. Under surface variable in colour, thorax coloured like the elytra. .24-.26 in...................*rubra*, Web.

AA. Elytral punctures in eight rows, costæ acute. Colour variable, usually with head dark, thorax and elytra pale with dark spots of irregular shape. .15 in................... ..*nervosa*, Panz.

Tribe XI.—CASSIDINI.

These are the "tortoise beetles" or "helmet beetles" found on morning glories and other convolvulaceæ. They are easily recognized on account of the peculiar form, which is circular or elliptical in outline, the upper surface convex, the margins of elytra and thorax explanate (to a varying degree), the head concealed. Some of them, notable *Coptocycla aurichalcea*, which, with its larva, is often abundant on the morning glory, are of most brilliant golden and greenish tints when alive ; these, however, being lost at or after death. The three genera found in Canada are as follows :

Size large (.38–.46 in.), form more elliptical.

Head partially exposed, thorax and elytra spotted.. *Chelymorpha*.

Head entirely covered, thorax spotted, elytra plain....*Physonota*.

Size small (.20–.30 in.), head entirely covered, antennæ longer than thorax...... ....*Coptocycla*.

COPTOCYCLA, Chevr.

Three species are recorded, one of which, *C. clavata*, Fabr., is easily known by its size (.30 in.), the brown elytra, which are roughened and gibbous, and the transparent spot on the middle of the outer margin. It occurs on the "ground cherry." The others have the elytra nearly even without gibbosities, and are closely allied. Mr. Crotch separates them by the fact that in *aurichalcea*, Fabr., the body beneath and the last four joints of the antennæ are black, while in *guttata*, Oliv., the sides of the body beneath are reddish and the last two joints of the antennæ are black. Both are of about the same size, a trifle under a quarter of an inch in length.

PHYSONOTA, Boh.

A rather large insect of a greenish or pale yellow colour, the thorax spotted, the principal and most constant spot being a large one near the middle. Two others are usually present near the base. Elytra not maculate. It is described by Say as *P. unipunctata.*

CHELYMORPHA, Chevr.

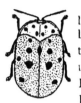

Represented by *C. argus*, Licht., of the size of the preceding species (.36–.48 in.), yellowish or reddish above, black beneath. Thorax with four black spots in a curved transverse row, behind which are often two others. Elytra usually with six black spots on each, arranged as shown in Fig. 11, and a common spot just posterior to the scutellum. Legs usually black. The prosternum is rather deeply longitudinally grooved and produced in front.

Fig. 11.

The following bibliography gives the names of the principal papers on the North American Chrysomelidæ; a few short articles have been omitted to economize space, since the genera have been treated in the more extended papers cited.

1845–1848.—Lacordaire, Th. Monographie des Coléoptères subpentamères de la famille des Phytophages. 2 Vols. Mem. Soc. Roy. Liege, Vols. III. and V.

1849.—Haldeman, S. S. Cryptocephalinorum Borealis Americæ diagnoses. Jour. Acad. Nat. Sci., N. S., I.; Proc. Acad. Nat. Sci., IV.

1852–1854.—Suffrian, E. Monographie und kritisches Verzeichniss der Nordamerikanischen Cryptocephaliden. Linnæa Entom.,VI. and VII.

1856.—Rogers, W. F. Synopsis of the species of Chrysomela and allied genera inhabiting the United States. Proc. Acad. Nat. Sci., VIII.

1862–1865.—Stal, C. Monographie des Chrysomelides de l'Amérique. Upsal.

1865.—Leconte, J. L. On the species of Galeruca and allied genera inhabiting North America. Proc. Acad. Nat. Sci., Philadelphia.

1866.—Leconte, J. L. Practical Entom., II., p. 9 [Prasocuris].

1873.—Crotch, G. R. Material for the study of the Phytophaga of the United States. Proc. Acad. Nat. Sci., Philadelphia.

1880.—Leconte, J. L. Short studies of North American Coleoptera. Tr. Am. Ent. Soc., VIII. [Cryptocephalini].

1883.—Horn, Geo. H. Chrysomelidæ, Hispini. Miscellaneous notes and short studies of N. A. Coleoptera. Trans. Am. Ent. Soc., X.

1889.—Horn, Geo. H. A synopsis of the Halticini of Boreal America. Tr. Am. Ent. Soc., XVI.

1891.—Leng, Chas. W. Revision of the Donaciæ of Boreal America. Tr. Am. Ent. Soc., XVIII.

1892.—Horn, Geo. H. Studies in Chrysomelidæ. Tr. Am. Ent. Soc., XIX.

1892.—Horn, Geo. H. The Eumolpini of Boreal America. Tr. Am. Ent. Soc., XIX.

1893.—Horn, Geo. H. The Galerucini of Boreal America. Tr. Am. Ent. Soc., XX.

1896.—Linell, M. L. A short review of the Chrysomelas of North America. Jour. N. Y. Ent. Soc., IV.

Since the note on the genus *Zeugophora* was printed (on p. 73 of the previous volume) two other species have been received from Mr. R. J. Crew, of Toronto : *Z. Kirbyi*, Baly (*Reineckei*, Grote), which is uniformly yellowish above, and *Z. scutellaris*, Suffr., in which the head and thorax are entirely yellow, while the elytra are black, with large punctures, separated by more than their own diameters. Collectors should be on the lookout for *Z. consanguinea*, Cr., which differs from *scutellaris* in having the occiput black, while the elytral punctures are close. It is known to me from Wisconsin, Illinois, and Manitoba.

Attention should be called to a clerical error in the table of *Chrysomela*. The name *labyrinthica* should read *pnirsa*. Dr. Leconte is said to have distributed it under the manuscript name of *labyrinthica*, and in thinking of it by this characteristic cognomen the error was committed.

ON THE MEXICAN BEES OF THE GENUS AUGOCHLORA.

BY CHARLES ROBERTSON, CARLINVILLE, ILLINOIS.

In the Transactions of the American Entomological Society, XX., 147, after notes and descriptions of five species of *Augochlora*, I gave the following note : "All of the species of *Augochlora* mentioned above agree in having the hind spur serrate with numerous fine teeth, and form a distinct section of the genus. Another section, to which belong *A. lucidula*, Sm., *A. sumptuosa*, Sm., and *A. humeralis*, Pttn., is characterized by having this spur provided with four or five long teeth."

In the January number of this journal, XXIX., 4–6, Prof. Cockerell makes use of these distinctions—under more obscure terms, however—and has given special names to these sections, and that, too, without referring to my note. I have no objections to his giving names to the sections, however, for I have had ample opportunity to do so, if I had desired. My note was intended for the use of students of these insects, and was given to call attention to the form of the hind spur, the importance of which was not indicated in the descriptions because all of these had the spur of the same form.

It is nothing new to me to hear that the males of *A. viridula* and *A. fervida* have the hind spurs different from the females. Indeed, I have never supposed that the spurs of the males of *Halictus* and *Augochlora* presented any important characters, though, as a rule, I have mentioned the form of the hind spurs in the descriptions of the females.

In Trans. Am. Ent. Soc., XXII., 118, I indicated *A. lucidula*, Sm., as a synonym of *A. viridula*, Sm. I intended to confirm Patton's view that the former was the female of the latter, and cited the place where he had expressed it. As regards the synonymy of *A. fervida*, Sm., and *A. humeralis*, Pttn., the description of the male of Patton's species is the only thing which leaves any doubt in my mind. I think they are the same, however. Two of my specimens have the tarsi pale testaceous, while a third has all except the basal joint dark, seeming to connect the typical *A. fervida* ♂ with the male described by Patton. I have no doubt at all about what I have identified as *A. humeralis* being the female of *A. fervida*, and that is all I have said about it.

The females of the first division do not have the spurs "ciliate or simple," but serrate with numerous fine teeth. The spurs are to be distinguished mainly by the number and length of the teeth, a fact which is obscured by the terms "ciliate" and "pectinate." The females of the second group have the spurs with only four or five long teeth.

It is one thing to use these characters in separating the species, and quite another to found named sections upon them before it is shown that they are valid indications of relationship throughout the genus. If we assume that *Augochlora* is a genus distinct from *Halictus*, or even a natural section of that genus, we must admit that the form of the hind spur is a case of parallel modification, and no proof of affinity. Otherwise, we must subdivide each genus and rearrange the species according to the form of the spur. In *Halictus*, I am satisfied that some species with few-toothed spurs are more closely related to species with finely serrate ones than to some species whose spurs are more like their own. Judging from analogy, we may expect to find the same thing in *Augochlora*.

NEW FORMS OF OSMIA FROM NEW MEXICO.

BY T. D. A. COCKERELL, MESILLA, N. M.

Osmia prunorum, n. sp.

♀.—Length, 9 mm.; shining dark greenish-blue, densely punctured with pale ochreous pubescence. Head subquadrate, face and front so densely punctured as to be cancellate; pubescence thin except on occiput ; clypeus punctured just like the front, with no central keel, the anterior margin broadly dark purple, the edge straight and entire, two converging brushes of orange hair projecting from beneath it. Mandibles with the two lower teeth long and pointed. Antennæ rather short, flagellum only feebly brownish beneath. Thorax very closely punctured, not very densely hairy ; basal triangle of metathorax minutely granular, its extreme base minutely longitudinally plicate. Tegulæ black, shining, sparsely punctured. Wings hyaline, faintly dusky beyond the nervures, nervures black. Legs black, with pale brownish or grayish pubescence, rufescent on inner sides of basal joints of tarsi ; hind femora quite broad at ends, basal joint of hind tarsi quite stout. Abdomen short, suboval, convex, shining, strongly but not very closely punctured, first joint covered with sparse long pale ochreous pubescence ; remaining joints with a sericeous pile, only noticeable in certain lights, when it will take more or less the appearance of bands. Apex with snow-white hairs. Ventral scopa black in middle and yellowish-white at sides.

♂.—A little larger ; face and clypeus densely covered with silky white pubescence ; pubescence of thorax a deeper ochreous, especially on scutellum. Antennæ long, flagellum rufous beneath. Colour of head and thorax a decided olive green. Wings not dusky beyond the nervures. Pubescence of last four legs sparse and black. Middle tarsi ordinary. Pile of second and third abdominal segments pale ochreous, that of the following segments black except narrowly along hind margin of fourth. Sixth segment with a shallow median depression ; its hind margin with a very distinct rounded emargination. Apex with two long black spines.

Hab.—Mesilla Valley, N. M.; 3 ♀, 1 ♂ at flowers of plum, College Farm, April 9 (Ckll.); 4 ♀, 1 ♂ at flowers of *Sisymbrium*, College Farm, April 12th (Ckll.). Resembles *O. distincta*, but easily known by the bicoloured ventral scopa. The ♂ seems to resemble that of *proxima*, which I have not seen. This species is apparently referable to the subgenus *Chalcosmia*, Schmeid.

Osmia phenax, n. sp.

♀.—Length, 9 mm. This so closely resembles *prunorum* that I had confounded it with it. It differs in the following particulars : Head and thorax olive green, clypeus strongly purple on the disc. Flagellum ferruginous beneath. Pubescence somewhat thinner, and entirely rather dull white. Tegulæ shining rufotestaceous. Wings faintly dusky all over. Abdomen duller, olive green with faint purple tints, punctures larger and closer. Ventral scopa thin and short, pale fulvo-ochreous, uniform in colour. Small joints of tarsi more or less rufescent.

Hab.—Mesilla, N. M., at flowers of honeysuckle, April 13, 1895 (Miss J. E. Casad). Also one taken some time ago at Las Cruces, by Prof. Townsend. Easily known by the colour of the tegulæ, which is very unusual for *Osmia*. A specimen was compared by Mr. Fox with the Cresson collection, and returned with the note : " Near *distincta*, colour paler, and wings clear throughout, tegulæ testaceous, punctures of dorsulum stronger."

Osmia cerasi, n. sp. or var.

♀.—Length, 9½ mm.; stoutly built, very dark blue, greenish on vertex and dorsum of thorax, purplish on clypeus. Pleura sometimes black. This agrees with Cresson's description of *O. densa* in almost every particular, and may be only a southern variety of it ; but it has the pubescence of the occiput and thorax above bright rust-red, as Cresson describes for *rustica*. The thorax is distinctly green anteriorly. The apical margins of the abdominal segments are dark blue, concolorous with the rest. Pubescence of pleura and face entirely black ; ventral scopa black. Tegulæ black. Pubescence of abdomen short, black, except that on first segment, which is longer and pale fulvous. The punctures of the head and thorax are large, and about as close as it is possible for them to be ; those of the abdomen are also close. Legs with black hairs.

Hab.—Mesilla, N. M., on flowers of cherry, April 14th, 1895 (Miss J. E. Casad); College Farm, Mesilla Valley, April 9th, 1895, on flowers of plum (Miss J. E. Casad). Also one taken at Las Cruces by Miss Agnes Williams (now Mrs. Herbert). The above three are all the species of *Osmia* observed in the Mesilla Valley.

NOTES ON EUPOEYA AND THE MEGALOPYGIDÆ.

BY HARRISON G. DYAR, NEW YORK.

I have had occasion to refer three times in these pages to the genus Eupoeya, placing it, with some doubt, in the Megalopygidæ. Very recently I have been so fortunate as to discover the larva in Florida on the mangrove. It is a true Eucleid, contrary to my expectation, but in confirmation of Dr. Packard's original statements. This genus, then, removed from the Megalopygidæ, renders it possible to define the family by the branching of vein 1 of primaries, instead of by the pectinations of the antennæ to the tip, which proves to be a fallacious character.

Megalopygidæ.

If the family be defined on this character, it appears unfortunate that Aurivillius does not refer to it, nor show that part of the wing in the figures in his recent paper on the group. Aurivillius would place the African genera Somabrachys and Psycharium in the Megalopygidæ, which is interesting, if well founded, as extending the geographical distribution of the family to the Old World. (Iris, Dresden, VII., 189, 1894.)

In Can. Ent., XXVII., 244 (1895), I referred eight genera to this family. Eupoeya may now be omitted, but *Alimera bicolor*, Möschl., may probably be added. Recently Grote doubted (Can. Ent., XXVII., 136) the correctness of Berg's union of Lagoa with Megalopyge. Möschler had previously expressed the same opinion (Abh. Senek. Naturf. Gesell., XVI., 122) and stated that *nuda*, the type of Megalopyge, differs in antennal characters. " Die Fühler von *nuda* sind kurz, kaum halb so lang wie die Vorderflügel, beim ♂ an der Spitze äusserst kurz gekämmt, während dieselben bei *crispats* länger als der halbe Vorderflügel, stärker und bis am Ende gekämmt sind."

If we accept these characters as diagnostic of the two genera, our species separate as follows :

Genus MEGALOPYGE : contains *nuda* (type), *lanata* and *opercularis*.

Genus LAGOA : contains *crispata* (type) and *pyxidifera*.

The larval characters confirm us in dividing our species into two genera, since the larva of *opercularis* has the hair crested and curled and is furnished with a terminal tail-like tuft, while those of *crispata* and *pyxidifera* are evenly and smoothly haired.

Grote states that Lagoa is preoccupied, but I do not find this to be the case in Scudder's Nomenclator. Pimela, Clem. is preoccupied by Pimelia, Fab. (Coleop.)

The genera of the Megalopygidæ at present are as follows :

Aidos Hübn., Carama Walk., Mesocia Hübn., Podalia Walk., Och-rosoma H.–S., Sciathos Walk., Alimera Möschl., Megalopyge Hübn., Lagoa Harr., Sombrachys Kirb. (?) and Psycharium H.–S. (?)

The larva of Eupocya.

The larva of *E. Slossoniæ* is flattened, green, with four dorsal red dots and fringed with a row of regular hairy appendages. They represent the subdorsal row, are detachable and furnished with heart-shaped basal pieces. There are no stinging spines. The form represents the same special adaptation as in Sisyrosea, but superimposed upon the phylo-genetic characters of Phobetron. Our larva is a green Phobetron with all the appendages of the same length and the lateral tubercles atrophied.

Dr. Packard states that Eupoeya is not the Cuban *Phryne immacu-lata*, Grote, but he has neglected to compare the forms listed as *Euproctis argentiflua*, Hübn., *E. fumosa*, Grt., and *E. pygmæa*, Grt., all from Cuba and one of which at least is a Eucleid as shown by Dewitz. (N. act. k. Leop.–Car. Deut. akad. nat., XLIV., 252).

It is curious that the Florida and Cuba forms of Eupoeya should be different species, while the recent description of a third form from Jamaica, by Schaus (Journ. N. Y. Ent. Soc. IV., 57), emphasizes this fact and renders it probable that still others will be found on other islands, possibly all mangrove feeders in the larval state.

FURTHER NOTES ON AUGOCHLORA.

BY T. D. A. COCKERELL, N. M. AGR. EXP. STA.

A portion of my table of Mexican species, on p. 4, should be amended to read as follows : —

5. Hind margins of abdominal segments broadly black, blue-green or more or less purplish-tinted species.

   (i.) Legs black ; only the coxæ, and front femora behind, me-tallic .... *Townsendi*, n. sp.

   (ii.) Legs metallic, blue or green ; nervures fuscous.

      a. Smaller, largely purplish, species .... *labrosa*, Say.

      b. Larger, green species, 5th abdominal segment basally purple .... : ...... *Binghami*, Ckll,

Hind margins of abdominal segments narrowly or not black, yellowish-green species...6.

Augochlora Robertsoni, n. sp.

This species had apparently been confounded with *pura*, but Mr. Robertson, who takes it commonly in Illinois, has pointed out good distinctive characters in Tr. Am. Ent. Soc., XX. (1893), p. 146, under the name of *labrosa*, Say. I possess a ♀ specimen from S. Illinois, sent by Mr. Robertson, and accepting his identification of it, had placed *labrosa* in my table of Mexican *Augochlora*, from the characters it presented. Say described his *labrosa* from Mexico, however, and suspecting later an error in identification, I compared Say's description. The result is, that I am convinced that Say did not have Mr. Robertson's Illinois insect before him, and that the latter stands in need of a name, being apparently different from other described North American species. It is accordingly named after the writer who first pointed out its characters, which are, principally, the evenly punctured, not roughened, mesothorax, the broad face and deep emargination of the eyes, in the female ; and the fourth ventral segment not broadly emarginate in the male. The stigma and nervures are brown, not very dark, the second submarginal cell is conspicuously longer (squarer) than in *pura ;* the legs are very dark brown, the front femora metallic behind. In most respects, the insect is like *pura*, and could easily be confused with it.

Say's type of *labrosa* is said to be a ♀ , while the allied *Binghami* is described from a ♂, but I do not think they can be the sexes of one species.

Augochlora Townsendi, n. sp.— ♂. Length, 10 mm.; head and thorax densely and confluently punctured, brilliant blue-green, pleura becoming very strongly tinted with blue ; but the face, especially the clypeus and supraclypeal area, yellowish-green, the latter with a coppery tint. Abdomen dark blue-green, not so blue as the thorax ; hind margin of first segment narrowly, and of the others broadly, black ; venter black, none of the segments emarginate, nor any trace of the fish-tail brush of *Binghami*. Face broad, emargination of eyes deep; clypeus shining, sub-cancellate with large punctures, its anterior edge very narrowly at sides, and the labrum, black ; labrum striate, mandibles dark, only very faintly rufescent about the middle ; antennæ reaching to scutellum, black, feebly rufescent beneath, not at all hooked at tip, first two joints of flagellum

broader than long, the first a little the shorter; third about as broad as
long. Prothorax with a very strong keel running to tubercles : mesothorax evenly and very closely punctured ; scutellum very finely and
closely punctured at the sides, the disc with a pair of small smooth
sublateral areas, a yellower green than the surrounding parts ; postscutellum very minutely punctured in the middle, coarsely subreticulate
at sides ; metathoracic enclosure distinct, shining, very blue, with numerous longitudinal ridges ; sides of metathorax and the ill-defined truncation
very closely punctured. Pubescence of head and thorax scant and pale,
rather conspicuous on upper part of face, the hairs beautifully plumose.
Tegulæ piceous, the outer edge hyaline, the base greenish and with
minute punctures. Wings dusky hyaline, stigma dark brown, nervures
piceous, second submarginal cell much higher than long. Legs black,
with thin white pubescence, coxæ in front, and anterior femora behind,
metallic blue-green ; anterior tibiæ in front, and anterior tarsi, rufescent,
remaining tarsi more or less rufescent within ; hind spur of hind tibia
minutely ciliate. Abdomen with first segment having rather large,
tolerably close punctures, and a small purple spot on each side ; second
segment with the punctures conspicuously smaller and closer ; third with
them still smaller, and much feebler ; remaining segments with them
minute and feeble. No hair-bands, but short pubescence, shining
brilliant silvery in certain lights.

Hab.—San Rafael, Vera Cruze State, last of June; collected by Prof.
C. H. T. Townsend on plant No. 31, which Dr. Rose says is a *Cordia*,
probably *C. ferruginea.* The coloration of this beautiful insect is
singularly like that of some new species of *Voluccella* taken by Prof. Townsend at the same locality, especially in the effect of the pubescence and
metallic colours on the abdomen. It resembles somewhat *A. urania*,
Sm., and *A. feronia*, Sm., from Brazil. On the same flowers, at the same
time and place as *A. Townsendi*, Prof. Townsend took both sexes of a
lovely *Temnosoma*, either *T. smaragdinum* or possibly a new species,
since it seems to differ from Smith's description, being larger, the head
hardly quadrangular, the wings darker, etc., but it differs so little that it
will be advisable to call it *smaragdinum*, Sm., var., until comparison of
specimens can be made.

Plant No. 4 (see p. 6) on which *A. Binghami* was taken, has been
identified by Miss Vail as *Calopogonium cæruleum* (Benth.) Britt.

A NEW PYRALID.

BY MARY E. MURTFELDT, KIRKWOOD, MO.

Titanio helianthiales, n. sp.

Alar expanse 15 to 16 mm.

Head small, with long, rather bristly scales, of which it is easily denuded, the colours mingled dingy white and buff; labial palpi projecting, elongate triangular, densely scaled, of a buff colour, indistinctly margined with white; maxillary palpi not in evidence; tongue slender, naked, eyes globular, large, purplish brown; antennæ silvery white above, pale brown beneath, the joints distinct and clothed with very short pubescence. Thorax buff with white median line, patagia buff, bordered more or less distinctly with white. Abdomen clothed with buff or fulvous scales, with indistinct bands of white at base. Legs shading from pale brown femora to yellowish-white tarsi. Wings broad. Fore wings, ground colour of black, buff and white scales intermingled, ranging from dark to light in proportion to the number of white scales, which is variable; a narrow, rather indefinite, white streak extends longitudinally from the base of the wing near the inner margin to about the middle; a more distinct white area has its base on the costa in the apical third extending obliquely backward about half across the wing; a narrow white line curves around the outer margin, diverging quite widely from the latter near the apical and the outer angles, most distinct near the costa, where it very nearly touches the base of the costal fascia, to this succeeds a dark band and a second narrower white line followed by a fine black marginal line; fringes white, variegated with two dusky bands. Hind wings yellowish-white at base, shading to dusky toward the outer margin, near which is an obscure whitish band; fringes similarly marked to those on fore wings. Under side of fore wings rather dark, silvery gray, except along the inner margin, where it is almost white; near the apical angle is a light spot, larger and of oblong form in the ♂, small and round in the ♀. Described from two ♂s and two ♀s. The combination of colours gives to the eye the general impression of pale purplish-gray, or "lavender"— to employ a milliner's term — and there is considerable variation in pattern and proportion of the silvery white scales, which makes an exact description difficult.

The adolescent stages of this insect are peculiarly interesting. It is a true leaf-miner and, so far as I have been able to learn, the only member of its family as yet discovered to have that habit. It works between the cuticles of the leaves of the Russian sunflower and probably of other species of *Helianthus*.

The mine is large, translucent, of irregular shape, but covering an area of from two to two and one half square inches. The black, granular frass drops to the lower margin. The mine and included larva bear considerable resemblance, on a magnified scale, to those of some *Lithocolletis*. Full-grown larva, 15 to 16 mm. long, 3.5 to 4 mm. in diameter across middle segments, from which it tapers gradually toward either end. Form cylindrical, with rounded segments and deep incisions, giving it a sub-moniliform appearance. General colour whitish green, often with a rosy suffusion. Head small, broadly triangular, polished, faintly mottled, dark brown on the lobes, with dingy white, triangular face. The corneous, whitish-green collar has two large, glossy, brown spots covering the greater part of its surface; or, it might perhaps be better described as brown, with broad, pale anterior and lateral margins and medio-dorsal line. Each of the other segments has the usual arrangement of conspicuous, round, dark brown, piliferous spots, from which proceed very fine, short hairs.

The pupation is irregular. In some cases the mature larvæ desert their mines and inclose themselves in oval cocoonets on the surface of the ground, but as a rule they spin up within the mine, in a nidus of loosely-webbed frass, with an inner, more firmly woven cocoon immediately inclosing the pupæ. The latter are short, and thick, of a golden-yellow colour, without marked characteristics.

The imago appears in eight or ten days after the larva spins up.

The mines were discovered August 2nd, 1896, and in all probability were those of a second brood. Another series of mines was found on the sunflower leaves September 5th to 10th, the moths from which issued shortly and probably hibernated — no later mines appearing. I am indebted to my friend, Prof. Fernald, for the generic determination of this interesting species.

Mailed March 4th, 1897.

The Canadian Entomologist.

Vol. XXIX.　　　LONDON, APRIL, 1897.　　　No. 4.

SYNONYMICAL AND DESCRIPTIVE NOTES ON NORTH AMERICAN ORTHOPTERA.

BY SAMUEL H. SCUDDER, CAMBRIDGE, MASS.

In a review of N. A. Decticinæ (CAN. ENT., XXVI.), I referred (p. 180) an undescribed Pacific Coast species provisionally to Drymadusa, an Old World genus of which I had not then seen specimens. Direct comparison shows that it differs from that genus in the lack of a humeral sinus on the posterior border of the lateral lobes of the pronotum and in the great posterior extension of the pronotum. I propose for it the generic name Apote (a-, $\pi\sigma\tau\acute{\eta}$). The species, which may be called *A. notabilis*, is testaceous, tinged on the pronotum with olive-green, the abdomen fusco-testaceous, much and minutely marked with black and light testaceous, the tegmina abbreviate but attingent, testaceous with black veins. The length of the body is 37 mm.; of the ovipositor, 28 mm. Oregon.

We have, however, another genus of Decticinæ not given in my table, consisting of long-winged species still more nearly allied to Drymadusa, but separable from it by the slender form, much narrower head and fastigium, narrower tegmina and the less incrassate base of the hind femora, and by the presence of spines on both sides of the under surface of the fore femora, though these are inconspicuous on the outer side of one species. It may be called Capnobotes ($\kappa a\pi\nu o\beta\acute{o}\tau\eta s$) in allusion to the smoky aspect of the insects.

To this belong two species described by Thomas and referred to Locusta, and which I had not determined when I published my former paper. Prof. L. Bruner has kindly sent them to me, as well as two other species, one of them from Lower California. The three United States species may be separated by the following table :—

*a.*¹ Outer margin of fore femora distinctly spined beneath; metazona considerably elevated above the prozona, so that the pronotum is subselliform.

$b.^1$ Metazona abruptly elevated above the prozona ; anterior sulcus of pronotum very deeply impressed ; ovipositor much shorter than hind femora ; tegmina marked with pallid spots and streaks ....*fuliginosus*, Thom.

$b.^2$ Metazona gradually elevated above the prozona ; anterior sulcus of pronotum distinct but not deep; ovipositor longer than the hind femora ; tegmina nearly uniform in coloration...*Bruneri*, sp. nov.

$a.^2$ Outer margin of fore femora very faintly spined beneath ; metazona scarcely elevated above the prozona, and the pronotum not sub-sellate... ....*occidentalis*, Thom.

Fuliginosus was described by Thomas from a male from Arizona, and the specimen, a female, sent me by Bruner is from the same territory; Bruneri comes from California and was sent me by Professor Bruner ; occidentalis was originally described from California, and the specimens I have seen come from Nevada and Utah. The sub-family Locustinæ to which Thomas thought these species belonged has not been recognized in the New World.

On different occasions I have received from the extreme south-western part of the United States specimens of a large speckled Acridian belonging to a generic type of Eremobiini very distinct from any known and differing widely from either of the two known genera of this group found in our territory. It may be called Tytthotyle (τυτθός, τύλη). It has a general Oedipodid aspect, not unlike Anconia or Hadrotettix. The body is not depressed, and but little compressed ; the head is normal, with rather large and prominent eyes ; the intraocular space, as seen from above, is narrower than the width of the eyes ; the vertex is carinulate ; the fastigium of the vertex sulcate, distinctly declivent, passing by a scarcely interrupted curve into the frontal costa ; the latter is not very broad, contracted and sulcate just below the ocellus, then disappearing. The antennæ are slender and shorter than the pronotum, at least in the female. The pronotum narrows rapidly from behind forward, is feebly carinulate with blunt lateral rugæ or shoulders, the lateral lobes of equal width throughout ; the metazona is a little longer than the prozona, sub-acutangulate posteriorly ; the prozona is twice cut by transverse sulci, and is a little tumid on the disk. The tegmina and wings are fully developed and much longer than the body. The hind femora are scarcely compressed, of general Oedipodid form, merely carinate above ; the inner and outer calcaria of the hind tibiæ are subequal, and the arolea minute.

I know of but one somewhat variable species, described by Bruner as *Thrincus* (?) *maculatus*. Mr. Bruner has kindly sent me types of this for examination.

The tribe Thrincini has not been found in America. The second species which Bruner has referred doubtfully to Thrincus, viz., *T. aridus*, belongs to Heliastus, a genus of Oedipodini in near vicinity to the Thrincini. The species described by Thomas under the name *Thrincus californicus* also belongs to Heliastus.

Among the Oedipodini, Mestobregma Scudder and Trachyrhachis Scudder are synonymous and the former has priority.

In Psyche (vi. 265) I pointed out that my *Leprus ingens* from California belonged to a new generic type, for which I now propose the name Agymnastus (ἀγύμναστος) in allusion to its clumsy inactivity. It is most nearly allied to Leprus Sauss., but differs from it in its more bulky shape, due largely to the exceptional breadth of the mesosternum, its abbreviated organs of flight, which do not wholly conceal the abdomen when at rest, and the presence of a subcostal taenia reaching the base of the wings from the transverse fascia common to both genera; the posterior process of the pronotum also in rectangulate instead of rounded subacutangulate, and the intercalary vein of the tegmina is more or less obscure proximally and only a little nearer the median than the ulnar vein; the upper and lower carinæ of the hind femora, and especially the lower, are subfoliaceous.

One of the genera of our Tryxalinæ has been very much named. It was first described by me under the name Aulocara, males only of which were seen. Very shortly afterward I redescribed it, from the female only, as Oedocara. A few years ago Brunner renamed it Coloradella, and recently McNeill has given it the name Eremnus; Aulocara of course has priority, and the species on which it was founded proves, as Bruner has already pointed out, to be identical with Thomas's *Stauronotus Elliotti*. The genus under the name Oedocara was included by Saussure in the Oedipodinæ and by Brunner (as Coloradella) in the Tryxalinæ, an excellent illustration of the difficult definition of these two sub-families.

Some years ago, in Psyche, V., I attempted to show that certain genera that had been referred to Tryxalinæ should really be placed in the Oedipodinæ. I now think I was mistaken, at least as regards all the genera found in our own country, and would follow Brunner in placing them in the Tryxalinæ. It was partly owing to my statements that Mr. McNeill has rejected them from his recent Revision of the Tryxalinæ.

The generic name Beta, proposed by Brunner in 1893 for two unnamed species in his collection from Texas and Colorado, is proved by a specimen sent me by him to be the same as my Phlibostroma (1875). His Pseudostauronotus, proposed at the same time and manner, is identical, as a specimen sent me shows, with my Stirapleura.

A REMARKABLE APPEARANCE OF CATOCALA INSOLABILIS.

On Friday, June 6th, 1896, the first Catocalas were noticed in this locality for the season. Three *Insolabilis* were taken. The weather was hot—87° in the shade at 1 o'clock. The Saturday following was also hot, and Catocalas were abundant. During the forenoon twelve were taken on trees near the house. In the afternoon twenty-one more were taken on trees at some distance from the house, and in the evening, at sugar, twenty-three more were captured. Of the entire number (56) fifty were *Insolabilis*, one *Nurus*, three *Ilia*, one *Uxor*, and one *Marmorata*. Sunday the weather was still hot, and on the way to and from church Catocalas could be seen on nearly every tree. The wind continued south-west. On Monday the wind had changed to south-east, and the Catocalas were still present, but resting higher up on the trees. This being a work day, I had but little time for observation or collection. After school hours, however, a few minutes were spent in the woods, and the Catocalas were found hard to capture. When startled they would light high up in the trees, sometimes fully twenty feet from the ground, and some would even alight upon the leaves of the trees. At dusk *Insolabilis* came to the sugar in abundance, and thirty were taken before it was dark enough to need a lantern. In all, fifty-seven were taken on Monday, all but five being *Insolabilis*. On Tuesday the wind was north-west, and not a Catocala was to be seen. Not one came to sugar that evening. The only Catocala that was seen on Tuesday was snapped out of a tree by a scarlet Tanager and immediately torn to pieces.

I have talked with other collectors of this vicinity, and all seem to have secured a goodly share of *Insolabilis*.

In the parks and suburbs of Chicago there were literally thousands of *Insolabilis* during the three days. Previous to this remarkable flight the species was not common, so far as I have been able to ascertain.

ARTHUR J. SNYDER, North Evanston, Ill,

DESCRIPTION OF THE STRUCTURAL CHARACTERS OF THE LARVA OF SIBINE FUSCA, WITH NOTES ON THE FOUR KNOWN LARVÆ OF SIBINE.

BY HARRISON G. DYAR, NEW YORK.

Stoll figures the moth of two species of Sibine. He also figures two larvæ of Sibine, but, owing to the unfortunate confusion into which his labels must have fallen, they are not attributed to the right imagines, but to two species of Dioptidæ. After Stoll, Sepp also illustrated two species of Sibine, with their larvæ correctly shown. One of Sepp's species is the same as one of Stoll's, the other is different in both larva and moth. This gives three species of the genus known in both larval and mature states, assuming only that the larva which Stoll figures as *micilia* (228 G.) really belongs to the moth *nesea**, which I think is probably the case.

The names of the species are *nesea*, Stoll ; *fusca*, Stoll (= *trimacula*, Sepp = *bonaerensis*, Berg. = ? *megasomoides*, Walker = ? *affinis*, Moeschler), and *vidua*, Sepp (= ? *fumosa*, Walk.). As a fourth species we have *stimulea*, Clemens (= *ephippiatus*, Harr.).

The larvæ have in common the following characters : (1) The shape of the body, which may be sufficiently described by a reference to the well-known *S. stimulea;* (2) the absence of subdorsal horns which bear stinging spines on joints 6 to 10 ; (3) the presence of a large patch of detachable spines above the horn on joint 13 and the lateral horn of joint 12 ; (4) probably the presence of skin spinules only without granules, though this can not be definitely asserted till the two species *nesea* and *vidua* have been microscopically examined ; (5) the coloration involves a square green patch on the middle of the back, variously modified. Other characters are shared by the whole group of spined Eucleids.

SYNOPSIS OF THE LARVÆ.

The subdorsal horns which are present, long.

Lateral horns long ; green, the horns all purple-brown, dorsal mark square, dark green, broadly edged with yellow..............*vidua*.

Lateral horns short.

•   Subdorsal horns and body green ; dorsal mark square, without a central dark patch, edged before and behind with yellow....*nesea*.

Subdorsal horns and body purple-brown ; dorsal mark elongate, projected below the posterior subdorsal horns, and bearing a central, elliptical purple-brown patch edged with white..*stimulea*.

---

* Stoll's so-called larva of *nesea* is an absurd error. It is a Notodontian with a long yellow horn on joint 6.

The subdorsal horns short; green, the dorsal mark much elongated, reaching the posterior end of the body and projected forward below the anterior subdorsal horns, edged with yellow... ....*fusca.*

The larva of *fusca* is evidently the most highly specialized. I have received a number of alcoholic specimens from Mr. G. Ruscheweyh, of Buenos Ayres, Argentina, under the name "*Streblota bonaerensis,*" but I am unable to find any differences in either moth or larva from Sepp's figures. The coloration is largely lost in my material through the effect of the alcohol, but the outline separating the two shades of green can easily be traced, and is exactly as shown by Sepp and Stoll.

Larva.—As compared with *S. stimulea,* Clem., the body is of the same shape, or a little more flattened, but all the horns are short. Subdorsal horns present on joints 3 to 5, 11 to 13, about .5 mm. long, alike, bristly with stinging spines ; absent on joints 6 to 10. Lateral horns on joints 3, 4, 6 to 12, even shorter than the subdorsals, sessile spined. A subventral row of two distinct pale setæ.

Dorsum broad, flattened, sides oblique, subventral space small, contracted. Segmental incisures deep, the depressed spaces (1) dorsal intersegmental paired, two lateral (4) and (6) all show as distinct black dots buried in the intersegmental folds ; addorsal spots (2) also present, small. A large, elongate patch of detachable spines above the lateral horn on joint 12, and a smaller one above the horn of joint 13. Caltrop patches present on the bare tips of the lateral horns of joints 6 to 12 apparently, but nearly all the caltrops are lost in my specimens. The caltrops and spines correspond with those of *S. stimulea* (Journal N. Y. Ent. Soc., Vol. IV., plate 1, figs. 5 and 6). Skin not very finely spinulose, the bases of the spinules enlarged, approximating granules, but still bearing the sharp tips. Colour largely green, a line of dark spinules joining the subdorsal horns of joint 5 runs forward on each side below the subdorsal horn on joint 4, turns down behind the lateral horn of joint 3, and runs backward just above the row of lateral horns to joint 12, turns up over the subdorsal horn of 12, and joins its fellow again just above the horn on joint 13 ; a detached ring also surrounds the subdorsal horn of joint 11. This line evidently marks the joining of the dorsal green with a different tint, which obtains over the horns, the stigmatal region and the dorsum of joints 3 to 5. Thoracic feet and venter as usual ; the spiracle on joint 5 moved up above the others.

Habitat.—If my synonymy is correct, *Sibine fusca* ranges throughout the eastern part of South America, from Guayana to Argentina.

DESCRIPTION OF THE LARVA AND PUPA OF AULAX NABALI.

BY THOMAS W. FYLES, SOUTH QUEBEC.

The tall White Lettuce, *Nabalus altissimus*, Hooker, is a striking and graceful plant. At Quebec it is found in glades and on the edges of woodland roads. Its wand-like stems rise sometimes to the height of six feet, and end in panicles of greenish-white or pale straw-coloured flowers. The stems are hollow, but have a lining or inner coat of white downy pith, which in the summer is sometimes found to be broken with discoloured warts. Late in the fall, when the stems of the plant have become indurated and the pith has dried up, the warts are found to have developed into galls of the size, shape and colour of grains of hemp. I have found them in the stems from about six inches above the ground up to a height of three feet or perhaps more. Sometimes they appear in clusters, sometimes in rows, and sometimes singly at intervals. The proper inhabitant of each of these galls is a footless, spindle-shaped grub, one-eighth of an inch long. In colour it is like white wax, with the mouth organs brown. It is more pointed at the head than at the other extremity. It lies curled round in the gall.

Towards spring the pupal change takes place. This change may be hastened by warmth; the specimens I have kept in my study are now (January 9th) passing through it. A week or two after the change the pupa is of compact form, white, waxen, with amber-coloured eyes. The head is small, the thorax large and convex, and the abdomen ovate and closely joined to the preceding part. The legs are drawn up by the sides of the thorax, and the tarsi are stretched backwards under the body. The antennæ (beautifully translucent) are turned under the head and extended between the tarsi, reaching nearly to the end of the abdomen.

The perfect insects were described by Dr. Brodie, of Toronto, in the 25th volume of the CANADIAN ENTOMOLOGIST, p. 12. I copy his description for the benefit of those who may not have the volume at hand :

"♀.—Length, 2.50 xx. Antennæ 13-jointed ; uniform brown ; head "and thorax black ; abdomen shining brown, with a large anterior dorsal "spot black ; all the tibiæ, femora and tarsi brown, a little paler than the "abdomen ; wings ample, veins well-defined, hyaline, iridescent at certain "angles."

"Abdomen of ♂ darker brown, and without the dark dorsal spot. "From numerous specimens."

Dr. Brodie discovered the galls in great abundance at the roots of the White Lettuce. I have not yet found them at the roots of the plant, and I am inclined to think that the insects are less numerous at Quebec than Toronto.

A NEW SPECIES OF ANCYLOXYPHA.

BY G. H. FRENCH, CARBONDALE, ILL.

Ancyloxypha Longleyi, n. sp.

Female.—Expanse 1 inch. Fore wings with the costa more straight from the shoulder to near the apex than in *Numitor*, in this respect approaching *Thymelicus;* apex rounded, but less than in *Numitor;* outer margin and hind wing rounded, much as in *Numitor;* antennæ reaching but little more than one-third the distance to apex of fore wings ; palpi as in *Numitor*, but the third joint longer ; abdomen surpassing hind wings, but less so than in *Numitor;* the whole insect more robust than *Numitor*.

Fore wings brown, darker than in *Numitor*, without the discal yellow patch, emitting a pale blue sheen in reflected light ; a few yellow scales below the costa between the venules, and a few scattered on the base of the wing, but in either case not enough to give a yellow colour ; otherwise the wing is uniform brown. Hind wings marked and coloured as in *Numitor;* yellow, with outer and costal borders and base brown, the brown along internal margin running to a point before reaching anal angle.

Under side differing very little from the under side of *Numitor;* the dark central and posterior area of fore wings a little darker brown, the costal and outer margins yellow, the yellow running to a point before reaching the posterior angle. Hind wings uniform yellow.

Antennæ black, annulate with white; club black, tipped with brown— the club of *Numitor* is tipped with black ; palpi white at sides, black above, terminal joint black ; thorax concolorous with fore wings, abdomen concolorous with hind wings.

The above description is drawn from a single specimen taken at Ridgeland, near Chicago, September 6th, 1896, by Mr. W. E. Longley, in whose cabinet it is and after whom I have named the species. In describing the species I have compared the specimen with *Numitor* because that species is so common all over this portion of our country. I hope the Chicago collectors will be on the lookout for this species the coming season.

THE COLEOPTERA OF CANADA.

BY H. F. WICKHAM, IOWA CITY, IOWA.

XXII. THE CERAMBYCIDÆ OF ONTARIO AND QUEBEC.

The size and beauty of the Longhorns are in themselves sufficient to render them objects of interest to a beginner ; adding to this the fact of the great abundance of certain species and the destructive work of their larvæ, we can readily understand their importance to all who are in any way interested in Entomology, whether as a pleasant recreation for leisure hours or a serious pursuit for gain. Although usually easily recognized by sight, the family is, as stated by Dr. Leconte, almost impossible to define. The tarsi are apparently four-jointed, the fourth joint being very small and connate with the fifth. The antennæ are usually very long, especially in the males, filiform or serrate, often borne on large frontal tubercles. The eyes are frequently deeply emarginate. Tibial spurs are present. The larvæ are grub-like, living in burrows or chambers which they excavate for themselves in the woody tissues or in the pith of plants, the pupa resting in a cell constructed by the larva in its gallery.

The collector will obtain many species of this family by carefully beating branches (especially if partially dead) and flowers, over a sheet or an umbrella. Dead logs should be searched, on both the upper and lower surfaces, and particularly freshly-cut timber or sawed lumber. A morning spent in a wood yard will often repay one richly in rare specimens. Some are to be found commonly under bark and may be trapped by loosely fastening pieces of bark to a tree over night and examining the under side of bark in the morning. A great number fly to lights after dusk. Dead twigs and branches may be sawed or cut off, preferably during the autumn months, and kept in large boxes or in an empty room until the beetles are disclosed through the development of the larvæ contained therein. While the activity of the Canadian collectors has already resulted in the recording of a great number of species, there can be no doubt that others will reward the efforts of explorers of the more remote districts.

Although mostly of at least moderate size, and after once identified easily recognized again, their classification presents considerable trouble owing to the fact that structural characters are so unstable and consequently of less than usual value for the separation of large groups. In the main, the arrangement adopted is that presented in the Leconte and

Horn " Classification," though the tables are constructed on a different plan and on account of the limits of the fauna it has been possible to do away altogether with the use of certain characters difficult of observation.

The prothorax in the Longhorns offers two principal types : that in which the lateral edge is sharp or thin for almost or quite the whole length, more or less toothed, giving us the form called *margined*, and that where it is cylindrical or rounded on the sides, which may, however, be either spined, tuberculate or plain. Thus we have a point of departure for sub-family separation, which may be aided by taking into account, among those genera in which the thorax presents the second form, a study of the palpi. These may have the terminal joint more or less compressed or subtriangular as in the Cerambycinæ, or this joint may be cylindrical and pointed at tip as in the Lamiinæ. The front tibiæ in the latter group have an oblique sulcus or groove on the inner surface, not always very distinct, but to be seen without difficulty in the larger species like *Monohammus;* once seen it may be used with some facility elsewhere. In the Cerambycinæ this groove is wanting.

Following the Classification, we may, then, throw the characters into tabular form, separating three sub-families, thus :

Prothorax margined, antennæ not pubescent, labrum connate with the
    epistoma .... PRIONINÆ.
Prothorax not margined, labrum free.

    Front tibiæ not grooved ; last joint of palpi not acute at tip, often
        subtriangular.. .... CERAMBYCINÆ.
    Front tibiæ with an oblique groove on the inner side ; palpi with last
        joint cylindrical, pointed at tip.................. LAMIINÆ.

The Canadian species of the first sub-family, the Prioninæ, are but three in number and represent as many genera. All of them are of rather large size, brown colour, and with elytra of a leathery appearance. The genera may be distinguished thus :

Sides of prothorax two- or three-toothed.

    Form elongate, parallel ; antennæ more slender, joints not overlap-
        ping.. .... *Orthosoma.*
    Form stout ; antennæ heavy, joints overlapping, especially in the
        male ....*Prionus.*
Sides of prothorax with one tooth, antennæ slender......... *Tragosoma.*

ORTHOSOMA, Serv.

Represented by *O. brunneum*, Forst. (Fig. 12), a large brown insect, .88 to 1.60 in. long, the elytra nearly parallel-sided, shining and rather thickly punctured. Prothorax more coarsely sculptured above than the elytra, each side with three sharp teeth. The head bears a deep, sharp impression between the eyes. The basal antennal joints are stouter in the males than in the females. I have found the larvæ in rotten pine timbers under sidewalks.

FIG. 12.

PRIONUS, Geoff.

The largest Canadian Longhorn is *P. laticollis*, Drury (Fig. 13). It varies in length from .88 to 1.88 in., and is of a brownish or blackish colour, the prothorax almost or quite as broad as the base of the elytra, sides with three teeth, of which the posterior is sometimes poorly marked. The elytra are much broader at base than at apex. Antennæ twelve-jointed in both sexes, much heavier in the male. The larva (Fig. 14) is said to injure the grape, poplar, apple, and pine, by boring in the roots.

FIG. 13.

TRAGOSOMA, Serv.

T. Harrisii, Lec. (now considered by some writers as identical with the European *T. depsarium*, L.), is a curious-looking beetle of elongate form and brownish colour. The antennæ are slender, the prothorax small in comparison with the elytra, very hairy and armed on each side with a single sharp tooth, in front of which the lateral margins are convergent. The elytra are shining, distinctly punctured and

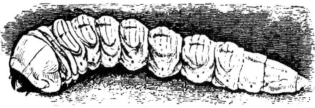

FIG. 14.

with numerous longitudinal raised lines. I have taken the species under pine bark. It varies in length from 1.20 to 1.40 in.

The next sub-family, the Cerambycinæ, is of great extent, and in consequence more difficulty is encountered in arranging the genera. In the use of the table considerable care must be exercised by those who are unfamiliar with the structure of the Longhorns. This is especially true of the first character involved, *i.e.*, the enveloping of the base of the antennæ by the eyes. In order to obtain a proper appreciation of this structure, the antennæ should be extended forward from the head: in this position it will be seen that in those genera where the " base of the antennæ is partially enveloped by the eyes " a line passing from the anterior or inner border of the upper lobe of the eye to a corresponding point on the lower lobe will pass through the antennal socket, whereas in the other genera this line would run behind the socket. Of course none of the genera in which the eyes are entire (*i. e.*, not emarginate) will belong to the former category, though those with emarginate eyes may belong to either. Comparisons of a few specimens ought to make this clear.* The remaining characters may be easily verified by careful examination of a few species the positions of which are already known to the student, and with these as a point of departure he should meet with no greater difficulty than is always to be expected in dealing with a group of large size, wherein colour and sculpture are inconstant and secondary sexual characters well marked. The following table is submitted for generic discrimination ; a short account of the method of using may be useful to some. Suppose on taking up our insect, which we have previously ascertained to belong to this sub-family, we examine the position of the base of the antennæ with regard to the eyes, since this is the first point of departure : ascertaining the antennal bases to be partly enveloped, we find ourselves referred to the number 12 at the end of the line. We now run down along the numbers at the *beginning* of the lines until we reach 12, which shows us where to recommence our analysis, with a scrutiny of the second antennal joint. Suppose we find this joint large, we are referred to the number 36, under which (on searching out its position at the beginning of a line) we are again confronted with a query, this time as to the relative proportion of the second joint to the fourth ; if these two joints are about

---

*Cases will, however, arise in which this point is in doubt. In such an event the choice will rest between the Callidioides and the Cerambycoides. The former have the second antennal joint larger (as a rule) than the latter, but I can find no hard and fast distinction which will serve the beginner as a sure test. A certain number of properly named specimens serving as a guide to tribal and generic facies is almost indispensable here. It should be stated that the table is based on the characters developed in the " Classification," but is intended to apply only to the Canadian fauna.

equal, our insect belongs to *Microclytus*. The generic sequence followed in succeeding pages is the same as that employed in the table and is hence slightly different from the Henshaw Check-list.

TABLE OF GENERA OF SUB-FAMILY CERAMBYCINÆ.

Base of antennæ not enveloped by the eyes ....2.

Base of antennæ partially enveloped by the eyes ....12.

2. Front coxæ transverse, not prominent (*Callidioides*)............3.
   Front coxæ conical, prominent (*Lepturoides*). .. ..37.

3. Eyes divided, apparently four in number...... .... *Tetropium.*
   Eyes not divided, often deeply emarginate................. .4.

4. Brown species, second antennal joint proportionately larger, often half as long as the third and sometimes twice as long as wide. Elytral costæ usually distinct ....5.
   Variously coloured, often ornate species, second antennal joint proportionately smaller, often much less than half the length of the third and never much longer than wide. Elytral costæ usually indistinct................................. ....6.

5. Eyes hairy, finely granulated .... *Asemum.*
   Eyes not hairy, coarsely granulated................. *Criocephalus.*

6. Elytra with narrow raised white lines, prothorax with very deep median groove, thighs strongly clubbed ....*Physocnemum.*
   Elytra without distinct raised white lines (traces are sometimes visible in *Merium*)...7.

7. Prothorax very short, strongly rounded on the sides. Upper surface entirely opaque, lustreless. Black, prothorax red....*Rhopalopus.*
   Prothorax not very short, the width not apparently much exceeding the length. Upper surface at least moderately shining........8.

8. Thighs more slender ; antennæ with the eleventh joint divided in the male. Colour above blackish, prothorax red........ *Gonocallus.*
   Thighs strongly clubbed, colour variable.....................9.

9. Anterior coxæ contiguous..... ....10.
   Anterior coxæ at least moderately distant................11.

10. Palpi unequal, the labial much the shorter...........*Phymatodes.*
    Palpi about equal................................. .... *Callidium.*

11. Dorsal surface of prothorax with narrow median and moderate or small lateral callosities........ ....*Hylotrupes.*
    Dorsal surface of prothorax with a very broad, smooth, shining median space, which bears a few large punctures. Elytra with more or less distinct raised lines of a yellowish or whitish colour.....*Merium.*

12. Second joint of antennæ large *(Cerambycoides)*...............36.
    Second antennal joint small.... ....13.
13. Eyes coarsely granulated.....................................14.
    Eyes finely granulated.........................:..............21.
14. Front coxal cavities open behind............................15.
    Front coxal cavities closed behind ; small pale species with the first
        abdominal segment very long............................20.
15. Scutellum acute, triangular, antennæ very long, prothorax with lateral
        spine... ...*Chion.*
    Scutellum rounded behind....................... ....16.
16. Elytra with elliptical elevated ivory-like spots, in pairs......*Eburia.*
    Elytra without raised ivory-like spots.........................17.
17. Femora not strongly clubbed, antennæ spinose...............18.
    Femora strongly clubbed.....................................19.
18. Large species ; metathoracic episterna narrower behind..*Romaleum.*
    Moderate-sized species, episterna parallel..............*Elaphidion.*
19. Antennæ bisulcate externally........................*Tylonotus.*
    Antennæ not sulcate nor hairy....................*Heterachthes.*
    Antennæ not sulcate but quite hairy..................*Gracilia.*
20. Prothorax much narrower at base than at apex............*Phyton.*
    Prothorax dilated at middle, but about equal at base and apex..*Obrium.*
21. Elytra either very short, not covering the abdomen, or rapidly narrow-
        ing behind and broadly dehiscent along the suture...........22.
    Elytra normal, not abbreviated nor notably dehiscent. ....23.
22. Elytra about as long as the prothorax................*Molorchus.*
    Elytra about twice as long as the prothorax......... .*Callimoxys.*
23. Scutellum rounded or (in *Cyllene*) broadly triangular...........24.
    Scutellum acutely triangular........ ....25.
24. Tibial spurs small, thighs suddenly and strongly clubbed. Form slen-
        der and cylindrical. Black, elytra and abdomen scarlet..*Ancylocera.*
    Tibial spurs large ....27.
25. Prothorax opaque, sides with spine or large tubercle ....26.
    Prothorax shining, sides unarmed ....*Batyle.*
26. Elytra coarsely punctate, sutural angle produced .....*Purpuricenus.*
27. Tibiæ strongly carinated, form slender. Elytra without narrow cross-
        bands of pubescence, punctuation sparse and coarse. Antennæ
        as long (♀) or longer (♂) than the body.........*Stenosphenus.*

Tibiæ not carinated, form usually stouter. Elytra in most cases with lighter coloured angulated cross-bands ; antennæ usually shorter than the body in both sexes. Punctuation fine.. .... 28.

28. Head comparatively small, front short, oblique ; legs hardly clubbed. Intercoxal process of first ventral rounded .... 29. Head large, front long, intercoxal process acute.............. 32.

29. Prothorax transversely excavated at sides near the base, prosternum perpendicular at tip......................... ..... *Cyllene.* Prothorax not excavated at sides, which are rounded and constricted at base. Prosternum declivous at tip.................. 30.

30. Antennæ filiform.. 31. Antennæ subserrate, compressed. Size large, colours strikingly contrasted with black and yellow bands.............. *Plagionotus.*

31. Large species, prothorax entirely black, much rounded on the sides. .... *Calloides.* Smaller, less robust ; prothorax with central black spot, the remainder clothed with g r a y pubescence, sides much less rounded .... *Arhopalus.*

32. Elytra plane ; moderate sized species...................... 33. Elytra gibbous at base ; small ant-like species .... 35.

33. Head with a carina of variable form .... *Xylotrechus.* Head not carinated....................................... 34.

34. Prothorax with transverse dorsal rugæ or ridges...... *Plagithmysus.* Prothorax without transverse ridges................. *Clytanthus.*

35. Elytra with a transversely oblique ivory-like band......... *Euderces.* Elytra without ivory band....................... *Cyrtophorus.*

36. Second joint of antennæ equal to the fourth.......... *Microclytus.* Second joint of antennæ less than half as long as the fourth.. *Atimia.*

37. Elytra short, not covering the wings................. *Necydalis.* Elytra normal .... 38.

38. Joints 3 to 5 of antennæ much thickened at their tips, inner angle sometimes much produced. Large insects, bright blue with an orange band across base of wing-covers .... *Desmocerus.* Joints 3 to 5 of antennæ normal, usually slender and never produced inwardly at tips. Elytra usually tapering to apex, sometimes more or less dehiscent....................................... 39.

39. Spurs of hind tibiæ terminal.......... .... :.......40. Spurs of hind tibiæ not terminal, but borne at the base of a deep excavation. Thorax tuberculate or spinose at sides..... *Toxotus.*

40. First joint of hind tarsi with the usual brush of hair beneath (except in certain *Acmœops*). Prothorax, with rare exceptions, distinctly tuberculate at sides or with heavy lateral spine. Head obliquely narrowed behind eyes..................................41.

First joint of hind tarsi without brush-like sole. Prothorax, with few exceptions, broadest at base, sides never spined and rarely tuberculate. Head suddenly constricted behind the eyes.........46.

41. Antennæ short, joints 5 to 11 wider. Prothorax with a heavy spine at sides, elytra strongly costate ....*Rhagium*.

Antennæ long or moderate, not thickened, elytra never strongly costate..42.

42. Eyes large or moderate. Thorax (except in *Pachyta monticola*) with sharp, strong, lateral spine................................43.

Eyes small, not emarginate, prothorax angulate or rounded on sides..45.

43. Eyes coarsely granulated, very prominent ; form of body parallel......... .. ..*Centrodera*.

Eyes finely granulated ; body narrowed behind.................44.

44. Eyes feebly emarginate............................ ..*Pachyta*.

Eyes more strongly emarginate........... ....*Anthophilax*.

45. Mesosternum not protuberant, body above more or less pubescent, sometimes moderately shining ....*Acmœops*.

Mesosternum protuberant, body above brilliant metallic green.................. ....*Gaurotes*.

46. Head constricted far behind the eyes, neck consequently very short. Form extremely slender, hardly tapering behind, prothorax with lateral tubercle ....*Encyclops*.

Head constricted close behind the eyes. Form variable, usually much narrowed behind, prothorax rarely bulging at sides and never with distinct tubercle..................................47.

47. Last ventral of male deeply excavated.......................48.

Last ventral of male not excavated........ ....49.

48. Antennæ without poriferous spaces, size large, sides of elytra deeply sinuate................................. ....*Bellamira*.

Antennæ with impressed poriferous spaces on sixth and following joints. Size moderate, sides of elytra sinuate, form very slender...... ....*Strangalia*.

49. Antennæ with poriferous spaces.....................*Typocerus*.

Antennæ without poriferous spaces....................*Leptura*.

THE LOST LEDRA AGAIN.

BY HERBERT OSBORN, AMES, IOWA.

The interesting note by Prof. Baker on *Ledra perdita (Centruchus perdita*, A. & S.) deserves notice on account of the mystery which it clears up, and it may also be worth while to add some testimony in the way of corroborative evidence.

A few weeks ago (Dec., '96) I had occasion to review the matter in an attempt to locate the *perdita*, and, in a critical examination of Amyot and Serville's figure and description, was struck by the resemblance to our common *Microcentrus caryæ*. On careful comparison, however, with this species and with the *Centruchus Liebeckii* of Goding, I concluded the figure and description must apply to the latter. It seemed so strange that a connection so obvious, when once seen, should have so long escaped the attention of Homopterists that I made a further search in the available literature, with the result of finding in a note by Dr. Goding, on "Fitch's Types of N. A. Membracidæ" (CANAD. ENT., Vol. XXV., p. 172), the statement that "No. 2152, labelled *Ledra perdita* and *capra*, Mels., is *Centruchus Liebeckii*, Godg." There is no comment to indicate that Dr. Fitch corrected the family reference from Ledridæ to Membracidæ, but considering his familiarity with the Homoptera in general, and the Membracidæ in particular, it is probable that he appreciated the full significance of his specific determination, and it is quite likely that his unpublished notes would show comments on this reference.

In any case, we have the testimony of Dr. Fitch in identifying his specimen as *Ledra perdita* and its recognition by Dr. Goding as *Centruchus* to confirm Prof. Baker's conclusion.

OCCURRENCE OF SCHISTOCERCA AMERICANA (DRURY) AT TORONTO.

Mr. C. T. Hills recently brought me a specimen of the large, handsome locust, *Schistocerca Americana*, Drury, which was taken about the 12th of October, 1896 (the exact date was not recorded), by Mr. H. Parish, while collecting at High Park. Mr. Parish found the insect resting on the trunk of a tree. The specimen is a female, in perfect condition, measuring 4.75 inches in expanse of wing, and is in every respect similar to examples of this species which I have from Tennessee. This is only the second time it has been taken in Canada ; Mr. J. A. Moffat having recorded it from London (CAN. ENT., XXVII., p. 52.).

E. M. WALKER, Toronto.

NEW COCCIDÆ FOUND ASSOCIATED WITH ANTS.

BY GEORGE B. KING AND T. D. A. COCKERELL.

[The species described below were all collected by Mr. King. The notes on the microscopical characters were prepared by Mr. King, but have been extended and rewritten from Mr. King's mounts by Mr. Cockerell, who is also responsible for the comparisons with allied species. The notes on the living insects, habitat, etc., are all by Mr. King.]

Lecanopsis lineolatæ, n. sp.

♀ (cleared and mounted).—Oval, length somewhat over 2 mm., dermis practically colourless, legs and mouth-parts tinged with sepia, anal plates a warm yellowish-brown, quite a different colour from the legs. The mouth-parts inclined rather to a madder-brown. Legs and antennæ small, hind legs not nearly reaching the anal plates, tip of femur of middle legs reaching extreme base of hind legs. Posterior cleft wide. Antennæ fairly stout, gradually decreasing in size distad, 8-jointed : 3 longest, not quite twice as long as broad ; 2 and 4 next, and about equal ; 5 and 1 of about equal length, but 1 much broader than long, 5 longer than broad ; then the last three subequal, but 8 the longer. Formula 3 (24) (15) 8 (76): 8 with several small hairs. Anterior tibia and tarsus as long as antennal joints 2 to 6, the tarsus about half as long as tibia ; femur very stout, not as long as tibia on its inner side, but a little longer on its outer; trochanter and coxa both very large. The legs are altogether noticeable for their stoutness, but the basal parts are especially enlarged. Claw large, moderately curved, digitules of claw stout, extending beyond its tip; tarsal digitules filiform, all but two broken off in the specimen. The claw-digitules are enlarged at ends to an obliquely truncate club, but the tarsal digitules with only an excessively minute club. There is the usual long bristle at the tip of the trochanter, and a short erect hair a little way up the femur on the inner side. Anal plates rather broad, the caudolateral side a little longer than the cephalolateral ; a large bristle near the tip and another at the extreme base ; these bristles are very large, and may possibly be dermal, beneath the plates ; especially as there is a corresponding pair on the skin laterad of the plates, that opposite the hindmost bristle being considerably shorter than it. Hairs of anal ring broken, but apparently they were stout and not numerous. Skin without any distinct markings ; marginal spines fairly large, pointed, simple, easily deciduous, a very little further apart than the length of one. Stigmatal spines in threes, one long, two much shorter but not very short.

Hab.—With *Cremastogaster lineolata*, two specimens in a nest at Lawrence, Mass., July 15th, 1894.

This is not a strictly typical *Lecanopsis*, but belongs apparently in the subgenus or genus *Spermococcus* of Giard. By its 8-jointed antennæ it resembles *L. formicarum*, Newstead, but it differs at once from that by the smaller (though still large) first antennal joint, the longer second joint, the much longer third joint, the femur decidedly stouter, the tibia not beset with numerous bristles, and the claw-digitules stout. *Lecanopsis* is simply a segregate from *Lecanium*, modified for underground existence. Maskell's *Lecanopsis filicum* hardly belongs here ; in some respects, but not in others, it seems to approach *Myxolecanium ;* it also recalls in some of its characters such forms as *Lecanium Urichi.*

Phenacoccus americanæ, n. sp.

♀.—When alive fusco-testaceous, smooth, soft, sticky, and free from any wax or down ; when put into alcohol its colour changes to a rufous-violaceous, and it becomes quite wrinkled, its general form is rounded, with a slice of nearly one-fourth cut off, making its under surface flat. Length (in alcohol) 3½ mm., width 3 mm.

♀ (cleared and mounted).—Oval, brown of a rather warm sepia tint, antennæ and legs very pale yellowish. The legs are slender, and although the insect is much larger, its legs are not so large as those of some of the ant's-nest species of *Ripersia ;* but at the same time they are well-formed and ordinary, not shortened or swollen as in the *Lecanopsis*. Trochanter with one long and at least two short bristles. Femur little longer than tibia, its inner margin straight, with four or five erect bristles; its outer margin very gently arched or bent, with a conspicuous erect bristle at the bend. Tibia slender, with eleven conspicuous bristles, tending to form three whorls. Tarsus rather over two-thirds length of tibia, with similar but finer bristles, no nobbed tarsal digitules. Claw large, little curved, with a minute denticle on inner side near the tip ; digitules of claw extending beyond its tip, slender, with hardly noticeable knobs. Antennæ slender, club not conspicuously swollen, formula 9 (123) (45678), or it might be written as well 9132 (87) (456), but the additional differences indicated by the latter formula are almost too slight to be accurately measured by the eye : 9 is very nearly as long as 7 + 8 ; 1 is cylindrical, its base not noticeably wider than the apex. The joints have sparse whorls of hairs, 9 having two such whorls. Eyes prominent. Mouth-parts small, mentum (so-called) very obscurely or not dimerous, broad and short, its apical half with three whorls of erect bristles.   Skin with sparse small round gland-spots.

Hab.—Andover, Mass., Oct. 27th, 1896, under a stone in the nest of *Lasius americanus*, Emery.   A small colony of five individuals captured, and only one herd as yet found ; they were not feeding on any roots entering the nest of the ants, but were altogether on the surface of the nest, and some of the ants were attending them.   It is to be presumed that they would eventually produce cottony matter.

Both by colour and habits this differs at once from *P. aceris*, Sign., which has been recorded from Massachusetts, and there is no species with which it is likely to be confounded.

Ripersia Blanchardii, n. sp.

♀.—Dark reddish-purple, segments prominent, much broader in front, pointed behind, subglobular or subelliptical, convex, antennæ short and thick.   Length, 2 mm.; breadth, 1½ mm.

♀ (cleared and mounted).—Skin quite thickly beset with round gland-spots, and also minutely hairy, the minute but abundant pubescence being a striking characteristic of the species.   So abundant are the hairs in the vicinity of the anal ring that it is impossible to be sure how many really belong to the latter, though there seem to be six, the usual number. The legs, antennæ and mouth-parts are tinged with ochreous, and are large for the size of the insect ; particularly the mouth-parts, which have at least twice the diameter, and many times the bulk, of those of the larger species *Phenacoccus americanæ*.   The mouth-parts are also much broader in proportion to their length than in *P. americanæ*, and the rostral fila-ments are quite stout.   The antennæ are stout, 6-jointed, just about as long as in *P. americanæ*, but very much stouter and quite different in appearance.   The formula is (36) 21 (45), but if anything, 3 is a little longer than 6 ; 3 about twice as long as broad ; 4 and 5 broader at apex than at base, so that the sutures between 3 and 4, 4 and 5, and 5 and 6, are very deep, the last two approaching a right angle.   The whorls of hairs are very sparse.   The legs are also peculiar ; fully a third longer than in *P. americanæ*, and very stout, with large coxæ and trochanters, they are tolerably thickly beset with small hairs.   The tarsus is somewhat over two-thirds the length of the tibia, and tapers quite rapidly from a broad oblique base, it shows a slight tendency to be jointed a little before the

end. Claw large, moderately bent, on one leg minutely notched at the
end. Digitules wanting ; there is a small bristle in the place of the claw-
digitule.

Hab.—Haverhill, Mass., Oct. 4th, 1896, in a nest of *Lasius claviger*,
Rog., under a stone with a small herd of another species ; only one found,
not feeding. Named after Mr. Blanchard, who has interested himself in
the Coleoptera associated with ants in the same region.

Of the Massachusetts species, this most resembles *R. lasii*, particularly in the antennæ, but it differs widely in its colour, hairiness and stout
legs. Still less does it seem to resemble any of the foreign species.

Reviewing the above three species, it seems that the *Lecanopsis* is
most modified for an underground life, the *Ripersia* somewhat, but the
Phenacoccus hardly or not at all. It is probable that the last will be found
in summer on some plant above ground.

ARGYNNIS IDALIA IN NEW BRUNSWICK.

On February 1st I happened to spend a few hours in St. John, N. B.,
and through the kindness of Mr. Herbert E. Goold, of Sussex, N. B., and
Mr. A. Morissey, of St. John, I was enabled to visit the very interesting
museum of the Natural History Society of New Brunswick. In looking
over the cases of insects I noticed two fine specimens of *Argynnis idalia*,
which Mr. Goold told me were taken by himself or his father at St. John.
I could not remember at the time any record of *A. idalia* having been
taken in New Brunswick, so asked Mr. Goold to enquire from his father
if he remembered anything of the capture. He has since written to me :
" In *re Argynnis idalia*—On my return home from St. John I asked my
father about the specimens. He remembered the circumstances of their
being caught distinctly, as he was perfectly familiar with the insect, having
been one of the most active members of the entomological branch of the
Natural History Society of Portland, Maine. In 1880 quite a number of
specimens of *A. idalia* appeared in the vicinity of St. John, and the
specimens you saw were taken at that time." It is to be deeply regretted
that at the present time very few members of the strong local Natural
History Society at St. John are studying entomology. The locality is one
of extreme interest scientifically, and very much requires working up.

J. FLETCHER, Ottawa.

[In the C. E. for March, 1896, Vol. XXVIII., p. 74, the capture of
a specimen of *A. idalia* at Windsor, Ont., is recorded.—ED. C. E.]

ON REARING DRAGONFLIES.

BY JAMES G. NEEDHAM, ITHACA, N. Y.

Field work in Entomology is full of delightful opportunities, and none, just at present, is more inviting, none more certain to repay well even a little effort, none more sure to yield discoveries of scientific value, than work upon the life-histories of Dragonflies.

Of the species occurring throughout the central tier of States, a majority perhaps has now been bred ; but of the Canadian, far western and southern species the known nymphs are few and far between.

FIG. 15.—AESCHNID NYMPH.

The nymphs (fig. 15), which are all aquatic, have an interesting distribution in depth. Those of *Agrionidae* and of most *Aeschninae* cling to floating or submerged vegetation. These at least every aquatic collector has seen. Those of *Libellulidae* sprawl upon the bottom amid fallen trash. Those of *Gomphinae* burrow shallowly along beneath the film of sediment that lies on the bottom, with the end of the abdomen turned up for respiration.

It is very easy to collect them, especially in spring. A garden rake with which to draw ashore the stuff to which they cling and a pail of water in which to carry them home is all the apparatus desirable at that season. Later, when a new growth of weeds is rooted fast to the bottom, the rake will have to be exchanged for a water-net. Withdrawn from the water, the nymphs render themselves evident by their active efforts to get back, and need only to be picked up. The number of species one will find will generally depend on the variety of aquatic situations from which he collects. The places apt to yield the best collecting are small permanent pools, shallow inlets in the shores of lakes, and the places where the trash falls in the eddies of streams.

They are quite as easily reared. I have found common wooden kits and pails half filled with water, with screen or netting covers, entirely satisfactory. A number of nymphs, if near one size, may safely be kept together (excepting only a few notoriously cannibalistic Aeschninas : *e. g.*, *Anax junius*), and if not grown may be fed upon such small insects as a net will gather in any pond. A good square meal once a week will keep them thriving. The water should be reasonably clean. Three things should be carefully observed. (1) There must be a surface up which they can climb to transform : if the sides of the kit are too smooth put in some sticks ; (2) there must be room enough between the netting cover and the water for complete expansion of their wings ; (3) they must remain out of doors where the sunshine will reach them. This last point especially is essential to success. But there is still an easier way to do it, and one which, when a species is very common, will prove entirely satisfactory. The several nymphal stages (excepting the youngest, not likely to be collected) are very much alike. I am in the habit of preserving the younger nymphs and putting into my kits only those well grown, as shown by the length of the wing-cases, which should reach the middle of the abdomen. But if, when a species is becoming common, one will go to the edge of the water it frequents, at the time of its emergence, one may find nymphs crawling from the water, others transforming, imagoes drying their wings, and others ready to fly, and may thus obtain in a few minutes the material necessary for determining nymph and imago. The time of emergence may be determined by noticing at what time pale young imagoes are seen taking their first flight, and then going out a little earlier. The unfortunate thing about it is that many of the larger species transform very early in the morning, and to take such advantage of them one must be on the ground between daybreak and sunrise.

Several imagoes should be kept alive until they have assumed their mature colours. It is most important that each imago and its cast skin should be kept together.

Eggs, also, are easily obtained. Every collector has seen the female of the species figured on the front of this magazine, or of related species, dipping the tip of her abdomen into the surface of the water, depositing eggs. If the ovipositing female be captured, held by the fore wings, leaving the hind wings free, and " dipped " by hand to the surface of clean water in a vial or a tumbler, an abundance of eggs will usually be liberated. Eggs of those species which possess an ovipositor and which place them within the tissues of plants may be obtained by collecting the stems in which they have been inserted.

Eggs and nymphs should be dropped in boiling water for a minute and then preserved in alcohol. Imagoes, if mounted, should have a wire or bristle inserted into the body its entire length to prevent otherwise certain breakage, or if placed unmounted in envelopes, these should be of soft paper, loosely packed, so that the eyes will not be crushed.

In my own field work upon Dragonflies I try to cover for each species the points of the following outline :

I. Imago.

(1) Name ; locality ; date ; occurrence ; etc.

(2) Haunts : places frequented ; places avoided ; the reasons, if discoverable.

(3) Flight : its hours ; its duration ; its directness ; average altitude ; places of rest : altitudes.

(4) Food : its kind ; how obtained ; where eaten.

(5) Enemies : what are they, and how do they destroy Dragonflies ?

(6) Oviposition : does the ♀ oviposit alone or attended by the ♂ .

(7) The eggs : where placed ; number in a place ; incubation period.

II. The Nymph.

Points 1, 2, 4 and 5 of above, and Imagination : hours ; places ; distance from water ; etc.

I shall have to admit at once that it is very difficult to determine all these points for a single species, but the effort will lead on into delightful intimacy with these beautiful insects.

At the kind invitation of the editors, I venture to say to the readers of this magazine that I am now engaged upon a semi-popular monograph of N. American Dragonflies, which, in so far as it includes accounts of habits and life-histories of the species, must of necessity be a co-operative work. And I have written this to invite co-operation. The foregoing simple methods are the very best. I will furnish (if desired) half a dozen named nymphs of typical genera to any one who will undertake to collect and rear others. I shall be very willing to determine nymphs or imagoes for any one, and to point out for description such as are new. But I especially desire that accurate field observations and notes be made on many of our species of which we now know only the names, and to such observers I will give all possible aid.

THE ANNUAL REPORT of the Entomological Society of Ontario for 1896 is now in type and will soon be ready for distribution.

Mailed April 1st, 1897.

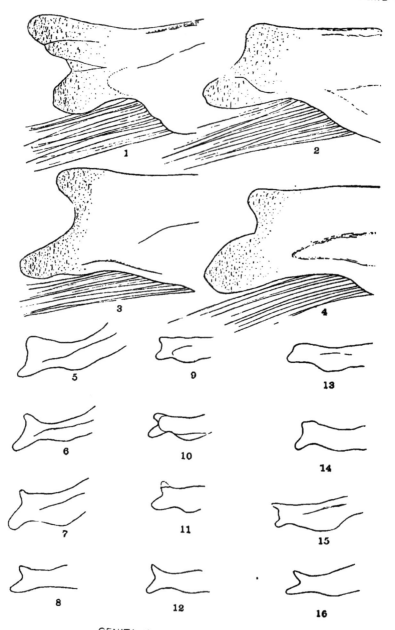

GENITALIA OF CALLIMORPHA.

The Canadian Entomologist.

Vol. XXIX.   LONDON, MAY, 1897.   No. 5.

CALLIMORPHA AGAIN.

Larva of Haploa fulvicosta and notes on the male genitalia.

BY HARRISON G. DYAR, PH. D., NEW YORK.

The difficulty of defining species in this genus is increased by the constancy of the local forms or races. I have elsewhere referred (*Ent. News*, VII., 218) to the race of *fulvicosta* which Mr. O. D. Foulks has discovered at Stockton, Md. Mr. Foulks was so kind as to send me over 100 hibernated larvæ, from which I bred a long series of moths. The type form is large, the size of *reversa* and *colona*, both wings immaculate yellowish-white, head, collar and the tips of the abdominal rings ochre-yellow.

In *var. A* the fore wings are nearly pure white, the hind wings much yellower, suggesting *conscita*, though never so dark as that form.

In *var. B* the ground of fore wings is white, marked faintly with ochreous bands in which the full pattern of *colona* can be traced ; the costa is narrowly brown-black ; the hind wings are pale ochreous. This looks like a washed-out *colona*, related to it in the same way as *var. A.* is to *conscita*.

Var. C is only slightly yellowish on both wings, the hind wings scarcely at all darker ; fore wings marked with various streaks and spots of brown-black, especially along the costa and margins, all more or less distinctly connected by ochreous shades, in which the full pattern of *reversa* can be read. This is a washed-out *reversa*, stained with the creamy yellow so characteristic of the Maryland race.

All these forms insensibly intergrade. I believe that this is practically the extent of variation in this Maryland race. There are no specimens that are true *colona*, *conscita* or *reversa*, but these forms are all strongly suggested. The view naturally presents itself that these names apply to local races rather than to distinct species. In his work on Callimorpha (Proc. U. S. Nat. Mus., 1887, p. 338) Prof. J. B. Smith describes the genitalia of *colona*, *Lecontei*, *contigua*, *reversa* and *vestalis*. The differences shown are at best slight, and Prof. Smith assumes the

forms which he figures to be constant.   In fact, they are not so.   I have
drawn the right side pieces of four males of the Maryland race of
fulvicosta.   They are shown in figures 1 to 4, viewed from within × 50,
the dorsal angles down.   These are not specimens selected for their
variation, but are all that were mounted.   The only selection applied
was in taking the poorest specimens for dissection.   Fig. 1 shows the
upper angle produced and rounded, the lower angle much more produced; fig. 2 shows the lower angle not produced, but simply rounded ;
fig. 3 both angles produced, the upper the most so ; fig. 4 both angles
produced, but the lobes of quite different shapes.   There is as much
variation in these specimens of *fulvicosta* as in all of Prof. Smith's
"species," and I am of the opinion that the genitalia are valueless as a
means of specific distinction in Haploa.   However, I add drawings of
most of the other forms and also reproduce Prof. Smith's figures.

It is possible that the larvæ, when fully known, will be of more help,
yet this is doubtful, as they seem to possess all the same habits and
hence are not markedly different in their colours.   Very full descriptions
are needed, especially of the mature larva, to test these points.   The
following observations were made on the larvæ sent by Mr. Foulks and
on the young ones bred out of the eggs from the moths.

Normal number of stages six ; hibernation in the fourth or fifth.
The young larvæ that were selected for observation passed two interpolated stages between the normal II. and III. and died before reaching
stage IV.

Egg.—Of the shape of two-thirds of a sphere, scarcely conoidal,
the base flat ; smooth, shining, rather dark yellow ; diameter .6 mm.
Reticulations obscure, visible in a strong reflected light, very narrow,
linear, irregularly hexagonal, the cell areas flat, uniform, no shadows.

Stage I.—Head high, bilobed, the lobes blackish brown, clypeus
pale, mouth brown ; width .3 mm.   Body pale yellowish, tubercles dusky
pearly ; hair short, stiff, white.   Setæ single, normal, no subprimaries ;
feet pale.   The larvæ grow considerably, becoming long and slender, the
tubercles surrounded narrowly by brown.

Stage II.—Head black, shining, clypeus whitish, jaws brown ;
width .45 mm.   Body whitish, warts rather small and with the shields
deep shining black ; hairs not numerous, but forming true warts, short,
bristly, black.   A wide space between tubercles i. suggests a dorsal band
Warts each narrowly edged with brown, most distinctly subdorsally, no
connected marks.   Subventral hairs pale.

Stage III. (interpolated)—Head shining black, clypeus and mouth brown ; width .55 mm. Warts large, black, hairs short, bristly, black and white. Body elongated, broadly whitish between warts ii., fading to smoky black in the region of wart iii.; below this another pale band, marked with yellow, transversely annulated streaks behind wart iv., two on each segment ; subventral region shaded with brown. Leg plates black. Later the appearance is more as in the next stage, though the bands are not really defined.

Stage IV. (interpolated) — Head black ; width .65 mm. Body black, a broad diffuse dorsal gray line, joining a narrower subdorsal one. Region of warts iii. and iv. yellow spotted, joining a substigmatal gray band and subventral gray marks. Warts black ; hair short.

Stage V. (normal III.)—Black ; head .75 mm. Pale whitish dorsal, subdorsal and substigmatal lines, the subdorsal faintest; bright yellow superstigmatal line, not perceptibly joined to the substigmatal one. Warts black.

Normal Stage V. (from Mr. Foulks; after hibernation) — Head shining black ; width 1.7 mm. Body black ; dorsal line broad, subdorsal faint, stigmatal broad, substigmatal fainter, yellow, traces of a line subventrally, all more or less white spotted. Essentially as in the next stage.

Stage VI.—Head and warts shining black, the latter bluish ; width 2.7 mm. Body deep black, the dorsal line broad, straight, narrowly broken in the incisures and centre of the segments, yellow, darker yellow or red in the centre of each segment, faint on joint 2. Traces of a subdorsal band, broken by wart ii., whitish, mottled. Lateral band broad, indented by warts iii. and iv., broken into three or four spots on each segment by transverse black lines, yellow, irregularly stained with darker yellow, connected inferiorly by mottlings and dots with a narrow substigmatal line which is yellow, mottled, broken and runs between warts iv. and v. Traces of a subventral line between warts v. and vi. on the base of each leg. Leg plates black. Venter broadly pale gray, blackish dotted. Hair very short, inconspicuous, black or black and white, stiff, pointed, not barbuled. In some individuals the subdorsal whitish dots are absent, and in some the dorsal band is distinctly marked with red ; otherwise there is very little variation. Corresponds well with Saunders's description of *reversa* (CAN. ENT., I., 20), and also with Riley's of *fulvicosta* (Third Report Ins. Mo., 134). The forms *colona* and *conscita* have not been bred.

Figs. 1 to 4.—Side pieces of male genitalia of *Haploa fulvicosta* seen from within ; four examples, specimens from Maryland.

Fig. 5.—The same, *H. clymene*, specimen from Kansas.

Fig. 6.—The same, *H. reversa*, specimen from Texas.

Fig. 7.—The same, *H. colona*, specimen from Texas.

Fig. 8.—Copied from Smith's figure of *H. colona.*

Fig. 9.—Side piece of male *H. lecontei*, var. *militaris*, specimen from Iowa.

Fig. 10.—Copied from Smith's figure of *militaris*.

Fig. 11.—Side piece of *H. vestalis*, specimen from Iowa.

Fig. 12.—Copied from Smith's figure of *vestalis*.

Fig. 13.—Side piece of *H. confusa*, specimen from Northern New York.

Fig. 14.—Copied from Smith's figure labelled *confusa* on the plate, but described as *reversa* in the text.

Fig. 15.—Side piece of *H. contigua*, specimen from New York.

Fig. 16.—Copied from Smith's figure of *contigua.*

SOME ANTS AND MYRMECOPHILOUS INSECTS FROM TORONTO.

BY GEO. B. KING, LAWRENCE, MASS.

During the summer of 1896 I received specimens of ants collected by Mr. R. J. Crew, of Toronto, in exchange for such Coleoptera as I could find for him in my locality. He writes me that he noticed no insects with the ants other than the Coleoptera and some aphids in a nest of ants, but did not capture any.

I have found, however, upon looking them over, they contain several very interesting species of various orders : some truly myrmecophilous, some occasional, while others were brought into the nests by the ants, to be used by them for food ; this will apply to a number of Hemiptera collected by *Formica subsericea*, Say.

It may appear to some who are collecting ants'-nests Coleoptera only that the finding of *Agonoderus pallipes*, Fabr., and *Otiorhynchus oratus*, L., is merely occasional. The position in which these Coleoptera are found with the ants here in Massachusetts, and the frequently finding them with various species of ants, lead me to believe that they are more than incidental or casual visitors.

I am not familiar with the scattered literature treating upon the

Formicidæ found in Canada. I will give, however, all that I know of, taken from Dr. Dalla Torre's Catalogue of Hymenoptera, Vol. vii., 1891 :

Stigmatomma binodosum, Prov.

Pogonomyrmex badius, Latr.

Leptothorax Canadensis, Prov.

Dolichoderus borealis, Prov.

Dolichoderus obliteratus, Scudd.

Formica arcana, Scudd.

Mr. Crew has not as yet found any of the above species at Toronto. The following is a list of those found by him :

Tribe CAMPONOTIDÆ.

Camponotus ligniperdus, Latr., var. pictus, For.

" herculaneus, L., sub-sp. pennsylvanicus, Deg.

" marginatus, Latr., var. nearcticus, Em.

Formica rufa, L., sub-sp. integra, Nyl.

" exsectoides, For.

" pallide-fulva, Latr., sub-sp. Schaufussii, Mayr.

" pallide-fulva, Latr., sub-sp. nitidiventris, Em.

" fusca, L., var. subsericea, Say.

" lasioides, Em., var. picea, Em.

Lasius niger, L., var. americanus, Em.

" niger, L., var. neoniger, Em.

" flavus, De G., sub-sp. myopes, For.

" claviger, Rog.

Tribe DOLICHODERIDÆ.

Tapinoma sessile, Say.

Dolichoderus plagiatus, Mayr.

" Taschenbergi, Mayr.

Tribe PONERIDÆ.

Ponera coarctata, Latr., sub-sp. pennsylvanica, Buckl.

Tribe DORYLIDÆ.

Solenopsis molesta, Say.

Myrmica scabrinodis, Nyl., var. sobuleta, Meinest.

" scabrinodis, Nyl., var. Schencki, Em.

Cremastogaster lineolata, Say.

The following are the miscellaneous insects found with Mr. Crew's collection of ants sent me.

Coleoptera.

CARABIDÆ.

Stenolophus conjunctus, Lec.—With Myrmica scabrinodis, Nyl., var. Schencki, Em.

Agonoderus pallipes, Fabr. — With Myrmica scabrinodis, Nyl., var. Schencki, Em.

OTIORHYNCHIDÆ.

Otiorhynchus ovatus, L.—With Formica fusca, L., var. subsericea, Say.

I have found this species in Massachusetts with :

Formica fusca, L., var. subsericea, Say ;

Aphaenogaster fulva, Rog.; and

Lasius americanus, Em.

STAPHYLINIDÆ.

Scopæus exiguus, Er.—With Formica fusca, L., var. subsericea, Say.

Aleocharini g. et sp.—With Solenopsis molesta, Say.

PSELAPHIDÆ.

Ctenistes piceus, Lec.

SCYDMÆNIDÆ.

Scydmænus bicolor, Lec.

These two last species were collected by Mr. Crew in company with ants ; but he did not at the time of capture deem it important to save any, so we cannot give the names of the ants. C. piceus was found March 23, 1895, and S. bicolor, Dec. 4, 1895.

Hymenoptera.

PROCTOTRYPIDÆ.

Proctotrypes californicus, Holmgr.—With Formica fusca, L., var. subsericea, Say. This, with a few other species of my own finding, are in the collection of the National Museum at Washington, by request of Prof. Howard.

ANDRENIDÆ.

♀ Halictus confusus, Smith.—With Formica fusca, L., var. subsericea, Say.

CYNIPIDÆ.

♀ Figitodes 5-lineatus, Say.—With Tapinoma sessile, Say.

I have found Aphaenogaster fulva, Rog.; Lasius flavus, L., and Lasius americanus, Em., to collect oak galls late in the fall. Two individuals came out of one lot of galls collected by L. flavus, L., in about two weeks after I collected them, and have been determined by Mr. Ashmead as Periclistus piratus, O. S. The ants lap the galls.

Diptera.

STRATIOMYDÆ.

Nemotelus globus, Low.—With Tapinoma sessile, Say.

MUSCIDÆ.

Ochthiophola polystigma, Meigen.—With Tapinoma sessile, Say.

Hemiptera.

CICADIDÆ.

Nymph of Tettigonia, sp. — With Myrmica scabrinodis, Nyl., var. Schencki, Em.

NABIDÆ.

Larva of Coriscus, probably ferus. — With Formica fusca, L., var. subsericea, Say.

LYGÆIDÆ.

Nysius thyus, Wolff.—With Formica fusca, L., var. subsericea, Say.

CAPSIDÆ.

Miris affinis.—With Formica fusca, L., var. subsericea, Say.

THRIPIDÆ.

A handsome species of Thrips.—With Camponotus nearcticus, Em.

ARANEINA.

Furolithus, sp.—With Tapinoma sessile, Say.

Quite a large quantity of a yellow seed unknown to me came in a mixed lot of ants in one vial. Mr. Crew states that he does not remember mixing any of the species found, but put each colony into separate vials. The following are the species from one vial, that contained the seeds :

Formica pallide-fulva, Latr., sub-sp. nitidiventris, Em.

Formica fusca, L., var. subsericea, Say.

Formica lasioides, Em., var. picea, Em.

Myrmica scabrinodis, Nyl., var. Schencki, Em.

The last species seemed to predominate greatly in numbers. So far as I know, this is the first time that any of the species here mentioned have been listed as being found in company with ants. In the determination of these insects I have received valuable assistance from Prof. Herbert Osborn, Prof. L. O. Howard, Mr. Ashmead, Mr. Coquillett, and Mr. Blanchard ; and not only for these, but for many others not yet published that I have found to inhabit ants' nests in Massachusetts.

ENTOMOLOGICAL SOCIETY OF ONTARIO.

We have great pleasure in announcing that a branch of our Society has recently been formed in the City of Quebec, with the following officers :

President—Rev. T. W. Fyles, F. L. S., Professor of Biology in Morrin College.

Vice-President—Miss Macdonald, Principal of the Girls' High School.

Secretary-Treasurer—Col. Crawford Lindsay.

Council—Messrs. D. H. Greggie, Richard Turner, J. E. Treffry, Miss Bickell, Miss Winfield.

With such an enthusiastic and experienced entomologist as the President, and such an able corps of officers, the Branch will no doubt do excellent work, and serve to unite together all those interested in this department of natural science in the neighbourhood of Quebec. We trust that the new Branch may have a long and useful career.

The Toronto Branch of the Society held its first annual meeting on Friday, April 2nd, in its room, 451 Parliament street. The election of officers for the ensuing year resulted as follows :

President—Mr. E. V. Rippon.

Vice-President—Mr. R. J. Crew.

Secretary-Treasurer—Mr. Arthur Gibson.

Librarian-Curator—Mr. T. G. Priddey.

Council—Messrs. C. T. Hill and C. H. Tyris.

The reports of the Secretary-Treasurer and the Librarian-Curator for the past year were read and adopted. They stated that twenty-four regular meetings had been held, at which papers relating to the study of insects were contributed by the members. The number of volumes in the library, including bulletins, pamphlets, etc., is 98, all relating to entomology, and all gifts to the Society. A fair collection of insects has already been formed through the kindness of members in presenting specimens, and will no doubt be largely increased during the coming season. The finances of the Society were shown by the Treasurer's report to be in a satisfactory condition.

The President, in his address, congratulated the members on the good work done during the year, and on the success which had attended the Society's operations. He hoped that during the coming season each member would take a special interest in some particular species of insect, and would endeavour to work out its life history ; he also trusted that much attention would be paid to the study of those species which are beneficial or injurious to mankind.

THE COLEOPTERA OF CANADA.

BY H. F. WICKHAM, IOWA CITY, IOWA.

XXIII. THE CERAMBYCIDÆ OF ONTARIO AND QUEBEC.—*(Continued.)*

TETROPIUM, Kirby.

This genus is easily recognized among its neighbours by the fact that the eyes are divided by a deep emargination into an upper and a lower portion, these parts being connected only by a narrow band from which the granulations or lenses have been lost. The Canadian *T. cinnamopterum*, Kirby, is brown, the wing-covers often much lighter than the head and thorax; the entire body is pubescent. Length .50 -.70 inch. The head and thorax are slightly shining, distinctly punctured, the punctures regular, usually close but distinctly separated. Elytra opaque or extremely feebly shining. Sculpture much finer than that of the prothorax. The sexes differ especially in the somewhat shorter antennæ and the broader and more strongly rounded prothorax of the female. The species occurs on or under bark of pine logs.

ASEMUM, Esch.

Two species are recorded from Canada. They are stout brown insects with short antennæ (from about one-third to one-half the length of the body), elytra sometimes yellowish. The thorax is about as broad, in its widest part, as the base of the elytra; the punctuation coarse and close on the pronotum, much finer on the wing-covers. The principal differences separating the two forms must be looked for in the prothorax, which is rounded on the sides in *mæstum*, Hald., and distinctly angulated near the base in *atrum*, Esch. The distinctness of the elytral costæ seems an evanescent character, since certain specimens of the former species approach the latter very closely in that respect. In length *A. mæstum* (fig. 16) ranges from .45 to .60 inch, while speci-

FIG. 16.

mens of *atrum* are known which slightly exceed the greater measurement and others which scarcely reach the lesser. In the larval stage *A. mæstum* is known to infest pine and spruce, and the beetles may be found on lumber piles.

CRIOCEPHALUS, Muls.

Contains larger species than the preceding genus, with coarsely granulated eyes which are not hairy. The prothorax is variably sculptured, sometimes roughened and with deep impressions on the disk.

Two are recorded from our region. They are both rather elongate brown insects and separate thus :

> Thoracic impressions deep, elytra finely punctured, third joint of hind tarsi two-thirds longer than wide, emarginate for about one-half its length. Sides of prothorax rounded, somewhat roughened. .90–1.10 in...................... ....*agrestis*, Kirby.
> Thoracic impressions fainter, elytra coarsely punctured, third joint of hind tarsi about as long as wide, cleft nearly to the base. Prothorax very finely punctured, sides rounded, hardly roughened. .94 in...*obsoletus*, Rand.

These insects are found about lumber piles in the northern and mountain regions of North America. *C. agrestis* is known to depredate on pine and spruce.

PHYSOCNEMUM, Hald.

P. brevilineum, Say, is .50–.75 inch long, black, somewhat shining, elytra sometimes bluish or with a faint reddish tinge along the suture. The upper surface is uneven, the prothorax with deep median longitudinal impression which is convex at bottom and limited on each side by an elevation, which is smoother than the external thoracic margin. Elytra distinctly closely punctured and ornamented with a few narrow, short, raised white lines ; the median region on each wing-cover is depressed and limited exteriorly by a smoother linear area, which extends from the humerus towards the apex. Thighs suddenly and strongly dilated near their tips. Hind legs very long. The larva is known as an elm borer.

RHOPALOPUS, Muls.

An easily recognized species, *R. sanguinicollis*, Horn, belongs here. It has been found on cherry trees. Length .62–.75 inch, colour black opaque, surface granulate ; prothorax red, tips of elytra sometimes brownish. The thighs are less suddenly clavate than in *Physocnemum*, and the tibiæ are stouter. The extreme shortness of the prothorax will separate it easily from most of its neighbours.

GONOCALLUS, Lec.

Differs from the adjoining genera by the slender thighs. *G. collaris*, Kirby, is black, shining, elytra sometimes with metallic lustre or clouded with fuscous, the prothorax red, legs sometimes reddish. The upper surface is punctate, the antennæ very slender. Length .35–.47 inch.

PHYMATODES, Muls.

Contains a number of species, all of rather small size and usually bright colour. The prothorax is rounded, usually sparsely punctured and shining. Elytral punctuation distinct, often rather coarse, surface usually shining. The following arrangement of species is taken from Mr. Leng's synopsis :

A. Elytra without narrow cross-bands.

    b. Thorax dark, elytra of lighter shade before the middle. .35–.50 in...........................*dimidiatus*, Kby.

    bb. Thorax rufous with broad black stripe. .25 in..*maculicollis,* Lec.

    bbb. Thorax yellowish ; surface metallic.

        c. Larger species, .50–.52 in., elytra and legs yellow, more or less marked with blue.............. *variabilis*, Linn.

        cc. Smaller species.

            Elytra blue, antennæ dark. .20–.32 in....*amœnus*, Say.

            Elytra piceous, thorax with more or less distinct dark lateral blotches. .34–.36 in........*thoracicus*, Muls.

AA. Elytra with two narrow white or yellowish cross-bands. Usually rufous, elytra dark, except at base. .25–.36 in. (Fig. 17.) ....*varius*, Fab.

These insects are usually to be met with in beating. *P. variabilis* has been recorded as depredating on oaks, while *amœnus* bores in grapevines. *P. varius* is believed to live as a larva in black oak, but I know of no breeding record.

Fig. 17.

CALLIDIUM, Fabr.

Two of the species are metallic green or blue, the other is brown or yellowish. They are mostly flatter than *Phymatodes*, and with heavier antennæ, especially in the male. The colour affords a primary means of separating them, *C. æreum*, Newm., being entirely testaceous or brownish, while *antennatum*, Newm., and *janthinum*, Lec., are metallic blue or green above. The last named has the thorax deeply punctured, not impressed, while in *antennatum* impressions are present and the thoracic punctuation is finer. All the species vary much in size, *æreum* from .34 to .50 inch, while the others run from .25 to .55 inch, *janthinum* averaging a little smaller. It is reported *æreum* has been bred from chestnut, while *antennatum* depredates on pine.

HYLOTRUPES, Serv.

The two species of this genus are very different in appearance. *H. bajulus*, Linn., is blackish, pubescent above, more thickly on the prothorax, where the hair is whitish, almost covering the surface except on the elevated median line and the two raised callosities, which are thus rendered very conspicuous. The elytra have two indistinct transverse fasciæ of whitish pubescence, one in front of the other behind the middle, the latter sometimes wanting. Length, .72 to .88 in. Depredates in pine and juniper. *H. ligneus*, Fabr., is extremely variable, the thorax usually black, less hairy than in *bajulus*, and with five callosities. Elytra yellowish or reddish, with a large blackish blotch occupying usually the apical third, and an elliptical spot of the same colour but varying in size between this blotch and the base. Bores in juniper in the larval state, perhaps also in pine, as the beetle is found on piles of lumber or on freshly constructed fences. Length, .30 to .45 inch.

MERIUM, Kirby.

M. proteus, Kirby, is .45 to .60 inch long, thorax metallic blue or violaceous, shorter than usual, varying in shape according to sex, densely punctured and rather opaque at sides, but shining and with only a few large punctures at middle. Elytra usually greenish metallic, densely and coarsely punctured, generally with two raised longitudinal yellowish lines before the middle, the side margin also yellowish in some specimens. Thighs reddish yellow, except at base and apex, which, with the tibiæ and tarsi, are blackish. Beneath shining black with a violaceous tint.

CHION, Newm.

Here belongs *Chion cinctus*, Drury, a large beetle of a brownish colour (fig. 18), sparsely clothed with whitish pubescence, each elytron usually with an oblique blotch of a yellowish colour near the base. The prothorax is nearly round, and bears a small spine on each side. The elytra are each bispinose at tip. The male antennæ greatly exceed the body in length. The species reaches a size of from .75 to 1.5 inch. It is known to breed in hickory. The name *garganicus*, Fabr., catalogued as a variety, refers to the spotted form.

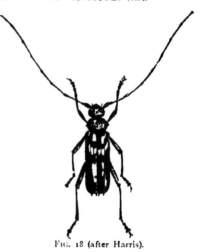

FIG. 18 (after Harris).

EBURIA, Serv.

The two pairs of raised white spots (looking like Ichneumon eggs) on each elytron will easily serve to distinguish this genus. The only Canadian species is *E. quadrigeminata*, Say, which is of a yellowish colour, the thorax with sharp lateral spine and two distinct discal callosities. The elytra are bispinose at apex, the middle and hind femora have each two long spines at tip. The ivory spots of the elytra are situated on the costæ, the outer one of each pair being the larger, this difference in size being much better marked in the posterior pair. Length, .90–1.20 inch. Breeds in hickory, ash, and honey locust.

ROMALEUM, White.

Contains two large species, among the most bulky of the Canadian Longhorns. Both are pubescent insects of robust build, the prothorax rounded at sides and without lateral spine, the elytra spinose at apex, tip of thighs unarmed, antennæ spinose internally. *R. rufulum*, Hald., is fulvous with uniform pubescence of the same colour. Length, .88–1.15 in. *R. atomarium* is darker, brownish, with irregularly mottled pubescence, and reaches a slightly larger size. It has been found under bark of walnut, while the larva has been bred up on hackberry.

ELAPHIDION, Serv.

The Canadian species of this genus are smaller and less robust than the preceding, and may be distinguished therefrom by that character alone. *E. villosum* is the well-known oak-pruner, and does, at times, considerable damage by ovipositing in twigs of oak trees, the larvæ then eating out the inner portion, so that the twig becomes weakened and may be blown off in a strong wind. Its depredations are not confined to oak, however, as Mr. Chittenden has recorded many other food plants. The table of species is an adaptation of the characters presented by Mr. Leng:

A. Antennal spines large, thighs spinose at tip, body above with irregular vestiture of gray pubescence. .60–.75 inch.

AA. Antennal spines small.

    b. Above clothed with mottled gray pubescence, elytra bispinose at tip.

        c. Sides of prothorax rounded. .70..........*incertum*, Newm.

        cc. Sides of prothorax hardly rounded ; nearly cylindrical.

            Prothorax scarcely longer than wide. .70 in. *villosum*, Fabr.

            Prothorax distinctly longer than wide. .70

            in ....*parallelum*, Newm.

bb. Above nearly glabrous, shining testaceous. Form very elongate, elytral spines long. .43–.45 in............*unicolor*, Rand.

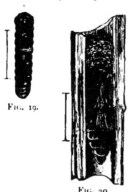

Fig. 19.

Fig. 20.

Fig. 21.

It is stated that *E. villosum* and *E. parallelum* are not distinct, but they are included in the above table, as their amalgamation has not yet been generally accepted. The figures 19, 20 and 21 represent the three stages of *E. villosum*.

TYLONOTUS, Hald.

Represented by *T. bimaculatus*, Hald.; of a brownish colour. .45 to .60 in. long. Each elytron with two rather large, somewhat rounded yellowish spots, one in front of the middle, the other sub-apical. The thighs are yellowish except at base and apex, rather strongly clubbed. The antennæ are bisulcate (more distinctly on the third and fourth joints), the thorax is thickly punctured with smoothish median line and two rather large dorsal callosities. The elytra are coarsely, rather sparsely, punctured. Pubescence thin, yellowish. The larva bores in ash; beetles have been found under bark of the white or paper birch.

HETERACHTHES, Newm.

Easily recognized by the elongate form, shining surface and extremely small second antennal joint. The thighs are strongly clubbed, the antennæ long and heavy. *H. quadrimaculatus* is .30 to .45 in. long, brown or testaceous with two paler spots on each elytron, one in front of and one just behind the middle. The pale specimens have the spots indistinct. Head closely, elytra and thorax very sparsely, punctured. Length, .30 to .45 inch. It has been bred from hickory limbs.

GRACILIA, Serv.

G. minuta, Fabr., does not occur on any of the Canadian lists, but has been described and figured (in the CANADIAN ENTOMOLOGIST, vol. xxiii., p. 102), by Mr. J. F. Hausen. His figure (fig. 22) and description are here reproduced. " It is of a uniform reddish-brown, the legs being somewhat lighter, with rather sparse cinereous pubescence, giving it a heavy appearance. The antennæ are ciliate, and the head, thorax and

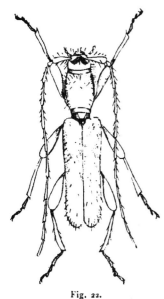

Fig. 22.

elytra furnished with flying hairs. Rather variable in size, .18–.27 in." It was taken by Mr. Caulfield, emerging from a barrel of some kind of dye. The species is supposed to have been introduced from Europe. It has been bred from white birch.

PHYTON, Newm.

A small pale insect, *P. pallidum*, Say, belongs here, and is perhaps doubtfully a true member of the Canadian fauna. It is a trifle under one-fourth of an inch long, of a yellowish colour, the prothorax broad in front of the middle, but narrowed in front and (much more so) behind, the surface with indefinite darker markings. Elytra with four oblique brownish bands, of which the one just behind the middle is broad, the remainder narrow. I have beaten it from palmetto blossoms in Louisiana. It has been bred from hickory and from *Cercis canadensis*.

OBRIUM, Serv.

The only Canadian form is *O. rubrum*, Newm., which is one-fourth of an inch in length, shining reddish-testaceous, the head broader than the prothorax, which bears an obtuse dilatation each side near the middle, and has the base and apex nearly equal. The elytra are more closely punctured than the thorax. Thighs strongly clubbed.

NOTES ON PHILÆNUS.

BY CARL F. BAKER, AUBURN, ALABAMA.

Philænus spumarius, L. — From various localities in the New England States I have large series of the typical form of this species, and also specimens representing the well-marked varieties, *leucocephala*, L., and *lineata*, Fabr.

Philænus abjectus, Uhl. — A portion at least of the material recorded under *Lepyronia angulifera* in the Prelim. List Hemip. Colo. belongs to this species. I have taken it at Fort Collins, Colo., and in the adjacent foothills, in May and June. The specimens from this locality are uniformly darker than the type.

Philænus lineatus, L.—I have a large series of this species from the North-eastern U. S., the specimens of which are identical with the typical European form. It seems probable that true *lineatus* is confined to the Eastern U. S. In American publications three distinct species have been confused under this name,—true *lineatus*, the *bilineatus* of Say, and a new species from New England which I shall call *americanus*.

The genus presents two types of elytral venation, one simple and regular, with three or four distinct apical cells, while in the other the elytra are distally irregularly reticulated. *Lineatus, spumarius, campestris, exclamationis*, etc., fall into the first group, while the second group appears to be strictly American, including *abjectus, bilineatus* and *americanus*.

Philænus bilineatus, Say.

1831. Say, Journ. Acad. Nat. Sci. Phila. VI., 304 (Aphrophora bilineata).
1872. Uhler, List Hem. Colo. and N. M., 472 (Ptyelus lineatus).
1876. Uhler, List Hem. region west of Miss. R., 347 (Philænus lineatus, var. bilineata).
1877. Uhler, Rep. on Ins. Coll. in 1875, 458 (Philænus lineatus).
1878. Uhler, List Hem. Dak., Ind., Mont., 510 (P. lineatus).
1895. Goding, Syn. and Cat. N. A. Cercopidæ (P. lineatus in part).
1895. Gillette & Baker, Prelim. List Hem. Colo., p. 70 (P. lineatus).

This is the very common western species, heretofore referred to *lineatus*. Say's colour description is a very good one. It is a larger, more robust species than *lineatus*, with the elytra broader in proportion to the width. The face is very much more strongly convex as viewed from the side than in *lineatus*. There are also other minor differences.

I have specimens of a small male variety from Northern Colorado in which the head and thorax are darker, and the elytra, except the usual costal markings, black.

Philænus americanus, n. sp.

Resembling *bilineatus* in size and form as viewed from above. It has the flatter face of *lineatus*, which it resembles very closely in colour. It, however, differs very markedly in the elytral venation, which is very weak and distally broken up into irregular reticulations. The vertex is longer in proportion to length of pronotum than in *lineatus*. Length, 6 mm.

I have before me nine specimens, all very uniform in characters, sent by Prof. A. P. Morse, of Wellesley College, from the following localities : Dover, Mass., June 26 ; Wellesley, Mass., Aug. 8 ; Thompson, Conn., Aug. 4.

TWO NEW PARASITES FROM EUPOEYA SLOSSONIÆ.

BY WILLIAM H. ASHMEAD, WASHINGTON, D. C.

The two new hymenopterous parasites described below were bred by Dr. Harrison G. Dyar, from the larva and cocoons of *Eupoeya Slossoniæ*.

PELECYSTOMA, Wesmael.

Pelecystoma eupoeyiæ, n. sp.

♀.—Length, 4.5 mm. Head, thorax and abdomen above brownish-yellow ; collar above, the middle mesothoracic lobe anteriorly, the lateral lobes and the metathorax, fuscous or blackish; head beneath mouth parts, pronotum, thorax at sides and beneath, tegulæ, legs and venter, white; the tarsi more or less and the femora toward apex with a decided yellowish tinge ; stemmaticum dusky, the ocelli pale ; occiput with two dusky spots. Antennæ 48-jointed, slender, much longer than the body, the scape and pedicel somewhat dusky, the flagellum pale brownish-yellow. Mesonotum smooth, trilobed, the metanotum shagreened. Wings hyaline ; the costa, stigma, poststigmatal and basal veins pale yellowish, the other veins dusky ; second abscissa of radius about three times as long as the first, the second submarginal cell, therefore, long, longer than the first and almost as long as the third. Abdomen as long or possibly a little longer than the head and thorax united. Segments 1–3 coarsely longitudinally striated, the following almost smooth, but opaque ; the first segment is scarcely as long as the second and third united, the third about half the length of the second, the fourth and following shorter, subequal ; ovipositor distinctly exserted, scarcely as long as the basal joint of hind tarsi, the tip black. Type, No. 3648, U. S. N. M.

Described from a single female specimen.

CRYPTURUS, Gravenhorst.

Crypturus Dyari, n. sp.

♀.—Length, 6.5 to 8.5 mm. Head and thorax marked with white, the abdomen black banded with white ; antennæ with a broad white annulus ; palpi and legs fulvous. Antennæ 29–30-jointed, black ; the apex of joint 6, joints 7–11 entirely, and base of 12th joint, white. Clypeus, a spot above, spot on cheeks, anterior orbits extending to back of eyes, collar above, large spot just before the hind angles of pronotum, two abbreviated median lines on mesonotum, spots on mesonotum ridges that extend to scutellum, the scutellum, the postscutellum, the tegulæ, a spot beneath, a large spot on mesopleura just above the mesosternal suture, spot at base of hind wings, the blunt but prominent metathoracic tubercles and rather broad bands at apex of all abdominal segments, white. Head sparsely punctate ; thorax punctate, the mesonotum medially somewhat rugoso-punctate, laterally more evenly and less closely punctate, the mesopleura medially with some coarse transverse striæ, just back of which is a smooth polished spot, but above and below closely punctate ; metathorax with only the basal transverse carina present and which is sinuate medially, the basal enclosure thus formed finely rugulose, but beyond it the surface is rather coarsely rugose ; the white metathoracic tubercles are short, blunt and wider or longer than high. Wings hyaline, the stigma lanceolate, brownish, the other veins black ; areolet quadrate in position but open behind. Abdomen shining, but under a strong lens exhibiting a very fine coriaceous punctuation.

♂.—Length, 7 mm. Agrees well with the female, except the face below the antennæ, including the semicircular labrum, is entirely white, the mandibles with a white spot at base, the antennæ entirely black, not ringed with white, 29-jointed, the front coxæ and trochanters whitish, while the hind tibiæ, except near base, their spurs and their tarsi, are black. Type, No. 3649, U. S. N. M.

Described from one male and three female specimens.

The two previous species known in our fauna were described by the writer and from the male sex only. The *males* of these three species may be tabulated as follows :

A. Head and thorax with rufous markings.

> Legs rufous ; hind tibiæ, except at base, and their tarsi, black ; tibial spurs red (Texas)...........(1) *C. texanus*, Ashm.

AA. Head and thorax with white markings.

> Legs rufous, the coxæ white with black markings ; second joint of hind trochanters, tips of hind femora, apical two-thirds of their tibiæ, black ; their tarsi, except extreme base of first joint and more or less of the last joint, which are black, white (Michigan)...........(2) *C. albomaculatus*, Ashm.

> Legs fulvous, anterior coxæ and trochanters white, hind tibiæ, except at base, their spurs and tarsi, entirely black, their femora not tipped with black........(3) *C. Dyari*, Ashm.

NOTES ON PREDACEOUS HETEROPTERA, WITH PROF. UHLER'S DESCRIPTION OF TWO SPECIES.

BY A. H. KIRKLAND, ASSISTANT ENTOMOLOGIST TO THE GYPSY MOTH COMMITTEE, AMHERST, MASS.

During the month of May, 1896, while making field observations in Malden and Medford, Mass., upon the insects known to attack the gypsy moth *(Porthetria dispar)*, I found that many of the common predaceous bugs upon emerging from hibernation greedily availed themselves of the food supply offered by the tent caterpillar and destroyed large numbers of this insect. *Podisus placidus, P. serieventris, P. modestus, Dendrocoris humeralis, Euschistus fissilis, E. tristigmus, E. ictericus, E. politus* n. sp., *Menecles insertus* and *Diplodus lividus* were often found feeding upon partially grown tent caterpillars. *Podisus placidus* and *P. serieventris* enter the tents and prey upon the inmates, but the other species generally attacked the larvæ while they were feeding. The species of *Euschistus* are the least predaceous and it is probable that they naturally feed more upon plants than upon insects.

When feeding, these Pentatomids insert the setæ only, and not the sheath, into the body of the caterpillar. I have watched them very carefully under a hand lens and my observations fully agree with those of Mr. Marlatt, as given in the Proceedings of the Entomological Society of Washington, Vol. II., p. 249. I have seen *P. placidus* extend its setæ beyond the end of the beak to a distance equal to the length of the last rostral joint. When the setæ are inserted in a strongly chitinized part the struggles of the larva often pull them from the sheath. In such cases the beak is drawn through the fore tarsi in the same manner that an ant cleans its antennæ, and thus the setæ are forced back into the sheath. I have also removed the setæ of *P. cynicus* from the sheath by means of a fine needle applied along the labrum and have seen them replaced in the same manner.

In the Report of the Massachusetts Board of Agriculture for 1896 I have published, with illustrations, notes on a part of the early stages and habits of some of these Heteroptera and the life history of *P. placidus*. This insect was first brought to the attention of entomologists through some very interesting notes published by Prof. Saunders in the CANADIAN ENTOMOLOGIST, Vol. II., p. 15. The nymphs of this species, at first thought to be *Stiretrus anchovago* (Fab.) *(fimbriatus*, Say), were found attacking the larvæ of the currant sawfly, *Pteronus ribesii* (Scop.). Walsh, on page 33 of the volume cited, corrects the identification and refers the insect to *Podisus spinosus* or *modestus* or to an allied species. Later specimens sent to Prof. Uhler (not Ulke) were found to represent a new species and were named *Arma placidum* (CAN. ENT., Vol. II., p. 93). Prof. Saunders also gives notes upon the predatory habits of this insect in the Report of the Entomological Society of Ontario, 1871, p. 31.

I have been unable to find the original description of the species. Through correspondence with Prof. Uhler I learn that he cannot recall the circumstances connected with the publication of the description, were such a description published, and he has very kindly sent me the following characterization of the species together with a description of *Euschistus politus:*

Podisus placidus, Uhler.—" Of a narrower and more oval form than *P. serieventris*, with a head somewhat tapering anteriorly, and rounded at tip instead of being truncated, and with the humeral angles rounded off and very moderately prominent. Colour pale testaceous, stained with pale brown and punctate with darker brown. Head much longer than wide, depressed, remotely punctate, the edge reflexed, brown ; each side of tylus is a slender brown line which is triangularly expanded on the base of the vertex ; occipital margin dark brown in the middle, pale and narrowly callous each side ; a pale callous line extends back from each ocellus ; throat whitish testaceous ; cheeks with a slender black line before each eye ; eyes brown, bordered with testaceous behind ; antennæ pale brown, paler at base and on the last two joints ; the basal joint testaceous, very short, the second longest, third scarcely more than half the length of the second, fourth about three-fourths as long as the second, fifth a little shorter than the fourth ; rostrum stout, pale testaceous, reaching upon the posterior coxæ, the apical joint narrow, about as long as the preceding one, brown. Pronotum with the sides straighter than usual, the lateral margin narrowly callous, pale ivory-yellow, and with a few indented points and small teeth before the middle ; the submargin with a brown line, surface with wavy, transverse pale lines between the pale brown marbling, more generally brown behind the middle ; posthumeral margins slightly sinuated ; anterior margins callous, having a small group of coarse punctures behind each eye ; punctures sunken, brown, mostly not close together in the transverse series ; posterior margin truncate. Scutellum long, bluntly rounded and margined with white at tip, punctures in short transverse series, grouped in about three spots at base. Corium slenderly bordered with pale testaceous, more broadly covered with brown at base and on the disk, the veins posteriorly yellow ; membrane pale bronze. Legs minutely speckled with red, the tibiæ and tarsi a little stained with brown. Under side finely punctate, the sternum with two series of black points. Connexivum depressed, punctate, the outer edge ivory white, callous and marked with two black points at each incisure of the segments ; the upper surface yellow, with the black points more linear. Length to end of abdomen, 8½ to 10½ mm. Width of pronotum, 4⅔ to 6 mm.

"Through the kindness of many friends I have had an opportunity to examine specimens from the Provinces of Quebec, Ontario, and Columbia, in British America; from nearly all of the New England States, besides Illinois, Iowa, Michigan, and Colorado. The genital segment of the male is deeply excavated, and with two short processes on the middle. The tergum is often bright red, which colour becomes brownish in more mature specimens. The humeral angle is usually more or less black. In some specimens there is a series of minute black dots each side of the venter, and a few obscure spots distributed over the ventral surface."

Euschistus politus. New sp.—" Pale dull fulvous, or rufo-fulvous, suboval, with the humeral angles almost rounded and very moderately prominent. Head narrow, as in *E. tristigmus*, Say, deeply and finely punctate, the tylus prominent at tip and a little longer than the lateral lobes, the lateral lobes deeply sinuated, with the outer margin blackish. A black line extends from the eye to base of antennæ ; antennæ clay yellowish ; the basal joint short, hardly reaching the apex of head, marked with a few black points ; second joint longer ; third a little longer than the second ; fourth longer, dusky at tip ; fifth a little longer than the fourth, fusiform, blackish excepting at base ; rostrum pale testaceous, slender, with the setæ piceous, reaching to the posterior coxæ. Pronotum much wider than long, polished, closely and finely punctate with brown ; the lateral margins very slightly sinuated, smooth, ivory white ; the submargins blackish ; humeral angles triangularly rounded ; posthumeral margins almost straight. An obsolete, callous, imperfect curved line extends between the humeral angles. Scutellum narrow and bluntly rounded at tip, where it is also slenderly margined with white ; the surface is less densely punctate in small spots. Wing-covers closely punctate ; membrane a little brownish, the veins and numerous dots darker brown. Legs pale yellow, remotely dotted with brown. Beneath pale greenish, finely punctate, highly polished, the pleura with a row of fine black dots, and an extra dot outwardly ; connexivum acute, the intersegmental sutures indented and marked with a black dot. Tergum black, the sutures, exteriorly, with a double black spot. Length to end of abdomen, 9 to 10 mm. Width of pronotum, 5½ to 6 mm. A pair of these insects taken in Massachusetts have been kindly given to me by Mr. A. H. Kirkland. Other specimens have been sent to me for examination from Rhode Island, Pennsylvania, and the District of Columbia. I have found it once, July 4, in a sandy pine woods district in southern Maryland. Only a few specimens have thus far been reported. It seems to be of rather uncommon occurrence."

GRAPTA INTERROGATIONIS.

BY ARTHUR J. SNYDER, N. EVANSTON, ILL.

Under the title " Notes on Vanessa Interrogationis," in the February number of CAN. ENT., Mr. W. F. Fiske gives some interesting statements corresponding to observations made here. I kept bait for moths on the trees in and near my yard from the beginning of the year 1896, and captured Noctuids during January, February, and March.

Diurnals came to the bait for the first time on April 12th. Vanessa Antiopa led the van, followed closely by the Graptas and Pyrameis Atalanta. In a few days *Interrogationis* and *Atalanta* were abundant. *Grapta Comma* appeared on the 17th of April.

April 24th I made the following note in my record : " Previous to this date all the *Grapta Interrogationis* were hibernating specimens and of the form *Fabricii*. This evening (my observations were made from four p.m. 'till dusk) all were of the dark form *Umbrosa*, but also all old hibernating specimens."

On the 25th both *Umbrosa* and *Fabricii* were seen. During the last of April and first part of May Graptas were exceedingly abundant.

On May 7th saw the first Grapta depositing eggs on elm. Captured the ♀ and found it to be *Umbrosa*. A single butterfly procured from these eggs was of the form *Umbrosa*.

Soon the eggs and larvæ of Graptas were abundant on the elm trees and shrubs, especially on the low branches of young trees. One could hardly turn over a bough of one of these without finding several larvæ.

Mr. Fiske came near proving a point concerning which many of us are interested, but the weak point is this : Did he examine the leaves of the branch of elm on which he netted the ♀ *Umbrosa?* If not, how does he know that there were no eggs upon the limb at the time of confining the ♀ there?

I have frequently found upon the same limb larvæ of Graptas in several stages of maturity, small ones just hatched, and others almost ready to pupate.

I am inclined to think that *Umbrosa* and *Fabricii* may be obtained from eggs laid by one ♀, just as Mr. W. H. Edwards has succeeded in raising imagoes of *Papilo Oregonia* and *Bairdii* from eggs laid by a single individual.

To prove these points just as we would have them, both sexes should be reared, each form paired with its kind, and *vice versa*, and the results noted. The second generation of specimens thus observed should settle the question.

While I cannot positively answer Mr. Fiske's question as to where the immense number of Umbrosa came from, the observations made here go to prove that the uncommon appearance of the species was not confined to one locality, but the "wave" probably extended over the entire eastern United States. It is my opinion that the preceding autumn was an unusually favorable one for the Graptas, for both *Umbrosa* and *Fabricii* were common here in August, 1895.

Grapta Comma was very abundant here in the autumn of 1892, but did not appear in great numbers again until the spring of 1896.

Papilio Ajax is very rare here in ordinary years, but in 1895 suddenly great numbers of badly worn specimens appeared and remained for some days. Every collector captured examples, I think, but hardly any one secured a perfect specimen.

The nearest point at which the food plant of *Ajax* is found, so far as I have been able to ascertain, is on the Michigan side of Lake Michigan. In this case the butterflies may have been carried from their usual haunts by winds.

Insects undoubtedly migrate, sometimes suddenly and in immense numbers, as has been noted of *Danais Archippus* and *Callidryas Eubule,* and sometimes slowly, taking years to reach a certain locality hitherto unknown to the species.

Chrysophanus Helloides is moving eastward. A few years ago it was considered a Rocky Mountain species, but lately specimens have been taken in Iowa, Illinois, and Indiana.

Another question is why the form *Fabricii* should appear before *Umbrosa* and then later on both forms appear at the same time?

The broods of *Interrogationis* seem very irregular as to time of appearance, but there are at least two annual broods here.

A NEW CŒLIOXYS FROM NEW MEXICO.

BY T. D. A. COCKERELL, MESILLA, N. M.

Cœlioxys menthæ, n. sp.— ♂ .   Length 9⅓ mm., black with the legs and base of abdomen ferruginous. Pubescence scanty, dull white, rather dense and tinged with ochraceous on face.   Head rather large ; vertex shining, with large, well-separated punctures ; mandibles bifid at ends, ferruginous except tips and extreme base ; antennæ black, flagellum faintly rufescent beneath towards the end ; mesothorax shining, with extremely large, well-separated punctures ; a band of dull white pubescence at base of scutellum and a patch above base of wings ; scutellum shining and sparsely punctured, without any trace of a keel, rounded behind, with a very small tubercle at the middle (representing the median tooth of *aperta*, etc.), lateral teeth large, flattened and rounded at tips ; enclosure of metathorax distinct, very finely granular, with a basal series of large pits ; tegulæ apricot colour ; wings dusky hyaline, the apical margin broadly smoky, nervures piceous, stigma fuscous, marginal cell more produced at tip than in *altilis* ; coxæ more or less darkened, legs otherwise entirely bright ferruginous, with the pubescence extremely scanty ; abdomen shining, segments 2–5 with transverse sublateral grooves ; punctures sparse, largest and densest at sides, rather small and numerous on dorsum of first segment, absent on dorsal middle of segments 2–5, except for an apical row and on 2 an imperfect basal one ; sixth segment with sparse minute punctures.   Hair-bands very narrow and interrupted dorsally, so as to be inconspicuous.   First segment except the extreme base entirely ferruginous ; second and third segments, and fourth more or less, ferruginous at sides ; venter ferruginous except apex. Apex with six teeth, of the terminal ones the lower are the longer.

Hab.—Deming, N. M., at flowers of garden mint in Mrs. Bristol's garden, July 9, 1896. (Ckll. B. 45.) Very distinct by the sparsely punctured (in parts impunctate) abdomen with its rufous first segment. Nearest, perhaps, to *C. texana*, Cr.

There is a *Cœlioxys* taken by Prof. Townsend on the Gila R. in numbers, which I could not definitely identify.   A specimen sent to Mr. Fox comes back marked " near *mœsta*."   Very possibly the species is new, but I do not at present care to give it a name, as there are several closely allied forms which I have not seen, and it may be one of them.

Mailed May 1st, 1897.

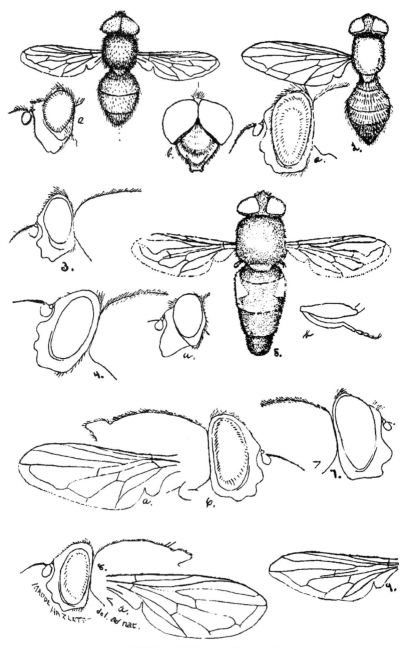

NORTH AMERICAN SYRPHIDAE.

The Canadian Entomologist.

Vol. XXIX.　　　　LONDON, JUNE, 1897.　　　　No. 6.

CONTRIBUTION TO THE KNOWLEDGE OF NORTH AMERICAN SYRPHIDÆ.—II.

BY W. D. HUNTER, INSTRUCTOR IN ENTOMOLOGY, UNIVERSITY OF NEBRASKA.

Plate V.

An interesting part of this paper will be found to deal with some material from Alaska. During the summer of 1896 Prof. L. L. Dyché, of the University of Kansas, the well-known taxidermist, made an expedition to Cook's Inlet, Alaska, and from there inland. A number of species of Syrphidæ were taken simply as a side issue, the expedition not being an entomological one at all. Unfortunately, during the long journey back many of these specimens were damaged beyond all hope of recognition. The material that came through, however, without damage, although consisting of only thirteen species, makes quite a contribution to the knowledge of the Dipterous fauna of that interesting region. Although some of the orders of insects, notably the Coleoptera, have been quite assiduously collected in Alaska, and extensive reports written upon them, in the Syrphidæ, as is the case in all of the families of Diptera, no collections of importance have been made. The whole of the literature of Dipterology contains the record of only seven species as occurring in Alaska. These are mostly from Loew's Centuries as follows :

Chrysotoxum derivatum, Walker, List, iii., 542 (Yukon River).

Platychirus pelatus, Meigen, Syst. Bschr., iii., 334 (Syrphus). Sitka, Loew.

The authority for this entry is Osten Sacken, Cat. 1872, 122, "Sitka according to Loew." I do not know where this record was made, nor indeed if it was ever made outside of letters.

Sphegina infuscata, Loew, Centuries iii., 23 (Sitka, Sahlberg).

Baccha obscuricornis, Loew, Centuries iii., 117 (Sitka, Sahlberg).

Sericomyia chalcopyga, Loew, Cent. iii., 20 (Sitka, Sahlberg).

Eristalis Meigenii, O. S., West. Dipt. 337 (Yukon River).

Xylota barbata, Loew, Cent. v., 40 (Sitka).

Since none of the species taken by Prof. Dyche duplicate those above, the total number of Syrphidæ known from Alaska is brought up to twenty.

These species are included in their systematic relation below. It has been thought best, however, to place them in ensembe form here.

Chilosia gracilis, n. sp.

Chilosia plutonia, n. sp.

Chilosia alaskensis, n. sp.

Melanostoma mellinum, Linn. I have also seen specimens of the species taken at Ft. Wrangel by Prof. Wickham.

Syrphus intrudens, O. S.

Syrphus mentalis, Williston.

Syrphus protritus, O. S.

Syrphus Lesueurii, Macq.

Syrphus umbellatarum, Schiner.

Eristalis occidentalis, Williston.

Helophilus latifrons, Loew.

Helophilus Dychei, Will.

Xylota ejuncida, Say.

The preponderance of *Chilosia* and *Syrphus* forms which are known to be mountainous is conspicuous; that the three species of the former genus are all new is not surprising considering the state of our knowledge of them in this country. The occurrence of three European species out of the relatively small total number is rather remarkable, and bears out the law of the occurrence of such forms in the West rather than in the East of this country, or at least that where they occur in the East they also occur in the West. That two species of *Helophilus* should be found is entirely as would be expected of such a northern genus; although that one of them should be new, and that in a restricted group of northern forms, which are of almost circumpolar distribution, is noteworthy.

All of this Alaskan material was placed in the form of a rough draft of a paper by Dr. Williston. In a most truly generous spirit he turned the paper with the specimens over to me, advising me to make any changes that I might see fit, and giving me full permission to incorporate it in the present paper. This has been done. The additions of mine are the preceding part, the descriptions of the three new species of *Chilosia*, and several notes.

1. *Microdon viridis*, Townsend, Dipt. Baja, California, in Proc. Cal. Acad. Sci., Series 2, Vol. III., p. 610 (April 8, 1895).

I have received from Prof. Aldrich a single specimen of this characteristic species.

This specimen bears the label " Knoxville, Tenn., 2nd July, '91." In reply to a letter in which I expressed some doubt as to the correctness of this label, Prof. Aldrich has assured me that the specimen was collected by Mr. H. E. Summers in Tennessee and that he has no doubt but that the label is authentic and perfectly correct. The specimen on which Mr. Townsend founded the species was from San José del Cabo Baja, California.

The remoteness of this locality from that of the type was a matter of no little surprise to me and has caused me to make an unusually diligent search of the literature to ascertain whether Mr. Townsend's species might not be the same as some previously described more widely distributed one. I have found, however, that M. viridis is entirely unique among the species of *Microdon*, although it approaches *M. devius*, Linn., of Europe.

I found recently in the collection of the Kansas State University another specimen of this species which I have ascertained was taken by Mr. Chas. Robertson at Orlando, Florida, March 16th, 1887.

2. *Microdon megalogaster*, Snow, Kansas Uni. Quart. Vol. I, No. 1, p. 34. Plate vii., Fig. 1 (July, 1892).

> *Microdon bombiformis*, Townsend, Trans. Am. Ent. Soc., Vol. XXII., p. 33 (March, 1895).

I have compared the types of these two descriptions in the collection of the Kansas University ; there is not the least doubt but that they are the same. The type of *bombiformis* is a female and that of *megalogaster* is a male of the same species. There is only a difference in size between these two specimens. Townsend states in regard to his species, " I can hardly identify this with *megalogaster*, Snow, from the differences in the wings." The wings in both specimens are fusco-hyaline, but in the female (*bombiformis*) they are perceptibly darker along all of the veins, precisely, however, as might be expected in that sex.

The locality of the specimen described as *M. bombiformis* is Dixie Landing, Va., and that of the specimen described as *M. megalogaster*, which Snow omitted to state, is Illinois,

3. *Chrysogaster pictipennis*, Loew, Centuries, iv., 58.

Numerous specimens of this species were taken by the writer at Cedar Bluffs, in Nebraska. The species has been recorded hitherto only as far west as New York. All of these specimens seem to differ from Eastern ones only in the fact that the wings are less distinctly marked.

TABLE OF THE SPECIES OF CHILOSIA INCLUDED BELOW.

1. Eyes bare..2
   Eyes pilose...3
2. Anterior cross vein distinctly bent at base, and parallel to vein at
     base of the discal cell ; thorax long pilose ; opaque black species ;
     wings very dark............................*plutonia*, n. sp.
   Anterior cross vein perfectly straight, not parallel to vein at base of
     discal cell ; thorax very short pilose ; shining olivaceous species ;
     wings hyaline... .... *gracilis*, n. sp.
3. Scutellum with bristles or bristle-like hairs on the margin........4
   Scutellum without bristles on the margin ; tibiæ largely reddish ;
     wings tinged with yellowish ; white pilose, bluish
     species...............................*punctulata*, n. sp.,
4. Face hairy ; thorax white pilose ; anterior cross vein oblique ; robust,
     shining brassy species......................*pacifica*, n. sp.
   Face not pilose ; thorax black pilose ; anterior cross vein rectangular,
     deep blue, shining species..................*alaskensis* n. sp.

4. *Chilosia alaskensis*, n. sp.   Plate V., Fig. 4.

Everywhere deep blue, shining, very short pilose. Eyes pilose, arista scarcely pubescent, incrassate on the basal half, scutellum with bristle-like hairs.

Female.—Eyes very short sparse pilose, appearing white from above. Front shining blue, sparsely punctured short black pilose, with a large sulcate swollen area above the antennæ. Face very prominent, deeply concave below the antennæ to the rounded, very prominent tubercle situated a trifle below the middle of the eyes, thence shortly but not very deeply concave to the epistomal tubercle which is only slightly less prominent than the upper, and is situated considerably above the lower eye margin. Below the lower tubercle straight, slightly receding. Cheeks narrow, lower border straight, epistoma not truncated at tip. Antennæ and margin of the antennal orifice reddish-yellow, first and second joints and the narrow upper margin of the third brownish. Third joint very large, circular with the upper outer margin slightly less convex.

Arista long, basal, brown, very indistinctly pubescent. Thorax shining blackish-blue, short black pilose. Scutellum with slender bristles on the margin. Abdomen oval, wider than the thorax, everywhere shining dark blue, almost bare. On the dorsum of the abdomen the pile is black, on the margins, especially anteriorly, it is white, and on the sides of the second segment rather long. Legs black, knees and narrow base of the tibiæ only lighter, short black pilose. Wings hyaline, the stigma and all the veins light luteous. L. corp. 8 mm., L. al. 8 mm.

One specimen : Cook's Inlet, Alaska ; Prof. L. L. Dyche, of the University of Kansas.

5. *Chilosia plutonia*, n. sp. Plate V., Fig. 7, 9.

Allied to *C. Willistoni*. Eyes bare, arista plumose, scutellum with bristles, legs black, second and third abdominal segments opaque except the anterior corners, thorax long black pilose, wings very dark.

Male.—Frontal triangle swollen, but little shining, long black pilose. Ocellar area similarly pilose. Face not pilose, very slightly pollinose, gently concave to the tip of tubercle which is round and distinct, thence only very slightly concave to the tip of the epistoma ; but little produced below the eye. Occiput long white pilose below. Antennæ small, black; third joint yellowish-red, a trifle longer than broad, rectangular with the lower basal corner bulging slightly outwardly. Dorsum of thorax and pleura subshining, long, erect, black (in all lights) pilose, finely punctured. Scutellum shining, quite distinctly punctured with bristles and coarse hairs on the margins. Abdomen not wider than the thorax, opaque. There are shining brassy triangles on the anterior angles of the second and third segments ; these spots extend about one-half of the width of the segment laterally and about the same distance inwardly. On the fourth segment there is a complete anterior shining band of metallic. Pile of abdomen sparse, long on the lateral margins, on the opaque portions black, on the shining portions whitish. Hypopygium shining, white pilose. Legs entirely black, long black pilose ; on the anterior and middle femora the pile is long and slender, forming loose cilia, on the inner side of the posterior femora it is short and spinous. Wings very dark, especially before the anterior cross vein. L. corp. 8½ mm., al. 8 mm.

In some lights the fourth abdominal segment seems almost entirely shining, and the anterior and middle legs seem whitish pilose.

One specimen : Cook's Inlet, Alaska ; Prof. L. L. Dyche.

6. *Chilosia Aldrichi*, Hunter.   Plate V., Fig. 8, a.

Several additional specimens have been received from the same locality as the type, Idaho.

7. *Chilosia gracilis*, n. sp,   Plate V., Fig. 3.

Eyes bare, arista plumose, scutellum with bristles on the margin, legs black.

Female.—Shining black, somewhat greenish, almost bare. Antennæ of moderate size, first and second joints piceous, third bright reddish-yellow, somewhat longer than broad, elliptical ; arista black, basal, long loose plumose. Front plane, short luteous pilose, longer black pilose near the ocelli. Face and cheeks bare, shining, lower anterior orbits very short white pilose. Face considerably obliquely produced below, with a conspicuous round tubercle below the middle moderately concave above ; between the tubercle and the tip of the epistoma there is a short deep concavity. Occiput white pilose. Dorsum of thorax shining, distinctly punctured, very short black pilose in the middle and yellow pilose around the margins, quite widely so anteriorly. Pleura more olivaceous than the dorsum, shining. Scutellum with two apical and three shorter lateral bristles on each side. Abdomen everywhere shining with a greenish tinge, much broader than the thorax at the apex of the second segment, with short white pile that appears to be arranged in bands on the segments ; the lateral margins of the first, second and third segments have longer erect pile. Legs entirely black, the knees, especially the anterior pair, lighter ; the pile is very short, sparse, and in most lights white. Wings uniformly grayish hyaline, veins black. Tegulæ white, fringed with somewhat yellowish. L. corp. 6 mm.; al. 6½ mm.

One female specimen : Cook's Inlet, Alaska, 1896 ; Prof. L. L. Dyche, of the University of Kansas.

This species is very closely allied to *C. Willistoni*. It differs, however, as follows : The tubercle is much more distinct, and between it and the tip of the epistoma there is a short deep concavity. In *Willistoni* the tubercle is so indistinct that between it and the epistoma the outline is almost perpendicular. The face is produced quite distinctly, more downwardly in this species. The pile of the dorsum is black ; in *Willistoni* it is luteous. The pile of the abdomen is also much more sparse and finer ; in *Willistoni* it is quite uniform and not arranged in bands.

8. *Chilosia pacifica*, n. sp. Plate V., Fig. 2, a.

Male —Eyes pilose, scutellum with bristle-like hairs on the margin, arista bare, abdomen largely opaque, robust, thickly white pilose.

Female.—Shining brassy, abdomen broad, entirely shining, antennæ brown, third joint reddish.

Male.—Eyes long dense, whitish pilose. Front swollen, sulcate, long black pilose. Face uniformly lightly white pollinose and short, sparse, white pilose below, extending only moderately below the eyes, obliquely truncate at the apex, the lower border of the cheeks straight. In outline the face is almost straight below the antennæ to the inconspicuous, obtuse, nasiform tubercle, thence distinctly concave to the tip of the epistoma, which forms a second tubercle almost as large as the upper one. Antennæ of moderate size, the second and third joints black, third brown on the upper half, yellowish-red below, very slightly broader than long, almost square, the lower outer angle rounded. Arista, bare, basal, black. Dorsum and pleura shining greenish, densely, long, erect whitish pilose. Scutellum with long, bushy, white pile, intermixed with slightly strengthened black hairs on the margin. Abdomen but little broader than the thorax, everywhere long erect whitish pilose, first segment shining, second entirely opaque, third ópaque, except a narrow posterior margin and lateral triangles reaching from the anterior margin two-thirds of the width of the segment, of shining green, fourth segment entirely shining greenish. Legs black, long white pilose. The basal third, and the narrow apex of the middle and anterior tibiæ and the basal third of the posterior are dull testaceous. The colouring of the posterior pair is very inconspicuous. Wings grayish hyaline, veins brown, stigma luteous. Tegulæ white. L. c. 10 mm., al. 8 mm.

Female.—Front shining brassy, coarsely punctured, pitted above the base of the antennæ, short white pilose. Along the eye margins, midway between the antennæ and the ocelli, there are short elevated ridges. Face shining greenish-black, short, sparse, whitish pilose below on the sides ; the orbital margins densely short, white pilose. Face considerably concave below the antennæ to the conspicuous tubercle, thence with a short, deep concavity to the epistoma, which forms another less conspicuous tubercle, not produced below the eyes nor obliquely at the apex, which is broadly truncate. Lower border of the cheeks slightly concave. Occiput long yellowish pilose. Eyes much shorter pilose than on the male. Antennæ moderate in size, deep brown ; third joint reddish-brown, blackish above, slightly longer than broad, the lower corner slightly less convex than the upper. Thorax shining greenish, short white pilose. Scutellum fringed with rather short white pile, with six slender black bristles arranged as follows : two on each side near the apex, one more slender on each side near the base. Abdomen considerably broader than the thorax, everywhere shining brassy and rather short, dense, short, appressed pilose ; on the lateral margins the pile is longer and erect. Legs short, whitish pilose, femora entirely black, except the extreme apex, anterior, and middle coxæ reddish, all the tibiæ obscurely reddish, except a broad subapical band occupying more than a third of the width of the tibiæ, front and middle tarsi reddish, the two apical joints blackish, posterior tarsi entirely blackish. Wings subhyaline, the basal half slightly coloured with yellowish, veins brown, stigma luteous, stumps of veins at the bases of the apical and posterior cross veins.

L. corp. 9 mm., al. 8 mm.

Two specimens : one bearing the label "Cal., R. W. Doane coll.," and the other, "Palo Alto, California, March 29, 1895."

This species is allied to *C. occidentalis*, Will., also from California. It may be distinguished, however, from that species among other characters by the colour of the pile and the presence of stumps of veins at the bases of the apical and posterior cross veins.

9. *Chilosia punctulata*, n. sp.   Plate V., Fig. 6, a.

Eyes pilose, arista bare, scutellum wholly without bristles. Everywhere profoundly punctured ; wings uniformly, distinctly yellowish.

Female.—Front deeply punctured, wholly without swollen processes, but little shining, pile short, dense, in some lights blackish, from above white. The orbits on the lower part of the front and the upper part of the face expanded as a narrow band just below the base of the antennæ, white pollinose. Face bare, shining black, deeply concave below the antennæ to the conspicuous round tubercle, thence shortly and deeply concave to the oral margin, which is obliquely truncate. Cheeks narrow, bare, shining, lower border straight. Antennæ situate above the centre of the eyes, second and third joints bright reddish-yellow (sometimes more brownish), first and the narrow orifice brownish. Third joint moderate, a trifle longer than broad, regularly elliptical. Arista bare, basal, yellow at apex. Eyes very short, sparse, white pilose. Mesonotum densely punctured but little shining, pile short, whitish, on pleura below the base of the wings longer and white. Scutellum without bristles, deeply punctured like the mesonotum, with a loose fringe of fine white pile showing from below the margin. Abdomen broadly elliptical, everywhere deeply and conspicuously punctured and subshining. Pile rather abundant, white. When viewed from above and at one side the pile of the third and fourth segments seems to form broad arcuate bands curving from the apical corner of the segment inwardly. Legs white pilose ; all the femora except a narrow tip black ; tibiæ reddish-yellow with an indication of a brown median band, more pronounced on the posterior pair. Tarsi yellow, two apical joints darkened. Posterior femora with several short spinous bristles below near the apex. Wings short, broad, uniformly tinged with yellow ; veins yellow.

Length, 8½ mm.; al., 6½ mm.

Two specimens : West Point, Nebraska, September 9.

This species is very closely allied to *C. sororia*, Will., from Mexico and to *C. petulca* from Washington State. In the shape of the antennæ and outline of the face it agrees precisely with *petulca* but differs in the absence of the scutellar bristles. This is the only character mentioned by Williston in the Biologia C. A. as distinguishing *sororia* from *petulca*. The character, however, in this species which leads me to consider it very distinct is the deep punctuation. The front of *C. sororia* is described as "shining metallic," and the mesonotum as "metallic green," which would certainly indicate that these parts are not deeply and closely punctured. In this species the front and mesonotum are very deeply and conspicuously punctured, so that they have a roughened, granulated appearance and are subopaque. The wings in this species are much more yellowish than the description of *C. sororia* would seem to indicate they are in that species, and there are several other differences.

10. *Melanostoma mellinum*, Linn.

Two specimens : Cook's Inlet, Alaska ; coll. L. L. Dyche.

11. *Platychirus chaetopodus*, Williston. Synopsis N. A. Syrphidæ, p. 59, 1896.

Four male specimens were taken on the Pine Ridge in North-western Nebraska by the writer during July, 1896. The species was described from the State of Washington and Snow has recently recorded it from Colorado. The abdominal markings are larger than the description seems to imply.

12. *Syrphus intrudens*, O. S.

Four specimens from Cook's Inlet, Alaska ; coll. L. L. Dyche. The legs are darker coloured than Osten Sacken describes them. This species has been recorded from California and Colorado.

13. *Syrphus mentalis*, Will.

Two specimens : Cook's Inlet, Alaska ; coll. L. L. Dyche. This species is known from the State of Washington. These specimens show a considerable variation from the description and from each other. However, the points in which one specimen differs from the description are the very points in which the other specimen agrees. I am thus led to believe that the species is variable.

14. *Syrphus protritus*, O. S.

One specimen : Cook's Inlet, Alaska ; coll. L. L. Dyche. This species was described from California.

15. *Syrphus Lesueurii*, Macq.

Two specimens : Cook's Inlet, Alaska ; coll. L. L. Dyche. This species has been recorded from New England and once from the Pacific Coast.

16. *Syrphus umbellatarum*, O. S.

Two specimens : Cook's Inlet, Alaska ; coll. L. L. Dyche.. This species has been recorded from New Hampshire to Arizona, but never from the North-west. This record gives the species an immensely increased range.

17. *Baccha clavata*, Fabr.

One female specimen taken on flowers of Aster multiflorus, Sept. 28th, 1896. It differs from the description in having two small yellow spots on the first abdominal segment corresponding to those on segments two and three and in lacking the white pile on all except the first. It is without doubt, however, this species. This is the second occurrence of this species at Lincoln, Nebraska. It was previously taken in 1895 under similar circumstances.

XANTHOGRAMMA, Schiner.

The astute Prof. J. Mik (Wien. Ent. Zeit., 1897, p. 65,) has discovered a character that will separate this genus from Syrphus as far as the European species of these genera are concerned. He states : " Als ein bezeichendes Merkmal für die Gattung Xanthogramma habe Ich Form und Farbe der Umwallung (das sind die Klappen) des Metathoracicalstigma (neber den Hinterhuften) gefunden. Diese Umwallung ist bei allen Arten nicht sehr hoch ; sie ist schwarz und trägt auf dem freien Rande kurze, feine, schwarze •der braune Wimperhärchen."

I have sought in vain to apply this character to the North American species of these genera. I have had, unfortunately, the opportunity of examining but one species of Xanthogramma, X. flavipes, Loew, and it is quite possible that the other species of the genus may differ from in it precisely this respect. However, it is important that the character that will separate all of the European species of these genera finds its exception in this one North American species at least.

In the absence of a positive illustration of the character used by Prof. Mik, I have had some difficulty in conceiving exactly what he means. I take it, however, that the " Umwallung " is the elevated orifice of the metathoracical spiracle and the " Klappen " are the lids fitting over them and bearing on their free edges cilia of the fine black or brown bristles. If I am right in this, the character does not apply at all to X. flavipes. The orifice of the spiracle is not in the least elevated more than in any of the fifteen species of Syrphus which I have examined with special reference to this character, and the cilia is not black or brown, but only slightly yellowish,

18. *Baccha lemur*, O. S.

Four specimens : Colorado Springs, Colo., Aug. 1896 ; Prof. Bruner.

These specimens show no variation among themselves, nor differences from the description. The posterior femora uniformly have only an indication of a preapical ring.

19. *Volucella apicifera*, Townsend, Trans. Am. Ent. Soc., 1895, p. 40.

One male specimen, Las Cruces, New Mexico ; coll. Townsend, April 8, now in the collection of Prof. Aldrich I have examined. The tpye of this species, which I have also examined in the collection of the Kansas University, which was taken at Las Cruces, N. M., April 17, and this specimen agree throughout. This species is certainly, as Mr. Townsend states, very closely allied to *V. isabellina*, Will. It differs in some respects in precisely such points as a tenental form of that species would be supposed to differ. However, the markings of the legs and abdomen are exactly the reverse of what would be expected if this were an external form of *V. isabellina ; i. e.*, they are darker and more extensive. I am inclined to think, with Mr. Townsend, that there are here two distinct though closely related species.

Pyritis, nov. gen. [πυρίτις, a precious stone].

Large black, thickly pilose species, without lighter markings. Marginal cell open, anterior cross vein in middle of discal cell, third vein straight. Antennæ shoit, third joint very broad ; arista basal pilose. Eyes long pilose, widely contiguous in the male. Femora and coxæ simple, without spines or tubercles. Face very broad, the diverging eye margins form an angle of at least 80 degrees ; the apex is just above the antennæ, swollen. (In *Sericomyia* and *Arctophila* the eye margins are almost parallel.)

Type of genus, *Pyritis montigena*, n. sp., North America.

This genus falls naturally into Williston's tribe Sericomyini, which contains the genera *Sericomyia* and *Arctophila*. From both of these it may be easily separated by the peculiar formation of the face and the pilose eyes. There is one genus in the Volucellini, Phalacromyia, which has the marginal cell open. From this it differs in having the outline of the face rounded and not produced conically downward, and also in having the third antennal joint circular and not elongate. The distinctive character of this genus, however, is the remarkably wide and swollen face.

20. *Pyritis montigena*, n. sp. Plate V., Fig. 1, a, b.

Male.—Black opaque, thickly pilose. Eyes long, dense black pilose. Face and front shining, sparsely clothed with yellowish pile, intermixed with black. Front very distinctly sulcate. Face swollen, perpendicular to below the eye margins, thence receding and very slightly concave to the oral margin. Antennæ black, third joint reddish, broader than long ; arista long, loose pilose on the upper side, much less so below. Thorax long, dense, whitish-yellow pilose, the margins and three narrow indistinct central lines shining. Scutellum shining, dull testaceous. Abdomen covered with long, erect, dark yellow pile ; first and second segments opaque ; third with a shining band on the anterior margin, becoming more opaque towards the middle, where it is broadly interrupted ; fourth segment entirely shining, except a subopaque band widely interrupted in the middle. Legs entirely black ; long yellowish pilose intermixed with black on the anterior pair. Posterior pair somewhat arcuate. Wings subhyaline, with black clouds on the cross veins, and at the furcation of the second and third veins. Third vein perfectly straight. L., 12 mm.

One specimen : Moscow, Idaho ; coll. Prof. J. M. Aldrich.

21. *Eristalis Meigenii*, Wiedemann, Ausseurop-Zweiflg.—Ins., ii., 177, 35, tab. x. b., f., 15 (1830) ; Williston, Proc. Am. Phil. Soc., xx., 322 (1882) ; ibid Syn. N. A. Syrphidæ, p. 165 (1886) ; ibid Trans. Am. Ent. Soc., xiii., 318 (1886) ; F. Lynch-Arribalzaga, Anales, d. l., Soc. Cien. Argentina, xxxiv., p. 38 (1892).

Eristalis foveifrons, Thomson. — Eugenies Resa, Dipt. 419, 78 (1878); Williston, Trans. Am. Ent. Soc., xiii., 318 (1886).

Eristalis Androclus, Osten Sacken.—Western Dipt. 337 (1877); non-Walker, List, 612 (1849); ibid Cat. N. A. Dipt., note 223, p. 249 (1878); Williston, Synopsis N. A. Syrphidæ, 165 (1886).

Eristalis Brousi, Williston.—Proc. Am. Phil. Soc., xx., 319 (1882) : (Brousii), ibid Synopsis N. A. Syrphidæ, 165 (1886); Snow, Kans. Uni. Quart. Vol. i., p. 38 (1892); ibid idem., Vol. iii., p. 243 (1895); Townsend, Trans. Am. Ent. Soc., xxii., 48 (1895); Hunter, CAN. ENT., XXVIII., p. 98 (1896).

This species was described by Wiedmann in 1830 from specimens from Montevideo in South America. Thirty-eight years later Thomson, in his work on the Diptera of the Eugenies Resa, redescribed it under the name of *Eristalis foveifrons*, basing his description on specimens from Buenos Ayres.

For some time previous to 1877 Osten Sacken and Loew had been sending out specimens of a species which they identified, however not certainly, as the *E. Androclus* of Walker's List, iii., 612, to their correspondents under that name. Osten Sacken has a note in his Western Diptera (1877) concerning this species which he still at that time considered as Walker's species, *E. Androclus*. Between this time and the time of the publication of Osten Sacken's catalogue in 1878, he had examined the type of Walker's species in the collection of the British Museum and found that it was a *Helophilus*. However, he retained the name *E. Androclus*, O. S., (non-Walker) to avoid confusion.

Now, strangely enough, Dr. Williston, in Proc. Am. Phil. Soc., xx., 319 (1882), recognized the male of this species as *E. Meigenii*, but at the same time described the female as *E. Brousii* (sic). In the synopsis this was corrected and the name *Brousi* given to replace *Androclus*. It was only the immense difference in localities that prevented Dr. Williston's identification of this species with *E. Meigenii*, as he states that the full description applies almost perfectly. He is now of the opinion that they are the same, and it is at his suggestion that the investigation which has resulted in the above arrangement of the names was undertaken.

22. *Eristalis occidentalis*, Will.

Five males and three females from Cook's Inlet, Alaska ; coll. Prof. Dyche. Some of the males agree quite well with the description, except that the basal joints of the middle tarsi are not yellowish, which was an error in the description, and there is no yellow posterior margin on the second and third abdominal segments. The pile of the median segments may be yellow, or mixed with black, or chiefly black. In the female the third and fourth segments are covered with dense deep black pile, and there is no posterior opaque margin on the third or else a very narrow one. This species has elsewhere been recorded only from the State of Washington. [Williston.]

23. *Eristalis montanus*, Will.

Several specimens of this species were taken during July on the Pine Ridge in North-western Nebraska. They all have the black on the second abdominal segment as broad on the posterior margin as on the anterior ; some of them have an indication of an opaque cross band on the posterior part of the third, and in others the posterior part is entirely shining. The pile of the eyes is entire.

These specimens were captured hovering over a small, shallow pond, at an elevation of a trifle over 4,000 feet.

24. *Helophilus latitarsis*, n. sp.

Male.—Antennæ black ; arista yellowish at the base. A spot directly above the antennæ, a broad facial stripe ending abruptly before the base of the antennæ ; cheeks and narrow oral margin shining black, the facial stripe may be more brownish. Front, except the vertex which is opaque black and black pilose, and face densely yellowish pollinose and yellow pilose. Face in profile not at all conically produced below, gently concave below the antennæ to half way to the epistoma, thence perpendicular to the notched epistoma. Lower border of cheeks forms with the plane of the occiput only a very little more than a right angle. Dorsum of the thorax opaque black, everywhere short yellow pilose, complete lateral margins yellow, two median moderately broad uninterrupted silvery white stripes which reach the scutellum. Scutellum entirely testaceous pile black, on the very narrow posterior and anterior margins yellow. Abdomen, first segment opaque black, the extreme angles yellow. Second segment opaque with a very narrow posterior margin shining, bright yellow with a broad central stripe of deep opaque black not reaching posterior border and expanded on the anterior border so as to cover three-fourths of the width of the segment ; posterior band ferruginous, very narrow at the lateral angles and increasing in width to the centre of the segment, where it unites with the central stripe ; pile short, yellow except on the posterior margin. Third segment yellow, tinged with reddish posteriorly, where there is a complete narrow reddish cross-band, pilose as in the preceding segment, the black markings consist of a triangular spot, the base of which extends two-thirds of the width of the segment behind, the sides of which are concave and the apex of which is expanded unto a small elliptical spot extending less than one-third of the width of the segment, touching the anterior margin. Fourth segment with the lateral margins narrowly and the posterior margins more widely yellow, for the rest black with a broad subinterrupted pollinose band, leaving a narrow anterior band and a posterior triangle shining black, pile of posterior third black. Femora black, apical third of anterior and middle pairs yellow, an obscure reddish spot near the apex of the posterior pair; anterior tibiæ yellow on basal half, intermediate entirely and posterior with only an apical band. All the tarsi except the intermediate metatarsi black; the anterior tarsi, especially the metatarsi, are very evidently widened and swollen. The posterior femora slightly thickened, their tibiæ slightly arcuate, unarmed. Wings cinereous hyaline. Length, 11 mm.

One specimen : Minnesota.

This species belongs to the *groenlandicus* group. It is easily separable from *glacialis* and *borealis* by the only gently concave face. From *groenlandicus* it differs: (1) The median dorsal stripes are not very narrow, but broad, distinct, and reach the scutellum, the lateral thoracic stripes are not obsolete posteriorly ; (2) the femora are more extensively yellow at the apex ; (3) the pile of the thorax is everywhere yellow. From *H. Dychei*, Will., it differs in the less robust and less pilose body throughout, in the outline of the face, which in that species is obtusely conically produced and which is perpendicular below the middle in this species, in the fact that the facial stripe ends abruptly before the base of the antennæ ; the median dorsal stripes are wider and reach the scutellum, the scutellum is largely black pilose (in *Dychei* the pile of the scutellum is entirely yellow), the apical femoral bands are wider, and there is a much greater extent of yellow on the third and fourth abdominal segments.

Mr. W. A. Snow has a note concerning a specimen of a species of *Helophilus* of the *groenlandicus* group (Kansas Univ. Quart. iii., 243) which in some respects differs from *groenlandicus* precisely as this species does. From the short note given by Mr. Snow I am not certain that his specimen belongs to a species distinct from mine. In his specimen the pile of the dorsum is entirely yellow, the median stripes reach the scutellum, and the middle femora are yellow at the apex. In all of these characters it agrees with mine. On the other hand, his specimen has the dorsal stripes narrow, and the lower border of the cheeks forms with the plane of the occiput an obtuse angle. In these characters it differs from mine. A comparison of specimens is necessary to clear up the difficulty.

I have noticed in the anterior metatarsus of this species a formation that I think is of considerable value as a specific character, and which apparently has not been observed in any other species of the genus. In this species it is widened and swollen considerably more than in any of the species of *Helophilus* with which I am acquainted. My autoptic knowledge of the species of this genus includes almost all of the species except those of the *groenlandicus* group, and of that group I know only *H. Dychei*, Will. Since that species has the tarsi somewhat wider than the other species of the genus, and since the authors have paid no attention to the tarsi, I am not certain but that in the *groenlandicus* group this character is not of much value. I am certain of this, however, that the swollen anterior metatarsi of the male of this species will separate it from any member of the genus outside of the *groenlandicus* group.

25. *Helophilus Dychei*, Williston, Ms.

Male.—Face below the antennæ only lightly concave ; on the lower half nearly vertical and straight in profile ; the lower line of the head forms with the plane of the occiput an obtuse angle ; face on the sides yellowish-white, the median stripe black ; cheeks black. Antennæ black. Front on the lower part yellowish-white with yellowish pile ; on the upper constricted part more brownish and with longer blackish pile. Mesonotum opaque black, with two slender, yellowish or yellowish-white stripes, sometimes narrowly interrupted at the suture and reaching only about half way from the suture to the scutellum, pile abundant, dusky yellow. Scutellum light yellow with yellow pile. Abdomen black, the second segment with two large yellow triangular spots, extending the whole width of the segment ; third segment with the anterior angles yellow, the black of the second segment is opaque or subopaque with the narrow hind margin metallic ; that of the third segment is opaque on the anterior half or a little more ; third segment wholly shining, no whitish lunulate spots ; pile erect, yellowish. Legs black, the immediate tips of the femora and the base of the tibiæ yellow. Hind femora moderately dilated ; hind tibiæ arcuate. Wings cinereous hyaline ; sixth vein sinuous. Length, 1–2½ mm.

The female scarcely differs. There is an indication of a gray lateral stripe on the mesonotum at the humeri. The species is closely related to *H. groenlandicus*, Stæger, but differs in the extent of the shining colour of the abdomen, the absence of the pollinose spots on the abdomen, and the colour of the pile of the mesonotum and abdomen in part. There are no yellow markings whatever on the fourth abdominal segment.

Four specimens : Sitka, Alaska ; Prof. L. L. Dyche.

The above description is Williston's. The manuscript containing it was most generously turned over to me by Dr. Williston, with the permission to change it in any manner I might see fit. I have not found any change or addition to be necessary.

26. *Helophilus mexicanus*, Mcq.

I have a specimen of this species from Custer in the centre of the black hills in South Dakota. The description applies exactly. This species has not previously been recorded except in Mexico and on the Pacific Coast.

27. *Helophilus pilosus*, n. sp.

Female. — Pile everywhere, including the face, long and rather abundant. Antennæ reddish-yellow. Front opaque black, clothed with yellow pollen on the lower half or more, everywhere black pilose. Face entirely yellow, rather deeply concave below the antennæ, thence almost perpendicular, produced downwards so as to form a short, regular, sharply-pointed cone. Cheeks black, their lower border forming with the plane of the occiput a very obtuse angle. Dorsum of the thorax opaque black, with four broad, complete yellowish-white stripes, the central black interval without whitish line ; pile short yellow. Scutellum yellow, with a blackish cast ; apical margin more yellowish, pile yellow. Abdomen a trifle broader than the thorax, the sides almost parallel, pile everywhere yellow, short except on the margins of the second segment. First segment whitish pollinose, a rather large spot on each side yellow. Second segment opaque, the posterior margin shining black, on each side with an L-shaped spot of yellow extending three-fourths of the width of the segment, their inner side concave : these spots leave a very broad interval of black between them. Third segment with a broader posterior margin of shining, with two small arcuate spots beginning at the anterior angles, not approaching each other, of yellowish pollinose. Fourth segment shining on the apical half, with two similar but almost straight spots separated at their inner ends by only as much as their width. Fifth segment entirely shining. Legs yellow ; anterior and middle femora with a wide black stripe, attenuated on the apical part on upper side, reaching into the apical third of the femora. · Posterior femora with a very broad median band and a very small apical spot black. All the tibiæ yellow, the posterior pair more brownish, especially at the base and apex. Tarsi yellowish, the posterior pair somewhat brownish. Wings hyaline. Length, 9 mm.

One specimen : British Columbia.

This species differs from *H. hamatus*, Loew, in the broad abdomen, although it agrees rather closely with the description of that species in coloration. From *H. divisus*, Loew, it differs in having the face not broadly truncate on the lower portion, but sharply conical ; in the absence· of black markings on the apical portion of the anterior tibiæ, the absence of a light stripe in the median dorsal black one, and very greatly in the maculation of the abdomen. From *H. integer*, Loew, it differs in not having complete abdominal bands, in the darker femora and the absence of the facial stripe ; and from *H. obsoletus*, Loew, in the distinct markings of the thorax and the darker legs.

28. *Helophilus latifrons*, Loew.

Several specimens : Cook's Inlet., Alaska ; coll. L. L. Dyche.

29. *Helophilus divisus*, Loew ; Centur. N. A. Dipt., iv., 78.

I have a male and a female specimen of this species which were taken in coitu. They bear no locality label. Another male bears the label "Westville, N. J." The males differ in the following respects from the females in the maculation of the abdomen. Second segment yellow except an opaque black anterior band not reaching the lateral margins by its own width, about one fifth of the length of the segment, a similar posterior band narrowed at the ends but reaching the lateral margins, and a broad median longitudinal band connecting the two, the posterior margin yellow with a small pollinose spot in the middle. Second segment yellow with the black markings similar but less extensive, the anterior band only a third of the width of the segment, the posterior one with the sides slightly arcuate, the interior corners of the yellow spots and a large median posterior spot between the arcuate bands pollinose. Fourth segment entirely yellow pollinose except a slender inverted Y-shaped mark, the base of which touches the anterior margin of the segment, and the broadly divaricate segment entirely pollinose.

There is in the male near the base of the posterior femora below an obtuse tubercle covered with very short black bristle-like hairs. In this character this species shows a relationship with *H. chrysostomus*, Wied, of this country, and a stronger relationship with *H. frutetorum*, Fabr., and *H. versicolor*, Fabr., of Europe.

30. *Helophilus integer*, Loew ; Centur. iv., 78.

I have a female specimen of this species taken at Newark, New Jersey. I would make the following additions to Loew's rather short description :

Face and cheeks yellow, front black pilose, below yellow and above black pollinose. The middle, as well as the anterior and posterior femora, have small black spots on the inside at the base. These spots consist of a dense mass of minute spinous bristles. The black colour at the base of the scutellum is visible only when viewed from in front, as is the case in the related species of the genus.

31. *Helophilus aureopilis*, Townsend, Trans. Am. Ent. Soc., xxii., p. 51 (1895), is the same as *H. lætus*, Loew, Centur., iv., 77 (1863).

I am unable to see any differences between the description given by Mr. Townsend and Williston's description (Synopsis N. A. Syrphidæ, 189) of *H. lætus*, Loew. I have also examined the type of *H. aureopilis* in the collection of the Kansas State University, and compared it with specimens of *H. lætus* from New York and Colorado and find not the slightest differences between them. Mr. Townsend describes his species as " H., n. sp., aff. *flavifacies*, Bigot." *Helophilus flavifacies*, Bigot, Ann. Soc. Ent. Fr., 1883, 344, must certainly be a distinct species that will be very likely recognized in time. It differs especially in the coloration of the posterior legs, which are described by Bigot thus : " Avec trois anneaux bruns, l'un, sis à l'extrémité des cuisses les deux autres sur les tibias," thus lacking the broad conspicuous black median band on the femora. Besides this the bases of the anterior and middle femora are presumably at least yellow in Bigot's species since he does not mention that they are black, and there seems to be a difference in the maculation of the abdomen.

32. *Pterallastes perfidiosus*, n. sp. Plate V., Fig. 5, a, b.

Front and cheeks black, the former with long erect black pile, intermixed below with yellow. Face yellow, pilose, slightly concave to tip of the inconspicuous tubercle, thence straight and slightly receding to the epistoma, which is truncate at the apex. Antennæ and arista yellow, third joint a trifle broader than long. Thorax opaque black, with narrow yellow lateral borders, rather short, sparse, yellowish pilose. Scutellum translucent yellow, with an apparent black in some lights. Abdomen shaped like that of *P. thoracicus*, but a little more elongate, short yellow pilose, first segment black, somewhat shining ; second opaque black, except a complete posterior cross band, and with elongate lateral yellow triangles, which reach from the anterior angle to just before the posterior shining band, and the inner angle of which extends towards the middle of the segment about a fourth of its width ; third segment shining, except a large, square, opaque spot with deep indentations on the sides, situated on the anterior part of the segment ; fourth segment with a similar much smaller spot. All the femora black on the basal half, the anterior pair more extensively so ; tibiæ yellow, the posterior pair more or less tinged with brown at the base and apex, posterior tarsi black. Posterior femora considerably thickened with short spinose bristles below, the femora arcuate. Wings hyaline, third vein very deeply bent, marginal cell wide open, last section of the fourth vein straighter than in *P. thoracicus*, anterior cross vein in the middle of the discal cell. L., 10 mm.

Described from two female specimens bearing the label " British Columbia."

The very great differences between this species and the only other described species of the genus *P. thoracicus* has caused me no little trouble in ascertaining its generic position. The extreme looseness of the definition of the genera of the Syrphidæ makes it impossible in many cases to locate a given species in its proper genus, except by a process of finding where better than elsewhere it may be placed. The present is by a great deal the best illustration of this fact that I have so far discovered. Its location in the Eristalini is without any doubt whatever. But as between *Triodonta*, *Teuchocnemis*, *Mallota* and *Pterallastes*, it seems to fit into one about as well as into another. Of these we may more easily throw out of consideration Mallota, on account of the formation of the face and general great pilosity, although the venation is precisely as in that genus. We may next dispose of *Teuchocnemis*, in which the third vein is only moderately bent, although we are here approaching differences that are only of specific value. As between *Pterallastes* and *Triodonta*, as far as the female sex is concerned, I know of no distinction sufficient to be called generic. In the male sex there are, however, good and sufficient grounds for generic separation. What has led me to place this species in

Pterallastes rather than in *Triodonta* is simply the general habitus. I think that too much importance has been placed on the presence of pollen on the thorax as a generic character.

33. **Criorhina verbosa**, (Harris) Walker. List iii., 568.

I have one male specimen bearing the label "St. Anthony Park, Minn." that I am quite certain must be this species. The description applies exactly except as to the median facial stripe. A thick coating of grayish pollen covers the face uniformly throughout; the cheeks, however, are shining.

34. **Pocota bomboides**, n. sp.

Black, but little shining, face black, first three abdominal segments black pilose.

Male.—Very much like *P. grandis*, but legs unarmed and much smaller. Antennæ and arista reddish-yellow, the basal joints brownish. Face black, indistinctly white pollinose, a broad stripe and the cheeks shining. Dorsum of the thorax long yellow pilose before the base of the wings, the remainder and the scutellum black pilose. Abdomen—First three segments black pilose; all except the first with indistinct posterior margins. Fourth segment more shining than the others, with a band of dense long yellow pile occupying the anterior half, the remainder of the segment black pilose. Legs simple, without spines or tubercles, black pilose; all the femora black except the extreme apex, on the posterior femora the apex is more broadly reddish; anterior tibiæ on the basal half, middle except an indistinct broad band, posterior entirely dark reddish brown, tarsi all reddish, two apical joints black. Wing strongly tinged with reddish, forming a large spot extending from the stigma to the base of the second posterior cell. L. corp., 12½ mm.: al., 11½ mm.

One specimen: Summit Sierra Nevada, California.

This species must resemble *P. apiformis* of Europe even more than *P. grandis*, Will., does. It differs from that species in not having yellow pile on the third abdominal segment and in the face being entirely black. It is very striking in general appearance as a miniature of *P. grandis*, Will. It is, however, easily separable from that species by the unarmed femora, coloured wings and black face.

The above is a manuscript name by Dr. Williston which I found attached to the specimen in the collection of the Kansas State University. The manuscript containing it had been misplaced. I thus continue the name although the description is my own.

35. *Brachypalpus inarmatus*, n. sp.

Very similar to *Brachypalpus frontosus*, Loew, but differs in the fact that the coxæ, femora and tibiæ of the male are entirely unarmed.

Male.—Antennæ dark reddish-brown, third joint slightly darker on the lower basal corner ; first joint shining ; arista yellow, its apex fuscous. Face front and cheeks bluish-black, somewhat shining, covered, except a broad oblique stripe on cheeks, with silvery pollen, more dense on the front, which in some lights obscures the ground colour. Occiput below with long yellowish pile. Face in profile concave, but the concavity not receding nearly as low as the lower border of the eyes nor as far back as the eye margin. Dorsum of thorax light shining green with four cupreous stripes, the median ones more slender and all abbreviated behind the middle ; the pile yellow and rather abundant. The scutellum and an irregular, poorly defined area in front of it on the dorsum cupreous. Abdomen shining purplish-black, with yellow pile longer on the sides of the second segment and on the posterior margin of the fourth, where it forms a conspicuous fringe. In the middle of the second segment there is a small, slender, opaque spot not reaching the posterior margin. Legs black : femora long golden pilose, the extreme apex of the femora, the narrow base of the tibiæ, and the tarsi, except the last two joints, black. Posterior femora and coxæ without spurs or protuberances, the former moderately incrassate. Wings distinctly infuscated on the anterior half.

One male specimen : Vollmer, Idaho, May 30th, 1896 ; Prof. J. M. Aldrich.

There are differences between this species and *frontosus* in the face, which is uniformly pollinose, but bare and shining below the antennæ in that species, in the presence of a golden fringe on the posterior margin of the fourth abdominal segment, in the pile everywhere being golden and not gray as in that species, the posterior femora are less curved and the tibiæ are darker than in my specimens of *frontosus*.

It has occurred to me that this might be simply a dimorphic form of *B. frontosus*, holding the same relation to that species as the form *Bautias* holds to *Mallota cimbiciformis*, Fall. From the differences enumerated above, however, it does not appear that such can be the case.

36. *Xylota barbata*, Loew.

A single male specimen [Santa Cruz Mountains, California, 18th April] agrees so well with the description of this species that I am constrained to think it is this species, although it lacks the posterior coxal spurs. It has occurred to me that possibly where Dr. Williston, in the Synopsis, p. 234, says "hind coxæ unarmed," he meant to state exactly the reverse. The second and third abdominal segments are opaque, but have obscure yellowish. shining spots ; the fourth segment is entirely shining bluish-black. The thoracic dorsum and scutellum are brilliant purplish-metallic.

37. *Xylota analis*, Will. Synopsis N. A. Syrphidæ, 226.

I possess a male specimen of this species taken on the Pine Ridge in Nebraska in July. This specimen agrees exactly with a specimen from San Pedro, California, Aug. 1896. This species has not been recorded outside of New Mexico and California.

38. *Xylota fraudulosa*, Loew. Centur. v., 41.

I have specimens of this species taken in North-western Nebraska.

39. *Xylota ejuncida*, Say.

One specimen : Cook's Inlet, Alaska ; coll. L. L. Dyche.

40. *Mallota facialis*, Hunter.

This species was described from a single male specimen from Pine Ridge, Nebraska. This season's collecting includes another specimen from the same region that is in every way a verification of the views I held at that time.

41. *Triodonta*, sp.

I have a female specimen of a species of this genus from Palo Alto, California, which undoubtedly is a species distinct from *curvipes*, Wied. It is, however, so closely allied to that species that I hesitate to describe it from only the female. Doubtless in the male there are abundantly sufficient characters for specific separation. This specimen differs from the female of *T. curvipes*, Wied., in having the thorax almost bare and shining, not densely brownish pollinose. The abdomen is bare and shining black with the narrow posterior margins of the segments yellow, with only very slight indications of pollinose spots on the segments laterally. It is also much smaller, 8 mm. in length.

42. *Tropidia montana*, Hunter; Ent. News, 1896, p. 215. (Change of name from T. nigricornis, which is preoccupied. See Ent. News, 1896, 305.)

Since writing the description of this species I have examined a female specimen of *Tropidia incana*, Townsend (Trans. Am. Ent. Soc., 1895, p. 53), from Colorado, as well as the type of that species in the collection of the Kansas State University. From this examination I am enabled to give further differences between these two species which are very closely allied.

The face in *incana* in the female is distinctly more concave than in *montana*. In *incana* the face recedes from the apex of the antennal callosity to half way to the epistoma ; from that point the outline of the face projects outwardly at the same angle that the upper half recedes inwardly. In *montana* the outline of the face on the upper half is exactly the same as in *incana*, but on the lower half the outline is an almost perpendicular line. Besides this the front is somewhat narrower in *incana*, the spots of the abdomen are much larger and the pile is considerably shorter.

43. *Tropidia mamillata*, Loew, Centur. i, 68, 1861.

Four male specimens of this species were taken by the writer at Cedar Bluffs, Nebraska, in April, on flowers of Prunus virginicus. This is, I believe, the only record of the capture of this species since the publication of Loew's first Century in 1861. The locality given in that case was Illinois.

LIBELLULA DEPLANATA OF RAMBUR.

BY JAMES G. NEEDHAM, CORNELL UNIVERSITY, ITHACA, N. Y.

In December, 1896, Mr. Adolph Hempel sent me from Orange Co., Fla., some full-grown dragonfly nymphs which were apparently not to be referred to any of our known genera. At my request he undertook to breed some of them, and soon had imagoes of the species named above. In the letter which accompanied his bred specimens he recorded some careful observations, which are so interesting and valuable I deem them worthy of permanent record. The following account of the habits of this species is from Mr. Hempel's letter :

This species frequents small ponds and the borders of adjacent woods. Imagoes fly, when undisturbed, quite leisurely. They will hover over one spot, then dart a few feet aside and hover again and again. The males are often found in low places about ponds, resting on the ground with wings aslant downward and forward. Sometimes they rest on reeds or snags in the water ; sometimes out in the pine woods several hundred yards from water ; they may be found resting on the sand warmed by the sun, on logs or on trees.

The female deposits her eggs while hovering over the water, descending to dip the tip of her abdomen repeatedly. She is generally interrupted in her peaceful occupation and soon driven away by the too importunate males. The females remain for the most part in the woods and come from the woods to the ponds to oviposit, but hardly has one shown herself over open water before several males are in pursuit and she quickly disappears again. The difference in the haunts of the sexes is so marked that males would seem largely in excess to one who collected only beside the water, females so to one who collected only in the woods.

The nymphs are quite active. When in the water they rest with the long abdominal appendages widely spread apart ; but withdrawn from the water, these are brought together so that the abdomen seems to end in a long point. When picked up they have a habit of curving the abdomen as if to strike with the terminal spines. Their transformation takes place in the early part of the forenoon, and imagoes leave their empty old skins generally clinging to stumps and logs fallen in the water.

The full-grown nymph measures 23 mm. ; abdomen, 16 ; hind femur, 5.5 ; width of abdomen, 6 ; of head, 4.5. Body slender, not depressed ; abdomen smooth ; thorax and legs clothed with tawny hairs.

Colour fulvous, yellowish beneath and on sutures ; eyes black ; sides of thorax indistinctly marked with black ; apical third of abdomen reddish, with two broad black lateral stripes.

Head wider than long ; eyes not remarkably prominent ; vertex roundly elevated. Rear of head straight or very slightly concave.

Labium moderate ; mentum without raptorial setæ ; median lobe prominent ; its border crenulate, with single spinules between the crenulations. Lateral lobes ample ; movable hook nearly straight to the short, abruptly incurved tip ; raptorial setæ 6 each side ; teeth of opposed margins crenate, each ending in a sharp, incurved hook, and armed with a stout spinule.

Meso-thoracic stigmata separated by less than the width of one of them. Wing-cases reaching well upon the 6th abdominal segment.

Abdomen lance-oval, with sharp lateral margins. Long, straight, sharp, lateral spines on 8 and 9. Dorsal hooks on 4 to 8, the first erect spine like the others directed backwards, the hindermost with their dorsal margin forming a straight line to the base of the segment ; 9th abdominal segment hardly longer on ventral than on dorsal side ; 10th segment a little shorter than 9th, conical. Abdominal appendages very long (13 mm.) and sharp, longer than segments 9 + 10 ; superior and inferior appendages equal ; laterals one fourth as long.

Libellula deplanata, Rambur, is but a smaller southern variety of *Libellula exusta*, Say, as was pointed out by Mr. P. P. Calvert in 1893 (Trans. Amer. Ent. Soc., XX., 258). But in recent repeated dismemberment of the genus *Libellula* no part of it has been left to bear that name in America. As genera go the European *Libellula depressa* of Linnæus is certainly worthy to stand alone, and by all the recognized codes it has the right to the original generic name. So that our N. American species belong to *Leptetrum*, Newman ; *Plathemis*, Hagen ; *Belonia*, Kirby, or *Holotania*, Kirby ; and Kirby (1890 a Synonymic Catalogue of Neur. Odon., London) has distributed our species rather freely among all these genera. I now have nymphs of species referred by him to all the genera, and, unfortunately, they do not confirm his arrangement of the imagoes. The unknown nymphs still in the majority would doubtless lend the best aid to drawing the lines where they belong.

As implied at the outset, the nymphs described above differ by good generic characters from all others known to me. They differ from all Libellulid nymphs which I have seen by the entire absence of raptorial setæ from the mentum of the labium. They are distinct from the nymphs representing the four genera named above by several additional characters : by hooked teeth on opposing edges of the lateral labial lobes ; by the extreme elongation of the abdominal appendages and especially by the shape and relations of the 9th abdominal segment which is not longer on the ventral than on the dorsal side, and consequently does not at all appear to enclose the 10th segment. The following characters of venation taken together appear to clearly segregate the imago : (1) The sectors of the arculus are not stalked in either wing. (2) The sub-triangular space consists of three areoles. (3) A short sector, which may be called the *apical sector*, arising beneath the stigma from the principal sector and extending to the apex in both wings, in this species arises under the proximal fourth of the stigma. This *apical sector* develops from a tracheal branch, is very constant in position, and may readily be recognized even when somewhat irregular if taken in connection with another which may be called the *sub-apical sector* which (in Libellulidæ) lies just posterior to it, parallel with it, and separated from it, except at the proximal end, by a single row of areolets. Hagen, describing *Libellula deplanata*, Rambur, in 1861 (Syn. Neur. N. Amer., p. 154), questioned whether it belonged to the genus. The nymph supplies an emphatic negative, which the venation and doubtless other adult characters corroborate, and which is equally applicable to the more recent subdivision of the genus. I therefore propose a new genus *Ladona* with *L. exusta*, Say (= L. deplanata, Ramb.), for its type. And for this interesting and locally common species, which ranges from Florida and Maine to the Columbian River basin, because of its very distinctive white humeral stripes, I would suggest the common name, " the Corporal."

NAKED AND COCOON PUPÆ OF ANTS.

BY GEO. B. KING, LAWRENCE, MASS.

Ordinarily the tribe Camponotidæ can be separated from the other tribes of ants by its habit of having cocoon pupæ in which their young go through their transformation period; whereas those of the (so-called) aculeate genera remain naked and do not spin a cocoon, it will appear, however, if diligent search be made, that several species of this tribe (Camponotidæ) do have naked pupæ, mixed with their cocooned ones, Latreille seems to have been the first to discover that *Formica fusca*, L., had naked and cocoon pupæ. He could not, however, understand why this should be, and indeed it remains one of the dark mysteries of the present day. So far as I am aware no other species of ants have been listed, other than *Formica fusca*, L., having this habit. During my researches in the study of the ants of Massachusetts, I have found that other species have acquired the same habit. And to satisfy myself that no mistake was made on my part in the determination of the larva, cocoon or imago, I sent samples of them to my friend and co-worker, Mr. Ernest Andra, of France, for his opinion, and at the same time enquired of him if any of the ants of Europe had been discovered with naked and cocooned pupæ, other than *F. fusca*. In his reply he stated that *F. fusca* is very frequently found with these two forms, and occasionally *Formica sanguinea*, Latr.; *Lasius niger*, L.; *Lasius fuliginosus*, Latr., and *Polyergus rufescens*, Latr., have been found in Europe having naked and cocooned pupæ, the last four species being very rarely met with in this condition. The species having this habit thus far found by me in Massachusetts are:

Formica fusca, L., var. *subsericea*, Say.; June 8.
" " " sub. sp. *subpolita*, Mayr.; June 20.
" *lasioides*, Em., var. *picea*, Em.; July 31.

This list may be extended after further research; they are, however, not very frequently met with. The season of the year in which they are to be found being hot and dry, and the ants much more active at this time, as their usual custom is, they will hasten off with their young very rapidly to the underground retreats of their nests, making it quite difficult to obtain samples of either. Furthermore, I might possibly have found more with similar habits if this were the only work which I am investigating, but as I am studying all the insects living with ants, it is quite possible that in many instances their cocoons and pupæ are overlooked.

THE COLEOPTERA OF CANADA.

BY H. F. WICKHAM, IOWA CITY, IOWA.

XXIV. THE CERAMBYCIDÆ OF ONTARIO AND QUEBEC.—*(Continued.)*

MOLORCHUS, Fabr.

Easily recognized by the very short elytra which are divaricate and separately rounded at apex, about equalling the prothorax in length. *M. bimaculatus*, Say (fig. 23), is somewhat variable in colour, but is ordinarily black except a large testaceous blotch on each elytron. The thorax is rather broad, roughly punctured, the sides irregularly rounded, Length, .20–.32 in. Usually found on flowers, but has been bred from hickory, maple, ash, and dogwood.

Fig. 23.

CALLIMOXYS, Kraatz.

Distinguished from *Molorchus* by the shape of the elytra, which are longer and drawn out nearly to a point at tip. The sexes differ in colour, the males usually having a partially red thorax. *C. sanguinicollis*, Oliv., is blackish (except as stated above), punctured, the elytra more or less fuscous with clear punctuation. Anterior and middle legs entirely blackish, the posterior yellow except the tips of the joints, which are black. The hind tibiæ are long and curved in the males, the exterior margin with numerous teeth. Length, .33–.40 inch. Found on flowers in June and July.

ANCYLOCERA, Serv.

It may be that the Canadian record for *A. bicolor*, Oliv., is incorrect, since the species is said to be a resident of the Southern States from North Carolina to Texas. It is unknown to me in nature, but is said by Mr. Leng to be "a very dainty insect, black with scarlet elytra and abdomen and with slender legs and clubbed thighs. The body is slender, head short and prothorax very long as compared with the cylindrical elytra. The antennæ are serrate, one-half as long as the body in the female and longer than the same in the male. The hind pair of thighs is armed with a terminal spine." Length, .50–.70 inch.

BATYLE, Thoms.

B. ignicollis, Say, is from .28–.52 in. long, black, the prothorax bright red. The elytra are densely rugosely punctured, with blackish pubescence. The prothorax is rounded, unarmed, the pubescence longer than on the elytra. *B. suturalis* is smaller (.28–.36 in.), red, the legs more or less black, the elytra often with a black line along the suture which may be dilated behind so as to extend over the greater part of the apical third. The prothorax is said to be occasionally black, but such specimens have never come under my notice. These beetles are often abundant on flowers on the Western plains of the United States.

PURPURICENUS, Serv.

Contains one species, *P. humeralis*, Fabr., a large insect, .50-.74 in. long, black, except a large triangular humeral spot on each elytron. Sides of prothorax spinose. Entire upper surface coarsely punctured, rugosely on the thorax, the elytral punctures distinctly and rather widely separated.

STENOSPHENUS, Hald.

Here belongs *S. notatus*, Oliv., a rather elongate beetle of nearly parallel form, the elytra slightly tapering behind. In colour it is black, the head beneath and the entire prothorax except a large central dorsal black spot, reddish. The punctuation is rather coarse but sparse and each puncture gives rise to a gray hair, those of the elytra being subseriate in arrangement. The antennæ are spinose, equalling or exceeding the length of the body. Length, .35-.48 inch. Adults of this species have been cut from hickory wood.

CYLLENE, Newm.

The two Canadian species of this genus are difficult to separate since they agree almost exactly in colour. The numerous cross-bands of yellowish (or rarely grayish) pubescence on the velvety black prothorax and elytra give them a very characteristic appearance. Dr. Horn has distinguished them as follows :—

Second joint of hind tarsi glabrous at middle, antennæ of male longer than the body. .42-.89 in.............*pictus*, Drury.
Second joint of hind tarsi densely pubescent, antennæ not longer than the body. Prosternum as wide as the coxal cavity. .40-.80 in................................... ...*robiniæ*, Forst.

The species differ in their times of emergence, *pictus* often appearing on its principal food-plant (hickory) early in spring, or even in winter if firewood of this sort be stored in a warm room. I have on one occasion seen several specimens copulating and ovipositing on felled honey-locust early in April at Iowa City. It also bores in butternut. *C. robiniæ* infests living black locust, often ruining the trees. It appears in late summer or early fall and may be found in great numbers on blossoms of golden-rod,

PLAGIONOTUS, Muls.

The soft-maple borer, *P. speciosus*, Say (fig. 24), is a most gaudy insect of large size (about an inch in length) and with heavier antennæ than most of its neighbors. The ground colour is

black or nearly so, the legs reddish; but owing to the dense clothing of yellow pubescence very little of the black is visible. Almost the entire under surface is thus rendered yellow, as are also the legs, the greater part of the head, two short bands on each side of the prothorax, and several cross-bands on the elytra.

Fig. 24.

CALLOIDES, Lec.

Includes another large insect, *C. nobilis*, Harr., black, pubescent, usually decorated on the elytra with a few small detached yellow spots, which may, however, be absent. Length, .80–.92 in. It is thought to breed in the chestnut.

ARHOPALUS, Serv.

A. fulminans is said to breed in oak, butternut and chestnut. It is .48 to .72 inch long, black with whitish pubescence forming irregularly defined bands on the elytra and leaving on the prothorax a large central black spot with a smaller one on each side. The thoracic marking alone will thus serve as a ready means of recognition.

XYLOTRECHUS, Chevr.

Includes several species which have the front of the head variably carinate; they are, for the most part, ornamented with transverse bands of lighter coloured pubescence, somewhat as in *Cyllene*.

A. Prothorax with four spots of (usually yellow) pubescence. Elytral markings indistinct and not conspicuous.   .32–.48 in............. ....*quadrimaculatus*, Hald.

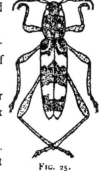

AA. Prothorax not spotted (except by breaking up of bands), sometimes fasciate with pubescence.

    b. Elytra obliquely truncate at apex, the outer angle spiniform. Sides of prothorax regularly arcuate.   .60–.72 in...... ....*sagittatus*, Germ.

    bb. Elytra obliquely truncate at apex, but not spiniform.

Fig. 25.

Thorax without apical and basal pubescent band.
Elytral bands about as broad as their intervals.
.32-.64 inch. (Fig. 25.)............*colonus*, Fabr.
Thorax with apical and basal pubescent band. Median
elytral bands angulated or undulatory. .44-.84
inch.........................*undulatus*, Say.

The above table is, in the main, taken from Mr. Leng's synopsis.
He adds, regarding *undulatus*, that there may, for convenience in cabinet
arrangement, "two names be retained: *fuscus*, Kirby, for the form with
the sides of the thorax entirely covered with pubescent blotches and the
elytral bands wavy, and *interruptus*, Lap. & Gory, for the form with
the bands greatly obscured by the sprinkling of white hair." As to food-
plants, *colonus* is known at attack oak and maple, while *undulatus* has
been beaten from spruce. The latter is often very abundant on freshly
cut pine logs or sawed timber.

PLAGITHMYSUS, Motsch.

This name is substituted for the *Neoclytus* of the Check List. The
prothorax is transversely rugose, and by this character the genus may be
readily distinguished from other Canadian Clytini. Mr. Leng separates
the species substantially as follows :—

A. Middle and hind femora spinose at apex.

   b. Thorax with a longitudinal elevated ridge, rugose at apex,
      antennæ filiform. Thorax with basal and apical bands only
      of pubescence ; colour reddish brown, posterior two-thirds
      of elytra and parts of thorax often darker. .28-.76
      in..................... ....*fuscus*, Fabr.

   bb. Thorax with a few distinct transverse rugæ, antennæ thickened
      toward apex. Blackish ; head, thorax and legs reddish,
      elytra with straight transverse bands of yellow pubescence.
      .20-.70 in.....................*erythrocephalus*, Fabr.

AA. Femora not spinose, antennæ filiform, thorax with many strongly
    elevated but more or less confused transverse rugæ.

   c. Elytra rounded at apex, the bands yellow (rarely whitish),
      forming an oval figure at the base of each, behind which
      are two slightly oblique fasciæ. Tip yellow. .48-.80
      in..................... ....*capræa*, Say.

cc. Elytra truncate at tip. Smaller species with long legs and
    whitish elytral bands.

      Thorax wider than long. .28–40 in. *muricatulus*, Kirby.
      Thorax longer than wide. .36–.44 in. . *longipes*, Kirby.

  P. erythrocephalus is known to depredate on elm, soft maple,
hickory and black locust ; *P. capræa* on ash, elm and hickory, while *P.
muricatulus* and *P. longipes* may be taken on freshly cut pine.

CLYTANTHUS, Thoms.

  C. ruricola, Oliv., is black, base of femora, the tibiæ, tarsi and
antennæ (except at tip) reddish. Pubescence yellow, forming a nearly
complete thoracic margin, a scutellar spot and elytral markings as
follows : A short oblique band near the base, posterior to which is a
hook-like (sometimes interrupted) figure the shaft of which is nearly
parallel with the suture, and behind this a rather broad, nearly straight
but oblique band. Beneath, the meso- and metathorax are spotted and
the apices of the abdominal segments more or less margined with the
same colour. Length, .28–.48 inch.

EUDERCES, Lec.

  Contains two small (Canadian) species which agree in their ant-like
form, the elytra gibbous at base and with an oblique ivory fascia. The
colour varies from black to almost entirely rufous, the tip of the elytra,
however, remaining black in the latter case. Mr. Leng separates them
by the following characters :—

    Eyes nearly divided ; prothorax uniformly rounded at sides. .20–.36
      in. . . . . *picipes*, Fabr.
    Eyes completely divided ; prothorax distinctly depressed each
      side near the anterior margin, laterally subangulate. .26–.36
      in. . . . .*pini*, Oliv.

  In my experience, *E. picipes* may be taken by beating hazel bushes.
When running up the side of the beating-net the resemblance to certain
black species of *Formica* (which are often abundant in the same thickets)
is truly striking. It has been bred from chestnut twigs.

CYRTOPHORUS, Lec.

  Until recently but one species has been recognized. Captain Casey
has of late described another form which he distinguishes from *verrucosus*
as follows :

Larger, pronotum compressed, prominent along the middle, basal elevation of elytra strong. Third antennal joint strongly spinose. .24–.40 in..*verrucosus*, Oliv.

Smaller, less convex, pronotum not compressed, basal elevation of elytra feeble, third antennal joint briefly spinoso - dentate within at apex. .24 in...................................*insinuans*, Cas.

These bear considerable resemblance in form to *Euderces*, but are without the ivory-like band of the elytra. In colour the former is blackish ; legs, in part, and basal three-fifths of elytra sometimes rufous, pubescence white or cinereous, arranged anteriorly in narrow oblique bands which follow the course of the basal elytral gibbosities. Behind these oblique bands is a very narrow cinereous one, nearly transverse in direction. Tip broadly covered with cinereous pubescence. I have not seen *C. insinuans*, which is described from a single male. Wild cherry is known to be a food-plant of *C. verrucosus*.

MICROCLYTUS, Lec.

M. gazellula, Hald., is found in the adult state on oaks. The genus differs from *Cyrtophorus* in not having the third antennal joint spinose at tip.* It is "a small insect, piceous or reddish-brown with the thorax above and the elytra, except about the middle of the suture, black and rather closely punctured, the legs and antennæ always paler. Elytral markings composed of long white hairs arranged as follows : An oblique line from the scutellum, a very short transverse or slightly arcuate line about the middle entirely distinct from the next, a broader band immediately behind and nearly transverse, a blotch covering the entire apical eighth of the length of the elytra." (Leng.) In the male the antennæ equal, in the female reach two-thirds the length of the body. In the former sex the elytral tips are very slightly truncate, in the latter separately rounded.

---

*Since publication of the table of genera I have come across the following note by Dr. Hamilton (CAN. ENT., XXIII., p. 63) :—" The characters separating *Cyrtophorus* and *Microclytus* were originally feeble, and have recently become more so by some one discovering that the relative lengths of the antennal joints in the male of the latter are the same as in the former, thus leaving in the males only the presence or absence of a small spine at the end of the third joint of the antennæ as diagnostic." By a clerical error the legend *Cerambycoides* is placed one line too high up on p. 86 of my table ; it should be on line 2, and embraces all the genera from *Chion* to *Microclytus*, inclusive.

NOTES ON RHOPALOCERA, WITH DESCRIPTIONS OF NEW SPECIES AND VARIETIES.

BY HENRY SKINNER, M. D., PHILADELPHIA, PA.

I have received beautifully fresh specimens of *Argynnis atossa* taken in the mountains near Tehachapi, Southern California, July 7th, 1895. The inner half of the superiors below is bright red, almost a blood red. The species was described by Mr. Edwards from a specimen taken by Mr. H. K. Burrison. It is quite distinct and ranks with *diana, idalia* and *nokomis* as one of our handsomest Argynnids.

Argynnis Snyderi, n. sp.— ♂. Expands three inches. Upper side : Superiors tawny as in other species, but dark and with considerable red. The black markings are distinct and sharply defined against the tawny background. The margin is distinctly but not heavily marked. The inferiors have the usual black markings, but they are unusually well defined and there are almost no black scales at base as in most species in the genus. Under side : Superiors have silver spots on outer margin, extending more than half way toward inner margin. There are two quite large subapical silver spots. On inferiors the silver spots are large and well defined, with wing-ground very light grayish-green with a distinct light buff intermediate border about one-eighth inch in width. Silver lunules on margin are large, well defined, and seven in number, the inner one extending up along inner margin as a line. The ground colour of wings on inferiors below is brownish in the female. This large species comes nearest *coronis,* and has been mistaken for it. I have specimens from Salt Lake City, Utah, taken June 23rd, 1895, and a female from Ogden, July 6th, 1895. All were taken by Prof. A. J. Snyder, after whom the species has been named.

Argynnis platina, n. sp.— ♂. Expands two and a half inches. Upper side : Rather light tawny or even light buff. Black markings dense and wide, with outer halves of wings looking rather clear or open, with row of round spots not very large ; marginal border light ; bases of wings not much obscured. Under side : Superiors have the two subapical silver spots and silver spots on margin well defined ; colour of inner half of wing rosy. Silver spots on inferiors are large and well defined and placed on a very light greenish-gray ground. The intermediate buff band is well defined, comparatively wide and very light in colour. Ground colour on inferiors below is reddish brown in the female. Described from specimens taken at Ogden, Utah, between July 18th and 24th, and Beaver Canon, Idaho, at nearly same dates. From Prof. A. J. Snyder.

The typical *Arg. nevadensis* comes from Nevada, and the types came from the valleys of the Sierra, near Virginia City. I have specimens from Reno and Verdi, Nevada. I mention this as I do not think the specimens from Colorado and Utah are typical but are var. *Meadii*, or more nearly related to that variety. I have females from Mammoth Hot Springs which are the colour of *leto* ♀. The species figured in Ent. News, pl. 2, 1892, is not *chariclea* but *polaris*. The other Greenland *Argynnis* brought back by the Peary expedition is *chariclea, var. artica*, Zett.

Melitaea Beani.—I propose this name for the Alpine form of *anicia* from the high elevations near Laggan, Alberta, the fauna of which has been so assiduously studied by Mr. Thos. E. Bean, and who has made known new species and interesting facts in regard to the butterflies of that region. This variety has quite a different appearance from the low valley form, being darker, smaller, and with markings apparently run together more and not nearly so bright in colour. Expanse of *Beani* 1⅒ inch. Expanse of low valley form 1½ inch. I have specimens of *Melitaea alma*, Strecker, from Coso Valley, Cala.; May. Types came from Arizona and South Utah.

Phyciodes Barnesi, n. sp.— ♂. Expands 1¾ inch. Shape and colour of *P. mylitta*. Superiors light tawny with less markings than any known species. Superiors have an eight-shaped mark in cell near base of wing; just below this is another better defined eight-shaped mark; in centre of cell is a small naught-shaped mark; below this on inner margin is a good-sized black spot; there is a black bar at end of cell and another black bar near angle of wing; the remainder of the wing is practically immaculate. Inferiors have a number of black lines extending out from base for about one-fourth inch; remainder of wing except margin is nearly immaculate, except that the markings on under side can be faintly seen. Under side: Superiors much as above. Inferiors have the markings as is usual, but are not so well defined and are quite light in colour. Specimens were taken at Glenwood Springs, Colo., May 8th to 15th, and June 1st to 7th, by Dr. Wm. Barnes, in whose collection are many co-types.

I have specimens of *Junonia cœnia*, var. *negra* (Feld. Reise Nov. Lep., 3, 399, n. 592, 1867) from S. E. Texas; Coleina, Mex.; Merchantville, N. J. (Kemp).

Cœnonympha (*Erebia*) *Haydenii*— ♀.   This differs markedly from the ♂ in being entirely different in colour.   Males are dark smoky-brown, and the females are nearly same colour as *Cœn. inornata* but not so reddish.   This species was found in numbers by Prof. Snyder at Beaver Canon, Idaho, last of July and first part of August, 1895.

Thecla damon, n. var. *discoidalis.*—Differs from typical form in having central area of both wings light greenish-yellow.   Round Mountain, Blanco Co., Texas, February 10th and August 16th.

Pieris ochsenheimeri, Staudinger (Stett. Ent., Zeit., 1886, p. 199). This species was described by Dr. Staudinger from Central Asia, and is beautifully figured in "Memoires sur les Lepidopteres" by N. M. Romanoff, 4, 220, pl. 14, f. 1 a, b, 1890.   Through the generosity of Dr. Herman Strecker, of Reading, Penna., I received two males and a female of a Pieris unknown to me from Mt. Wrangel, Alaska.   They prove to be the above-mentioned species.   As Romanoff's work may not be accessible to many, I append the following description :—

♂.—Expands 1¹⁰⁄₁₀ inch.   Upper side : Superiors white with costa blackish-gray ; apical costa, apical portion of wing and upper part of outer margin blackish.   There is a round black spot in the space between last costal and first discoidal nervure.   Neuration shows faintly gray scales.   Bars of wing black.   Inferiors white with only one spot and that on outer third of costa, round, black.   Base of wings black ; there is a very narrow, dark, submarginal line to both superiors and inferiors.   Under side : Superiors much as above except that apices of wings are yellowish, and there is an additional spot (not always well defined) below the third discoidal nervure.   Inferiors have mixed yellow and gray spots as in *Pieris napi bryoniæ.*   The female differs from the male in having the veins rather heavily marked with dark scales, as are also the apices of superiors and bases of all four wings.   It has an additional dark spot on superiors.   Below the veins are not as heavily marked and the ground colour of wings is white instead of yellow.

Systasea pulverulenta, Feld.—I have received a specimen of this species from Prof. T. D. A. Cockerell, who sends the following particulars : "Caught April 22nd at Mesilla, New Mexico, on flowers of *Biscutella Wislizenii.*   It is different from any Hesperid I have caught here.   When I saw it I thought it was a moth near to *Drasteria.*"

SOME NEW AND LITTLE-KNOWN DORYDINI (JASSIDÆ).

BY C. F. BAKER, AUBURN, ALA.

Spangbergiella vulnerata, Uhler.—There are two specimens of this species in the National Museum collection from New York, and another in Fitch collection from Arkansas.

Spangbergiella Lynchii, Berg.—Signoret quotes the description of this species in his Essai sur les Jassides and says: " This species might well be the *S. vulneratus*." Berg takes this suggestion as the final disposition of the species, and reduces *Lynchii* to a synonym of *vulnerata*. I have a specimen of what is undoubtedly this species, from the Herbert H. Smith collection taken at Corumba. While it is very near *vulnerata*, still I think it should retain its place as a good species. It differs from *vulnerata* in having the head more slender, vertex a fourth longer than width between the eyes, the red lines not reaching the middle. In North American specimens of *vulnerata* (and so figured by Signoret) the vertex is but little if any longer than broad between the eyes, and the red lines converge considerably beyond the middle—at the tip as figured by Signoret.

Spangbergiella mexicana, n. sp.— ♀. Length, 6.5 mm. Pale green, darker on vertex, pronotum, and bases of abdominal segments. Two oblique slender red lines on vertex, converging towards the tip, which they do not quite reach. Pronotum with two red lines extending its whole length, nearly in line with those on vertex, at its base with a median yellowish dash. Scutel immaculate. Elytra whitish towards the tips ; claval suture and all veins except apical, yellow. A black dot at end of claval suture, and one each at end of first and fourth apical veins.

Vertex triangular, obtusely angulate anteriorly, but little longer than breadth between the eyes, about a fourth longer than pronotum. Clypeus subrectangular, broadly rounded at tip. Pronotum twice as wide as long. Ovipositor two-thirds length of rest of venter, exceeding the elytra by ½ mm. Last ventral segment a half longer than preceding, hind margin truncate.

Described from a single female collected at Vera Cruz, Mexico, by Rev. H. Th. Heyde. This species is nearly related to *S. punctato-guttata* and *S. felix*, but is distinct from both as described above.

Bergiella, n. gen. Type, *Parabolocratus uruguayensis*, Berg.—The head is broader than long, somewhat angulate and sloping as in *Parabolocratus*. The frontal sutures are arrested at the antennal scrobes. The clavus has but a single longitudinal vein. A specimen of this species, collected at Chapada, is in the H. H. Smith collection. I name this genus in honour of the author of " Hemiptera Argentina."

Parabolocratus flavidus, Sign.—This species, described from North America, was omitted from the Van Duzee List. There are specimens in the National Museum from Texas. I have also collected it at Auburn, Ala.

Paraphlepsius, n. gen.—Head about the same width as the prothorax and considerably shorter, three and a half times as broad as long, anteriorly foliaceous, angulate, vertex level. Face of the normal Jassid type. Frontal sutures continued to the edge of vertex. Ocelli on the edge between vertex and face, somewhat rem)ved from the eyes. Elytra broad, slightly exceeding abdomen, bluntly rounded at tip, with a narrow appendix. Apical cells four, anteapical two, basal transverse vein entering radial cell. Clavus with two longitudinal veins. Wings with three apical cells exclusive of the closed costal.

This genus is nearest to *Psegmatus*, Fieber, from which it differs in having the head broader and much shorter than pronotum, and the frontal sutures nearly straight, instead of strongly bent inward, as in *Psegmatus*. Type :—

Paraphlepsius ramosus, n. sp.— ♀ ♂ . Length, 7 mm. Robust. Thickly marked with fine brownish dots and ramose lines. Face and below brown, the face marked with numerous yellowish dots. Legs yellowish, annulate with dark brown. Vertex and pronotum brownish, with numerous small, partly confluent whitish dots, which are larger on the latter. Elytra whitish translucent, with very numerous brown ramose lines adjoining the veins and in the cells ; in the female a large irregular clearer space towards base; in the male this clearer space is more pro-nounced, and there are small clear spots in several of the cells, the ramose lines becoming darker in a broad transverse band at middle of elytra.

Genæ broadly angularly emarginate below the eyes, the succeeding angle very obtuse, beyond attaining the tip of the clypeus. Loræ large, semilunar. Clypeus trapezoidal, narrower at base, truncate at tip. Front rapidly broadening above, apex rather abruptly bent forward, sides nearly straight. Width of vertex between eyes two and a half times the length ; length about two-thirds that of pronotum. Pronotum two and one-third times as wide as long, broadly rounded anteriorly, hind margin gently concave ; posteriorly the surface is rather coarsely, subobsoletely creased.

Last ventral segment of female twice the length of the preceding, shallowly trisinuate, the median sinus acute.

Described from two specimens from the Cornell University collection, kindly sent me by Mr. A. D. Macgillivray, collected at Ithaca, N. Y., the female on Aug. 3rd, 1889. This insect might readily be mistaken for a *Phlepsius*.

Dorydiella, n. gen.—Head broader than prothorax and somewhat longer, more than twice as broad as long, anteriorly foliaceous, angulate, and inclined upward. Face normal. Ocelli on the edge between vertex and face, adjoining the eyes. Elytra long and narrow, with a narrow appendix, somewhat exceeding abdomen, toward the apex narrowed to an acute point. Apical cells four, anteapical two, basal cross vein entering radial cell. Clavus with two longitudinal veins. Wings with three apical cells exclusive of the closed costal.

This genus is much like *Dorydium* in everything except the head, which is far shorter. Type :—

Dorydiella floridana, n. sp.— ♀. Length, 8 mm. Pale sordid whitish. Face variously marked with fine light brown dots, leaving portions below, and several indistinct transverse bands above, light. Vertex and pronotum with a number of very pale brownish indistinct longitudinal stripes. Anterior edge of vertex with five dark dots. Elytra with very sparse brownish ramose lines, densest about and extending back from the second apical cell. A dark spot at apex of clavus.

Genæ feebly emarginate below the eyes, then broadly rounded, slightly exceeding clypeus. Loræ large, semilunar. Clypeus somewhat narrower towards base, sides sinuate, apex truncate. Front with sides nearly straight, rapidly broadened above where it is bent somewhat back. Length of vertex three-fourths of width between eyes, somewhat longer than pronotum. Pronotal width nearly two and a third times the length ; anteriorly the pronotum is broadly rounded, the surface very sparsely punctate and posteriorly finely creased, the hind margin gently concave. Last ventral segment but little longer than preceding, hind margin with a broad, blunt, median projection having a small notch at its extremity and a black dot on either side.

Described from a single specimen in the National Museum collection, labelled " Fla." It is to be hoped that collectors doing miscellaneous sweeping in Florida will look particularly for further specimens of this rare and interesting insect.

CORRESPONDENCE.

BROTIS VULNERARIA AGAIN.

One of the many fine things secured by Mr. Bice at electric light during the season of 1896 was a specimen of that perplexing aberrant Lepidopteron, *Brotis vulneraria*, Hub.

In the CANADIAN ENTOMOLOGIST for 1886, Vol. XVIII., page 72, Mr. Ph. Fischer reports the capture of a specimen in Buffalo at electric light and gives some description of it and an account of the difficulty experienced by the various authors to decide its position in systematic classification. At page 136 the Rev. G. D. Hulst comments on that report and gives further information upon the subject, and quotes Walker as saying that " it does not seem to fit well anywhere."

Mr. Fischer identified his specimen by Hubner's figure. I had no difficulty in recognizing the London specimen by Guenée's illustration of it in his Lepidopteres Phalenites, plate 22, fig. 9, under the generic name *Sphacelodes*, but was indebted to Dr. J. B. Smith for a clue to its location in his List of 1891. I had forgotten these notices, where Dr. Hulst gives its generic synonymy, and the cause of it, although I read them with interest mingled with curiosity at the time, knowing nothing whatever of the moth referred to.

It is an interestingly anomalous insect. Whether in a tropical collection it has fitting associates with which it may harmonize and bear a resemblance, it certainly stands out conspicuously distinct in the Ontario one to which Mr. Bice has kindly donated it.   J. ALSTON MOFFAT.

PODISUS PLACIDUS.

In the May number of this journal Mr. Kirkland, of the Gypsy Moth Committee, publishes the descriptions of two Pentatomids by Mr. Uhler— *Podisus placidus* and *Euschistus politus..* Of *Podisus placidus* he says he was unable to find the original description, nor could Prof. Uhler at the time give him the reference. This description may be found in the American Entomologist, Vol. II., page 203.   E. P. VAN DUZEE.

Buffalo, N. Y.

ERRATUM.—On page 101, seventh line from bottom, for DORYLIDÆ read MYRMICIDÆ.

Mailed June 4th, 1897.

THE COLUMBINE BORER, HYDROECIA
PURPURIFASCIA, G. & R.

The Canadian Entomologist.

VOL. XXIX.   LONDON, JULY, 1897.   No. 7.

THE COLUMBINE BORER (HYDRŒCIA PURPURIFASCIA, G. & R.).

BY M. V. SLINGERLAND, CORNELL UNIVERSITY, ITHACA, N. Y.

In 1894, Mrs. J. J. Glessner, Littleton, N. H., called my attention to a "worm" which was feeding in the roots and stems of her columbines. It was not until July, 1895, however, that she succeeded in getting specimens of the "worm" for me. The "worm" proved to be a caterpillar which was unfamiliar to me, and in accordance with my usual practice in such cases, it was described and photographed. The photographs, giving dorsal and lateral views of the caterpillar, twice natural size, are reproduced on the plate.

The full-grown larva measured one and three-eighths inches in length. Its general colour is mars brown, much lighter on the venter of the first two thoracic and last four or five abdominal segments. The head is of a light russet colour, black about the eye-spots. Mandibles dark brown, black-tipped. Thoracic shield concolorous with the head on the dorsum, but merging into black on the sides and sometimes into a narrow black cephalic border; the shield is divided by a narrow whitish mesial line. Anal shield large, black, merging into brown mesially. The true legs are brownish-black, and the bases of the pro-legs are marked with blackish areas. Short light brown hairs arise from conspicuous, comparatively large blackish spots; the piliferous spots on the dorsum of the last two abdominal segments are considerably larger than the others. The spiracles are black. There is a continuous narrow white mesial stripe extending along the dorsum. A similar white stripe extends along the subdorsum on each side, but it is not continuous, being entirely obsolete on the first four abdominal segments, and sometimes on the last thoracic segment also. The discontinuance of these two white side stripes gives the larva a rather curious appearance, as the figures show.

One of the caterpillars, which was received in the latter part of July, 1895, pupated on or about August 8, and the adult insect (the beautiful moth shown twice natural size on the plate) emerged September 3, 1895.

The moth proved to be the one described by Grote and Robinson in 1868 as *purpurifascia*. Imagine the light spots in the figure to be of a delicate creamy white colour, the other shades as various shades of orange, purple, and russet brown, and you will have a faint conception of the rather uncommon but beautiful combination of colours presented on the wings of this insect.

I can find no reference in the literature to the early stages or habits of this insect. On account of its destructive work in Mrs. Glessner's columbines, it may be appropriately called "the Columbine borer." The moth has been recorded as occurring in Maine and Massachusetts in September and October, and in New York in August; it is also known to occur in Illinois and Colorado. Mrs. Glessner writes that she has found that rich soil, cultivation, and Fowler's solution of arsenic (diluted one-half with water) poured around affected plants seemed to check and control the pest.

THE RASPBERRY-CANE MAGGOT
(PHORBIA RUBIVORA, COQUILLETT).

BY M. V. SLINGERLAND, CORNELL UNIVERSITY, ITHACA, N. Y.

This new raspberry pest has been discussed in detail in Bulletin 126, issued in February, 1897, from the Cornell Agricultural Experiment Station. The life-history and habits of this Anthomyiian are fully illustrated in the Bulletin. At the time the Bulletin was written, however, the name of the insect had not been determined. In April, 1897, I reared several more of the flies and sent some to Mr. D. W. Coquillett at Washington. He soon reported that the insect was a new species of *Phorbia*, and sent me the following technical description of the fly, which he had drawn up from the specimens I sent him.

Phorbia rubivora, Coquillett, n. sp.— ♂. Ground colour black; sides of front and of face white pollinose, eyes sub-contiguous, more approximated to each other than are the two posterior ocelli, frontal vitta at the narrowest part linear; third antennal joint less than twice as long as broad, slightly over twice as long as the second, arista thickened on the basal third, the penultimate joint slightly longer than broad. Thorax grayish pollinose, marked with three black vittæ; three postsutural and three sterno-pleural macrochætæ. Abdomen quite thickly covered with suberect bristly hairs; narrow, subcylindrical, greenish-gray pollinose, marked with a black dorsal vitta ; tip of abdomen greatly swollen, bearing a subconical process in front of the hypopygium. Front and middle tibiæ each bearing two bristles on the posterior and one on the outer side below the middle besides those at the tip, hind tibiæ each bearing a single bristle on the inner side near the lowest third, three on the front side and two on the outer side, in addition to those at the tip ; under side of each hind femur bearing a row of bristles, those at the base the shortest. Wings hyaline, tinged with gray at the base and less distinctly so in the marginal cell, costa strongly arcuate along the costal cell, costal spine shorter than the small cross vein, the latter beyond the middle of the discal cell, hind cross vein nearly straight and subperpendicular, last sections of the third and fourth veins distinctly diverging ; calypteres whitish, halteres yellow, the extreme base of the peduncle brown.

♀. Front three-fourths as wide as either eye, frontal vitta destitute of a pair of macrochætæ, sides of front yellowish-gray pollinose ; abdomen ovate, pointed at the apex, almost bare, destitute of a black dorsal vitta ; costal spine slightly longer than the small cross vein ; otherwise as in the male.

Length, 4 to 5.5 mm. Two males and two females, bred by Mr. M. V. Slingerland, from larvæ boring in the stems of the cultivated raspberry at Ithaca, N. Y.

The male, will be easily recognized by the narrow abdomen and the arrangement of the bristles on the legs ; the female, by the absence of the usual pair of macrochætæ on the upper part of the frontal vitta.

D. W. COQUILLETT.

Mr. R. H. Meade, of England, sends me the following report upon some of the flies which were sent to him at the same time : " I have examined the flies carefully, and they seem to be an undescribed species of *Phorbia*. I cannot identify them with any European species that I know, and I think you may describe them as new. You might call them *P. rubi* or *P. ruborum*."

I shall be glad to send a copy of Bulletin 126 to anyone who may be further interested in this raspberry-cane maggot.

The annual meeting of the Association of Economic Entomologists will be held at Detroit, Mich., on Thursday and Friday, August 12th and 13th.

PRELIMINARY STUDIES OF N. AMERICAN GOMPHINÆ.

BY JAMES G. NEEDHAM, CORNELL UNIVERSITY, ITHACA, N. Y.

Examples of the emphasized importance of larval life better than that furnished by the subfamily Gomphinæ of Odonata are few even among insects. The nymphs live under the sediment (mostly organic debris) which falls to the bottom of ponds and streams. They are aquatic burrowers which live at such slight depth that their anal respiratory orifice is never beyond the reach of clean water. This thin stratum, which forms their home and which they only leave to transform, is one of great biologic richness. In it they have found room for development in enormous numbers and necessity for extreme specialization. They are, at least when well-grown, among the more powerful members of its teeming hidden population. The imagoes emerge, flit about under cover for a few days, lay their eggs and die. They emerge largely by daylight and are subject to great decimation of numbers at this time, and are sought later by numerous powerful enemies. The females which live to oviposit lay a very large number of eggs. A female of *Gomphus fraternus* laid for me in a watch glass of water over 5,000 at one time. The imagoes of the ancient genus Gomphus are regarded as a race of weaklings. Their nymphs, on the contrary, are splendidly equipped for the battle of life. And it is to the perfection of their adaptation that the prevalence of Gomphines with us is due.

These conditions have developed a large and very uniform series of imagoes, with one colour pattern, one plan of venation, one *habitus*, consisting of many very closely related species difficult to study. Specific characters, though slight, are yet constant. The slight specific variations of an ancient colour pattern long retained are unusually reliable. Secondary sexual characters reach here their maximum of importance and of specific individuality. This is as one would expect, recalling the vicissitudes of adult life and that its chief concern is with reproduction.

The real competition of life, however, is carried on by the nymphs, and the outcome of it is that they have become specialized. They have developed along several lines and have become segregated into well-marked natural groups which are not so obvious among the imagoes,

De Selys separated from the great genus Gomphus* as he found it three genera represented in our fauna, *Ophiogomphus*, *Herpetogomphus* and *Dromogomphus*, and divided the remainder into groups of species. My breedings of the nymphs during the past three seasons in the main confirm these groups and show that three of them at least are worthy to rank as genera.

One of the genuine surprises of this season was the finding here, at Ithaca, of nymphs like those described by Hagen from Rocky Creek, Ky., (Trans. Amer. Ent. Soc., XII., 281, 1885) and doubtfully referred by him to *Tachaptryx Thoreyi*, and the rearing from them of *Gomphus parvulus*, Selys. "This extraordinary nymph combines head and antennæ of *Hagenius* with legs and abdomen of *Gomphus*," wrote Hagen in the beginning of his very careful description. The length of the wing pads showed the nymphs not to be young, as Hagen supposed, and made it impossible to consider them as belonging to *Tachaptryx*, but that they should yield this dainty little Gomphine was still a surprise.

In June and July, 1896, I bred *Gomphus fraternus*, Say, in numbers at Havana, Ill. The nymphs are exactly described by Hagen (loc. cit., p. 262) as No. 13, *G. adelphus* (supposition). In May, 1895, I bred *Gomphus graslinellus*, Walsh, at Galesburg, Ill. These, especially the former, are very near to the typical *G. vulgatissimus* of Europe.

---

*Nomenclatural.—In the case of *Aeshna vs. Gomphus* I have examined the evidence and find it is as follows : Linne included all dragonflies known to him in one genus, *Libellula*. Fabricius (1775. Syst. Ent., pp. 420-426) divided the genus into three. *Libellula*, L., *Aeshna*, Fabr., and *Agrion*, Fabr., placing under Aeshna, among other species, *L. grandis*, L., and *L. forcipata*, L. It is worthy of note that he left *L. vulgatissima*, L., in *Libellula*. Illiger (1802. *Magazin für Insekten kunde*, p. 126) corrected the spelling to Aeschna, merely to accord with its etymology. Latreille was the first to designate types. He specifies (1802. Hist. Nat. Gust., Ms. III , 286) *L. depressa*, L , as the type of *Libellula* : *L. vulgatissima*, L., as the type of Aeshna, and *L. virgo* as the type of *Agrion*. With regard to the second, which alone concerns us here, *L. vulgatissima*, L., was described and figured by Latreille under the name "*Aeshna forcipata*, Fabr.," as was shown later by both Hagen and De Selys. Kirby's Catalogue of Neuroptera Odonata (1890) gives the correct synonymy and thus contains in itself the evidence which condemns the substitution it proposes. For if the type named by Latreille for *Aeshna* was *vulgatissima*, L., this species having been *excluded* by Fabricius when he founded the genus, cannot be its type. Leach (1815. Edinburgh Encycl. VIII. part 2, p. 726, of Amer. reprint) founded the genus *Gomphus*, with *L. vulgatissima*. L., for its type and placed under *Aeshna*, Fabr., the sole species *L. grandis*, L. However, Cuvier had previously (1798) characterized *Aeshna* (as pointed out by De Selys, *C. R. Ent. Soc. Belg.*, 1890, p. CLXI.) and described under it the sole species *grandis*, L. This usage has since been universally followed until 1890, and one is glad to find there is now no reason for change.

I follow De Selys in using the name *Ophiogomphus*, Sel., which seems to have been quite properly given.

In the *pallidus* group I find another type of nymph very distinct in the two species I have reared (*pallidus*, Ramb., at Galesburg, Ill., May 1895 ; *villossipes*, Sel., Ithaca, N. Y., May 1897).

The *plagiatus* and *notatus* groups of De Selys together present another type of nymph, already pointed out by Hagen (loc. cit., p. 269) as perhaps of more than subgeneric value. The bred nymphs of this group are of *plagiatus*, Sel., *notatus*, Ramb., *spiniceps*, Walsh, and *segregans*, n. sp. (vid. sub finem.)

Believing that the immature stages throw much light on the relationship of the imagoes, and that the study of this large and homogeneous group will be facilitated by the setting apart of distinguishable sub-groups, I propose three new genera which need here have no further characterization than that of the following tables : *Lanthus* (λανθανη contracted), type *G. parvulus*, Selys, *Orcus* (nomen proprium), type *pallidus*, Ramb., and Stylurus (στυλος and ουρα), type *plagiatus*, Selys. With these apart *Gomphus* is still somewhat polymorphic. The *dilatatus* group, characterized by extreme dilatation of the apex of the abdomen in the imago and correspondingly greater width to the 9th abdominal segment in the nymph, may yet, with advantage, be set apart. A clear line of demarcation, however, is not yet apparent.

I now hazard a table for separating these subdivisions of the *Legion Gomphus*, Selys. It is to be regarded as preliminary and tentative, the more so as I have endeavored to base it on characters common to both sexes. This legion is distinguished from others of Gomphinæ by the absence (normally) of cross veins from all the triangles and supra-triangular spaces.

Table for Imagoes of the Legion Gomphus, Selys.

1. Outer side of triangle of fore wing distinctly angulated at the origin of the cross vein between the two upper discoidal areolets...*2.*

   Outer side of triangle of fore wing straight or nearly so......*3.*

2. Inferior abdominal appendages of ♂ recurved upward in their apical half; vulvar lamina of ♀ shorter than half the 9th abdominal segment ....*Herpetogomphus*, Selys.

   Inferior abdominal appendages of ♂ recurved upward only at their extreme apices; vulvar lamina of ♀ almost equalling the 9th segment................................*Ophiogomphus*, Selys.

3. Upper sector of the arculus arising from its upper end ; *i. e.*, the part of the arculus above the sectors shorter than the part below them.......................... *Lanthus*, gen. nov.

   Upper sector of arculus arising from its middle ; *i. e.*, the part of the arculus above the sectors longer than the part below them .... *4*.

4. Hind femora with 5 to 7 long spines intermixed with smaller ones................................. *Dromogomphus*, Selys.

   Hind femora with numerous shorter spines................. *5*.

5. Ninth abdominal segment a little longer than 8th. Segments 7, 8 and 9 very little enlarged........ .. .. *6*

   Ninth abdominal segment not longer, generally shorter, than the 8th ; segments 7–9 more or less enlarged.... *Gomphus*, Leach.

6. Dorsum of thorax pale with darker stripes ; 8th abdominal segment cut obliquely at apex, longer on the dorsum than at the sides. abdominal appendages of ♀ hardly longer than the 10th segment...................................... *Orcus*, gen. nov.

   Dorsum of thorax dark with paler stripes ; 8th abdominal segment not longer on the dorsum than at the sides ; abdominal appendages of the ♀ at least one half longer than the 10th segment ....*Stylurus*, gen. nov.

Nymphs of four of our N. American Gomphine genera remain to be discovered. *Tachaptryx* and *Dromogomphus* of the eastern U. S., *Gomphoides* of Texas and *Octogomphus* of California. I venture now a preliminary table for our known nymphs. Doubtless many modifications of it will be necessary as the unknown nymphs still largely in the majority are discovered.

TABLE FOR GOMPHINE NYMPHS

1. Wing-cases strongly divergent............................ *2*.

   Wing-cases laid parallel along the back .... *4*.

─────────────────────────────────

*This will not apply to gaping *exuviæ* in which originally parallel wing-cases have been forced apart

2. One third or more of the length of the abdomen, formed by the
   10th abdominal segment................(supposition). *Aphylla.*

   Tenth abdominal segments not longer than the other segments..*3.*

3. Middle legs less distant at base than fore legs......*Progomphus.*

   Middle legs not less distant at base than fore legs.....(These
   apparently not separable) *Herpetogomphus* and *Ophiogomphus.*

4. Third joint of antennæ flat, circular ....*5.*

   Third joint of antennæ cylindric, at least twice as long as wide..*6.*

5. Abdomen flat, subcircular ....*Hagenius.*

   Abdomen ovate, twice as long as wide ....*Lanthus.*

6. Abdominal appendages longer than the 10th segment, front border of median lobe of labium straight (or in *Gomphus* occasionally very slightly rounded), with the usual fringe of flat
   scales, but without teeth. Abdomen not abruptly narrowed before 9th segment.....................................*7.*

   Abdominal appendages shorter than the 10th segment; front
   border of median labial lobe produced into a prominent rounded
   lobe which is generally armed with a conic apical tooth.
   Abdomen rather abruptly narrowed to the base of its 9th
   segment, more slowly tapering to the apex............*Orcus.*

   Body spindle-shaped, little flattened; fore and middle tibiæ with
   small external apical hooks or with none ....*Stylurus.*

   Ninth abdominal segment one half longer than the 8th, its lateral
   margins nearly parallel. A minute middorsal apical spine on
   the 9th segment only. Lateral lobe of the labium with a
   strongly incurved end hook and teeth on the inner margin
   increasing in size posteriorly.

   Body flat, lanceolate; fore and middle tibiæ with strong external
   apical burrowing hooks. Ninth abdominal segment hardly longer
   than the 8th, much narrowed posteriorly. Rudimentary dorsal
   hooks on some of the segments before the 9th. ....*Gomphus.*

   (To be continued.)

THE COLEOPTERA OF CANADA.

BY H. F. WICKHAM, IOWA CITY, IOWA.

XXV. THE CERAMBYCIDÆ OF ONTARIO AND QUEBEC.—*(Continued.)*

ATIMIA, Hald.

Represented by *A. confusa*, Say, the only Canadian species of the group Atimioides. Aside from the structural peculiarities given in the table of genera, it may be characterized by the blackish colour and the punctate surface clothed with rather long yellowish pubescence, which is irregularly disposed so as to leave abraded smooth spots. The elytra are broader than prothorax, truncate at tip. Length, .33-.40 inch.

NECYDALIS, Linn.

This genus, by the short elytra, bears some resemblance to *Molorchus*. The third and fourth antennal joints together are distinctly longer than the fifth. Our species is *N. mellitus*, Say, unknown to me, but described by Mr. Leng as being of variable colour, " usually rufo-testaceous, head, antennæ (base and tip tinged with rufous), thorax, scutellum and abdomen above black ; elytra punctate, more coarsely toward the margin ; reddish-brown, with paler spot at tip or entirely rufo-testaceous." The elytra are marked by an oblique impression which is not deep and does not reach the tip. Length, .60-.84 inch.

DESMOCERUS, Serv.

D. palliatus, Forst., is found on the elder *(Sambucus)* in July. It is a very showy beetle, with narrow head, deeply impressed above, bell-shaped prothorax, and faintly costate elytra. Colour blue except the base of the elytra, which is broadly orange or yellow. Length, .70-.90 inch. This insect can be mistaken for no other Longhorn.

TOXOTUS, Serv.

" This genus is sharply defined by the spurs of the hind tibiæ, which are inserted at the base of a deep excavation instead of at the extreme end."—(Leng.) This character is of easy verification, and is in itself sufficient for the separation of *Toxotus* from other Lepturoides. A modification of Mr. Leng's table may be used for the Canadian forms.

A. Elytra striped, black with marginal and discal yellowish vitta. .60-
    .68 in ....*trivittatus*, Say.

AA. Elytra unicolorous or nearly so.

 b. Third joint of antennæ much longer than the fourth.
  Larger species, legs bicoloured. .76–1.00 in.. *Schaumii*, Lec.
  Smaller species, legs unicolored. .40–.60 in.. *vestitus*, Hald.
 bb. Third joint of antennæ but slightly longer than the fourth.
  Tips of elytra obliquely truncate, sub-bidentate. .87–.90
  in...*cylindricollis*, Say.

The name *trivittatus* replaces that of *vittiger* in accordance with the synonymy proposed by Mr. Leng.

RHAGIUM, Fabr.

R. lineatum, Oliv., is often common under pine bark or in lumber piles. It has scarcely the appearance of being a Longhorn at all, the antennæ being so short as to usually fail of attaining the base of the elytra. The prothorax is much narrower than the elytra, armed on each side with a strong spine or acute tubercle. The elytra are narrowed behind, sharply costate. In colour the insect is black or nearly so, the prothorax appearing gray from the pubescence which clothes it, excepting a smooth stripe on each side (including the spine) and one on the median line. The elytra are marked by a few reddish or yellowish spots, and the pubescence is irregular, giving a mottled appearance. Length, .54–.80 inch. My small specimens are from the Lake Superior region, while the large ones came from the forests of the mountains of Arizona.

CENTRODERA, Lec.

A large species, *decolorata*, Harr. (Fig. 26), is our only representative. The head, prothorax, under sur-face and appendages are reddish-brown, the elytra lighter. The eyes are more prominent than usual, the prothorax shining, nearly smooth at middle, closely punctate and somewhat opaque at sides, lateral tubercle large and acute. Elytral punctuation coarse at base, becoming finer to tip, sides nearly parallel. Mr. Leng describes the antennæ as "about as long as the body," but they may fall one-third or more shorter. Length,  1.20 to 1.25 inch. Rather rare. Found on beech by Mr. Harrington.

FIG. 26.

PACHYTA, Serv.

 A. Elytra reticulate with raised smooth lines, the intermediate spaces coarsely punctured. Black, subæneous, antennæ, femora and base of tibiæ ferruginous. .51–.64 in............*rugipennis*, Newm.

AA. Elytra simply punctured.

Punctuation finer, surface of elytra finely pubescent, opaque or nearly so. Black, legs and antennæ often reddish, elytra yellow, four spots on each and tip black. .35–.38 in. *monticola*, Rand.

Punctuation coarse, surface of elytra glabrous, shining, colour black, elytra sometimes testaceous, or more or less distinctly maculate with black. .60–.72 in . . . . *liturata*, Kirby.

I am unacquainted with *P. rugipennis*, and the description is taken from Mr. Leng's synopsis. *P. monticola* is to be found on blossoms of wild rose, while I have taken *liturata* in numbers on piles of sawed pine lumber.

ANTHOPHILAX, Lec.

Three Canadian species are recorded, only one of which, *A. attenuatus*, Hald., is known to me. The others, *A. viridis*, Lec., and *A. malachitus*, Hald., are suspected by Dr. Horn to be respectively the ♀ and ♂ of one species. Following his table they separate thus : all belonging to that section of the genus in which the antennæ are slender, the third joint much longer than the fourth.

" Elytra coarsely punctate scabrous, more or less metallic.

Elytra greenish-blue, legs black. .70 in . . . . *viridis*, ♀ .

Elytra cupreo-æneous to blue, legs pale. .52 in . . . . *malachitus*, ♂ .

Elytra testaceous, irregularly maculate with piceous spots. Surface coarsely sparsely punctate and with small spaces which are distinctly pubescent ; median line of thorax distinctly impressed. .56 in . . . . : . . . . *attenuatus*."

Mr. W. H. Harrington has taken the last mentioned insect at Ottawa on beech.

ACMÆOPS, Lec.

Only two species, *pratensis* and *proteus*, are recorded in the Society's lists and additions. I have, however, seen *bivittata* with the label " Quebec," and recently Mr. Chagnon sent a specimen of *subpilosa* as coming from Montreal ; *longicornis* is known from the far north of Canada, and is included in the subjoined table, which is in the main equivalent to those prepared by Dr. Leconte and Mr. Leng.

A. Front and mouth much prolonged, body moderately robust, prothorax bell-shaped, sides sinuate but not tuberculate. Black, elytra variable, either blackish, reddish or clouded, occasionally indistinctly vittate. .24–.34 in . . . . *pratensis*, Laich.

AA. Front not greatly prolonged.

    b. Body short and stout, antennæ thicker, hind tarsi stout, the joints 1–3 equally pubescent beneath. Prothoracic tubercle distinct, elytra closely punctured. Colour varying from entirely black to almost entirely testaceous ; or the thorax may be yellowish while the elytra are black. Typical form has yellowish elytra, each with two black stripes. .24-.36 in..*bivittata*, Say.

    bb. Body more slender, antennæ more delicate, hind tarsi slender, pubescence wholly or in part lacking beneath on second and sometimes on first joint.

      c. Disk of prothorax convex, slightly channeled, densely punctured.

        Prothorax longer than wide, elytra rather sparsely punctured, pubescence short and scant. Colour extremely variable, black to testaceous, elytra often vittate. .36-.44 in.............*longicornis*, Kby.

        Prothorax broader than long ; blackish, pubescence very long. .36-.44 in.......... .. *subpilosa*, Lec.

      cc. Disk of prothorax flattened behind and prolonged into two dorso-lateral tubercles. Colour variable, blackish to testaceous, legs variable, but apparently with the base of the femora at least always rufous. .24– .36 in........................ *proteus*, Kirby.

While definite information is lacking, it is probable that *A. proteus* and *A. pratensis* breed in pine, since they are so frequently found on piles of pine lumber. *A. bivittata* (Fig. 27) is to be collected on flowers of *Anemone pennsylvanica*. Mr. Leng calls the punctuation "sparse," but it is rather close and coarse.

GAUROTES, Lec.

G. cyanipennis, Say, is readily known by its brilliant colour. The body is black, shining often with a purplish tinge, the elytra bright green, polished, the antennæ, legs and mouth-parts yellowish. The head is distinctly but sparsely punctured, the prothorax almost smooth except at sides, the elytral punctuation very distinct but widely separated. Length, .36-.40 in.

Fig. 27.

In Wisconsin I found this insect almost confined to Sumac blossoms. It is said to have been found ovipositing on butternut.

Encyclops, Newm.

E. cæruleus, Say, belongs here. It is smaller than most of the Lepturoides, and of slender parallel form, the elytra scarcely tapering to tip. The head is broad, squarish, the constriction far behind the eyes. Lateral thoracic tubercle distinct. Colour usually blue, varying to greenish, legs testaceous, antennæ with the bases of the joints (especially the distal ones) more or less testaceous. Punctuation strong, rugose. .28–.32. inch.

FOOD PLANTS OF THE SAN JOSE SCALE (Aspidiotus perniciosus) IN OHIO, EXCLUSIVE OF FRUIT TREES.

BY F. M. WEBSTER, WOOSTER, OHIO.

The following list includes forest and ornamental trees and shrubs, upon which the San José scale has been found breeding in Ohio*. Nearly all of these have been found either by myself or my assistant, Mr. C. W. Mally, in sufficient numbers to indicate that the insect might thrive on any of them. The Cotoneaster was sent for inspection, it having been recently received from a Long Island nursery firm, and when received was literally covered with the scale :

Grape, *Vitis labrusca.*
Linden, *Tilia Americana.*
European Linden, *Tilia Europæa.*
Sumac, *Rhus glabra.*
Japan Quince, *Pyrus japonica.*
Cotoneaster, *C. frigidum.*
Flowering Peach, *Prunus*, sp.
Flowering Cherry, *Prunus*, sp.
American Elm, *Ulmus Americana.*
Black Walnut, *Juglans nigra.*

Willow (imported), *Salix verminalis.*
Cut-leafed Birch, *Betula*, sp.
Lombardy Poplar, *Populus dilatata.*
Carolina Poplar, *P. monilifera.*
Golden-leaf Poplar, *P. Van Geerti.*
Catalpa, *C. speciosa.*
Chestnut, *Castanea sativa.*
Osage Orange, *Maclura aurantiaca.*
Snowball, *Viburnum opulus.*

To these must be added the several varieties of roses, currants, gooseberries and raspberries. The Early Richmond cherry I believe to be exempt from attack, as I have found trees whose branches interlocked with those of a pear that had been killed by the scale, yet the cherry was uninfested ; and in two cases that came under my observation, where this variety of cherry had been grafted upon mahaleb stock, and shoots had sprung up from below the graft, the shoots were badly infested with scale, while none at all could be found on the trees themselves.

*The determinations have been kindly verified by Dr. L. O. Howard, of the Division of Entomology, Department of Agriculture, Washington, and his assistants,

THE HIND WINGS OF THE DAY BUTTERFLIES.

BY A. RADCLIFFE GROTE, A. M., HILDESHEIM, GERMANY.

I wish to offer here a few remarks on the structure of the hind wings of the diurnals especially, in extension of my recent paper on the Butterflies of Hildesheim.*

The first point relates to the fact that the hind wings are more specialized as compared with the primaries. The probable explanation I offer is, that the hind wings bear more of the weight of the body (abdomen), and that they regulate the downward stroke of the fore wings. A parallel suggests itself with the vertebrates in which the hind legs are more specialized; and the cause is then, in both cases, a mechanical one. This specialization in the hind wings of the day butterflies manifests itself primarily in the inequality of the wings, of which the secondaries have the Radius 1 branched, the primaries 3 to 5 branched. In the second place by an advance over the front wings in the process of the absorption of the median veins, so that the radius or cubitus of the secondaries draws the branches nearer to itself than the corresponding vein of the primaries. Vein IV_2, in the case where its condition is not permanently generalized (*Lycaenidæ*, *Riodinidæ*, *Hesperiidæ*), is thus usually more drawn out of its original central position on the secondaries; it submits also first to degeneration (*Hesperiidæ*) on the hind wings, showing that here the cross vein has degenerated for a longer period than in the primaries, isolating the vein and depriving it of nourishment over a longer ancestral line. The cross vein itself vanishes first on the secondaries. Here the cell may be open, all trace of the scar vanished (*Araschnia*, *Melitæa*), while on the fore wings the degenerate vein is present, closing the cell.

The progress in the evolution of the neuration is evidently taking place in identical directions on both wings. The generalized condition of the radius (it being 5-veined) of the primaries in *Papilio* gives way to a specialized condition (4-veined) in *Parnassius*, with an intermediate 5-veined state in *Thais*, in which latter the upper branch of the median series, vein IV_1, which has left the cross vein to emerge from the radius in *Parnassius*, leaves the cross vein near the upper angle of the cell.

The absorption of the veins is everywhere attended by the same indications of a physiological process which, in its external manifestations, it is easy to trace. It is the same with veins II. and III. of the hind

*Mittheilungen a. d. Roemer Museum, No. 8, Feb., 1897.

wings. The greater the extent of absorption of II. by III. (the radius), from the base of the secondaries outwardly, the more specialized is the form. In the *Limenitini (Nymphalinæ)* the absorption is carried forward to the point of issue of the rudiment of I., so that the subfamily *Nymphalinæ* may apparently be separated from the *Argynninæ* by this character. While I have in various places in my paper correctly stated the change in . the position of II. and III., owing to this basal fusion of the two veins. I have in others written of a withdrawal of I. towards the point of junction of II. and III., which, in fact, is the reverse of what takes place, although the effect seems the same. I. probably remains constant, or nearly so ; in the cases where it is reduced to a mere scar it seems still to occupy the same relative position on vein II. It is extinguished by absorption. At the same time the fusion of II. and III. constantly changes in extent. In low forms, such as *Leptidia*, the two veins seem wholly separate at the base of the wing. In *Argynnis*, which is the lowest Nymphalid I have examined, the fusion at base is very limited, whereas in the highest Nymphalids the fusion is carried up to the point of issuance of I. In the Pierids the fusion is generally limited, and here, as I have pointed out in my essay, they lag behind the Nymphalids. The extent of the absorption is everywhere the measure of the specialization.

The last point to which I would here draw attention is the junction of the cross vein on hind wings with IV_3, or rather V_1. Here the Pierids have again lagged behind, the cross vein reaching IV_3, although the portion of the base of IV_3, between the junction of the cross vein and V_1, must be held to belong to the cross vein. In the *Pararginæ* and *Nymphalidæ* the cross vein is withdrawn to the point of issuance of V_1. The lower Meadow Browns agree with the *Limnadidæ* and *Pieridæ* in the position of the cross vein of secondaries. In the *Riodinidæ* (I have only examined the type) the cross vein is specialized as in the Nymphalids, while it is slightly removed outwards in the *Theclinæ* and *Lycæninæ*. Where the cross vein fails to meet the point of issuance of vein V_1, lying outside of it, we must describe IV_3 as issuing from the cross vein, to which the base of IV_3 morphologically belongs.

A study of both fore and hind wings shows that on both the same processes are repeated, but the initial impetus for the changes seems to be always given by the hind wings. It is as if a wave passed over the wings, coming from the hind pair and breaking over the primaries, carrying these frail creatures further along their airy paths into their unknown future.

FURTHER NOTES ON SECTIONS OF AUGOCHLORA.

BY CHARLES ROBERTSON, CARLINVILLE, ILLINOIS.

Since my note on the Mexican bees of the genus *Augochlora* was published I have been informed by Prof. Cockerell that he would not reply in this journal, but probably elsewhere. This conclusion seems to me to be remarkable, but I shall take this occasion to say what more I have to say on the subject and then leave it.

When I suggested two sections of *Augochlora*, in Trans. Am. Ent. Soc. XX., 147, I did not base my conclusion on the hind spurs alone, but because the two sets of species also agreed in other characters. I was too well acquainted with the characters of *Halictus* to suppose that a valid section of *Augochlora* could be maintained unless the spurs of a certain form were associated with other characters which indicated affinity. For example, *Halictus coriaceus* and *H. Forbesii* form a natural group of the genus and have finely serrate hind spurs. If I remember correctly *H. fuscipennis* belongs to the same group, but *H. parallelus*, which also has finely serrate spurs, does not. The sections of *Augochlora*, as I formed an idea of them at the time I mentioned them, might be defined as follows :

1. Slender species, having the sides of truncation of metathorax rounded above ; hind spur of ♀ finely serrate ; ventral segments of ♂ not metallic, or more or less metallic medially.

2. More robust species, having sides of truncation sharp ; hind spurs of ♀ with 4–5 long teeth ; basal ventral segments of ♂ metallic.

These characters belong to the species I indicated as coming in these sections, but it does not necessarily follow that other species with the same spur forms belong to either of them. Thus *A. splendida*, with basal fasciæ on second and third abdominal segments, may not belong to my second section.

That Prof. Cockerell did not know that the peculiar spur forms were secondary sexual characters of the females is shown by his failure to indicate the fact in the table ; by his insisting that *A. viridula* and *A. fervida* could not belong to the second group on account of their spurs ; by the use of the terms "ciliate or simple," which I think were taken from the males ; and by his comparison of types through Col. Bingham. Smith's male types were referred to the first group without regard to any except their spur characters, which were of no value. If the types of *A. aspasia*, *A. aurora* and *A. splendida* had been males these species would have been referred to the first so-called subgenus ; in other words, the author could not tell to which one of his own subgenera an *Augochlora* belonged. He failed to indicate valid characters of any natural group of *Augochlora*, and, in fact, showed that he had no idea of them.

SUCCESSFUL COLLECTING AT ELECTRIC LIGHT.

BY J. ALSTON MOFFAT, LONDON, ONT.

I herewith give a full list of the Lepidoptera new to the Society's collection, taken by Mr. J. W. Bice at electric light during the season of 1896.

Mr. Anderson and I picked out from amongst Mr. Bice's captures of about 2,000 mounted specimens of good material what seemed to be new to us ; and after comparison with named specimens, or illustrations, having failed to recognize them, they were laid aside for others more competent than we to decide upon them.

I am greatly indebted to Dr. J. B. Smith for the patient endurance, amidst his multitude of professional duties, with which he attended to and promptly returned a number of small lots sent to him by mail—the unreasonable demands of the U. S. customs officer at the boundary line prohibiting their being sent in bulk by express, and thus increasing the labour connected with it. And not only for the names of the specimens, but also for interesting and instructive remarks upon many of the species; Dr. Hulst also assisting me with the Geometers. Most of those new the collection were in single specimens of their kind, and Mr. Bice has generously donated them to the Society.

The names and their sequence are in accordance with Dr. Smith's list of 1891.

Protoparce carolina, Linn.
Cisthene unifascia, G. & R.
Lithosia bicolor, Grote.
Parorgyia parallela, G. & R.
Oedemasia badia, Pack.
Acronycta dactylina, Grote.
Acronycta impressa, Walk.
Cerma cora, Hub.   Upon this species, Dr. Smith remarks :   " Distinctly rare."
Semiophora tenebrifera, Walk.
Agrotis catherina, Grote.
Pachnobia salicarum, Walk.
Dicopis muralis, Grote.
Dicopis Thaxterianus, Grote.   Dr. Smith says : " Very good indeed, not in my collection."

Eutolype bombyciformis, Smith.

Eutolype Rolandi, Grote.

Mamestra assimilis, Morr.

Hadena passer, Guen.

Hadena indirecta, Grote.	Dr. Smith remarks: "Quite a new locality for this species. I have it from British Columbia and the Rocky Mountain region, but have never had it from anywhere near you."

Hadena diversicolor, Morr.

Taeniocampa vegeta, Morr.

Homoglœa hircina, Morr.

Cucullia florea, Guen.

Heliothis (Chloridea) rhexia, S. & A.

Galgula hepara, Grote. I took my first specimen of this insect in July, 1896, and sent it to Prof. Fernald, under the impression that it was a Tortricid, who kindly named it for me; Mr. Bice's specimen was so dissimilar that I did not recognize it.

· Homoptera Woodii, Grote.

Palthis asopialis, Guen.

Brotis vulneraria, Hub.

Semiothisa dislocaria, Pack.

Boarmia pampinaria, Guen.

Eubyia cupidaria, Grote.

Besides those altogether new, there were many interesting and unexpected varieties of common things brought to view by Mr. Bice's collection; which when disclosed were quite surprising to one not familiar with the extent and direction variation may go in some species, emphasizing with special force what Dr. Skinner gives in the subjoined extract as his experience with the butterflies:

"When I commenced my collection I was satisfied to have a single pair to represent the species, but now I cannot get enough individuals to represent all manner and kinds of variation brought about by natural causes. In the past I, therefore, knew this species or that, but now in many of our genera I nearly get brain fever in trying to determine where a species begins or ends."

BOOK NOTICES.

THE PARASITIC DISEASES OF POULTRY; by Fred V. Theobald, A. M., F. E. S.; 12 mo., pp. 120. Gurney & Jackson : 1 Paternoster Row, London, 1896.

It is encouraging to see a growing interest in applied entomology in England, and Mr. Theobald has given, in this handy little volume, a popular account of not only insect parasites but all other parasitic troubles likely to confront the poultry breeder. Not only is the little manual especially fitted for the wants of such, but it will doubtless find its way to the library of many other gentlemen who rely upon their estates to furnish fowls for their tables. The book is divided into several parts, relating to protozoan parasites, insect parasites, mite parasites, worm parasites, and vegetable parasites. Besides containing twenty - three illustrations, appendix I. gives a list of the parasites of *Gallus domesticus*, with the part of the fowl attacked by them ; appendix II. a quite full bibliography of the literature of the subject, which, with a very complete index, renders the volume of scientific as well as practical value, and Americans will find it of interest to them as well as Englishmen. We wish Mr. Theobald success in his efforts to add to the practical entomological literature of his people. F. M. W.

UBER DIE PALPEN DER RHOPALOCEREN. Ein Beitrag zur Erkenntnis der Verwandtschaftlichen Beziehungen unter den Tagfaltern ; mit 6 Tafeln : von Dr. Enzio Reuter. Acta Societatis Scientiarum Fennaciæ. Tom. XXII., No. 1. Helsingfors, 1896.

Entomologists in general, and lepidopterists in particular, will be interested in this work of Dr. Reuter's, occupying as it does a folio volume of 577 pages, the investigations, upon which the facts are chiefly based, requiring the examination of 3,557 palpi, belonging to 670 species, contained in 302 genera of the Rhopalocera. The work is divided into two parts : the first dealing with the direct microscopical examinations in descriptive form, while in the second is given the conclusions based on the same, as well as a discussion of other taxonomic characters allied to those brought out by himself, and their values. The plates are very fine, and the sixth of especial interest generally, as it presents, in the form of an evolutional tree, the relations of the various groups and genera to each other.

Dr. Reuter calls attention to the fact that at the base of the basal joint on the inner side of the palpi of butterflies is found a clearly distinguishable, naked spot, which he proposes to term the basal spot, on the surface of which are fine grooves and ridges as well as sparsely placed foveæ, and great numbers of peculiar, subconical, hairy rugosities. These last, though occurring normally in the Diurnals, and especially in the Nymphalidæ, and being clearly distinguishable with a low power lens, have formerly remained unknown, or if known have not been mentioned in entomological literature.

These ridges were by Landois considered as stridulating organs, and the two last structures in analogy with those observed by Kræplin, Forel, Hauser, and others, on the antennæ of various insects and looked upon as being sense organs ; but whether the peculiar structures in question served to convey the sense of smell, or, perhaps, some other and nearly related sense, is still an open question.

The Rhopalocera especially, of all the lepidoptera, have a special interest, because in them these cones present the greatest variations in form and are here the most highly developed.

Through further research, Dr. Reuter was convinced that a thorough study of the palpi, and especially of the basal spot, would afford a not inconsiderable basis for a knowledge of the family affinities of the individual genera and groups contained in the Rhopalocera, and he therefore determined to direct his especial attention to them, extending his studies over all of the families, and where possible over the smaller groups, as also to study the material at his disposal from a direct and thoroughly morphological point of view.   F. M. W.

Oviposition of Dorytomus Squamosus (Lec.).

This is a very common beetle upon cottonwoods in Colorado, but I have never known anything of its injuries until recently, when I had the good fortune to come upon a female preparing a burrow for her eggs in a terminal flower bud. When first observed she had her beak in the side of the bud up to her eyes. The twig was broken from the tree and carried in the hand without in the least disturbing the work of the beetle. After about ten minutes she removed her beak, turned quickly about and applied the tip of her abdomen to the hole she had made. After remaining in this position for about two minutes she ejected a small amount of a dark brown, thick liquid, which completely covered and hid the opening in the bud. This done she walked away.

The bud contained the catkin of a staminate flower which was nearly ready to burst forth, and immediately beneath the puncture in the bud scales, on the axis of inflorescence, were found three eggs lying close together.

The eggs were light yellow in colour, with a very thin, flexible shell, and although somewhat irregular in shape, measured about .85 mm. in length by .5 mm. in breadth.   C. P. Gillette.

Mailed July 8th, 1897.

PLATE 7.

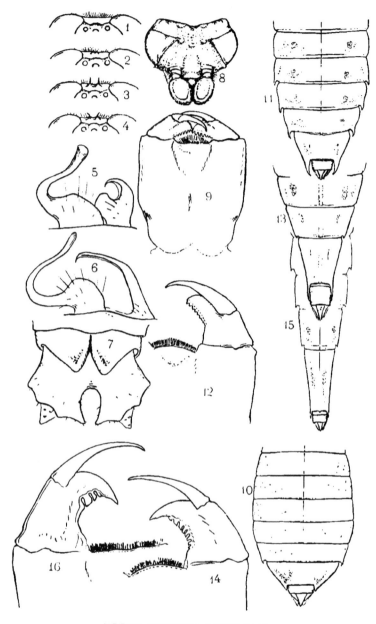

NORTH AMERICAN GOMPHINAE.

The Canadian Entomologist.

VOL. XXIX.   LONDON, AUGUST, 1897.   No. 8.

PRELIMINARY STUDIES OF N. AMERICAN GOMPHINÆ.

BY JAMES G. NEEDHAM, CORNELL UNIVERSITY, ITHACA, N. Y.*

(Continued from page 168.)

Herpetogomphus pictus, n. sp. Male.---Ithaca, N. Y.

Length, 49 mm.; abdomen, 35 : hind wing, 27.

Green and brown, varied with black and yellow.

Face and frons above entirely yellow ; a broad black band between the eyes, including the ocelli : antennæ black, the extreme rim of their cuplike insertions yellow. Occiput yellow, its border convex, ciliated with black. Rear of eyes brown, paler externally.

Prothorax fuscous, with a median twin spot greenish.

Thorax bright green, very thinly clad with brownish hairs, and faintly striped with brown. Dorsal and both lateral stripes subobsolete. Humeral stripe complete, irregular ; antehumeral, isolated above, and separated from the humeral by a narrow green line. Subalar and antealar carinæ brown.

Wings hyaline, flavescent at the base. Membranule minute, pale ; stigma brown ; veins black ; costa faintly yellow externally.

Femora straw yellow, lineated with black internally and each with a subapical incomplete ring of black. Tibiæ black, each with an external straw yellow line. Tarsi black : hind tarsi with a yellowish mark on the second and third segments superiorly.

Abdomen brown with transverse apical rings of black on segments 2 to 9 ; additional transverse lines of black on segments 3 to 7, at one-third the length of the segments. A middorsal yellow line, diffuse on segments 3 to 6, sharply bordered with black on 7 to 9. Apex of segment 10 and sides of 8 and 9 (except extreme lateral margin, which is black) and appendages yellow.

Superior appendages scarcely longer than the 10th segment, clad with blackish hairs Seen from above they are divergent half their length, then parallel to their blunt

---

*An unfortunate misarrangement of the table for nymphs crept in at the end of the last paper. The two paragraphs immediately preceding the last one on page 168 both relate to *Stylurus*. They should therefore be consolidated and preceded by 7.

Prof. T. D. A. Cockerell has promptly and very kindly called my attention to an oversight in proposing the name *Orcus*, which is preoccupied. I replace it with *Arigomphus*.

tips. Seen from the side they are thickest at the base and are gradually thinned and slightly declined to their truncate tips, beneath which are three or four rows of minute black denticles, extending more than half way to the base. Inferior appendage bifid for nearly half its length, the branches slightly divergent, truncate a little obliquely on tip, bent up at an angle with the declined basal portion, and bearing on each supero-lateral margin a broad quadrangular elevation just before the obtuse apex.

The appendages of the 2nd segment are very similar to those of *O. carolus*. (See plate.)

Two ♂ s from Ithaca in the Cornell University collection. One ♂ collected by Mr. J. O. Martin, at Ithaca, June 7th, 1897. A handsome species.

The occurrence of a *Herpetogomphus* at Ithaca was quite unexpected. All other species of the genus are from the extreme west and southwest. This one is related to *H. elaps*, Selys, of Mexico.

Ophiogomphus johannus, n sp. Male.—Wilmurt, N. Y.

Length, 43 mm.; abdomen, 30 ; hind wing, 26.

Black and yellow.

Labrum pale with a narrow brown margin which is broadened laterally. Face yellow ; rear of frons and vertex except the rear black. Occiput yellow, its margin ciliated with black.

Thorax yellow, with thin brownish hairs ; a narrow middorsal fuscous stripe subobsolete anteriorly, forking above with the carina to unite on either side with the fused humeral and antehumeral stripes, which are separated by a narrow yellow line only in their middle portion. Sides yellow, with an incomplete fuscous stripe on the 1st and a complete narrow one on the 2nd lateral sutures, and with fuscous markings above the bases of the legs.

Wings hyaline (immature).

Legs fuscous ; front femora paler below.

Abdomen black, marked with yellow as follows : Sides of segments 1 and 2, except behind auricles ; two lateral spots on 2 to 7; sides of 8 and 9, except the inferior margin ; apical half of 10 ; a maculose middorsal line reduced to very narrow basal spots on 5 to 8, wanting on 9.

Superior appendages slightly longer than 10, yellowish, darker at the tip, cylindric, moderately divaricate and equally narrowed in their apical fourth to an acute tip. Seen from the side they are a little angulated near the base and beyond this point irregularly denticulate beneath to a point just before the tip, where they are suddenly contracted from below upward, leaving the point at the upper side.

Inferior appendage bifid almost to its base, its branches straight, cylindric, about as long as superiors and twice as stout, almost as divergent, each apparently forked by reason of a very large external upturned tooth at two-fifths of its length ; at the extreme apex another stout upturned tooth. (For genital hamules see plate.)

A single ♂ , with its cast skin, from Wilmurt, N. Y., in the Cornell University collection.

Ophiogomphus carolus, n. sp. Male and female.—Ithaca, N. Y.

Length, 40-42 mm.; abdomen, 28-31 ; hind wing, 24-26

Greenish-yellow and blackish-brown.

Face greenish-yellow, paler toward the mouth. Rear of frons and vertex except the rear, black. Occiput yellow, its slightly convex margin ciliated with long black hairs. In the female there is generally in front of the margin a pair of black-tipped spines, whose various development is shown in plate, figs. 1 to 4. These sometimes occupy the margin which then becomes notched between them. Rear of eyes black above, mottled with paler below.

Prothorax blackish, its hind lobe with a median twin spot and a lateral spot each side yellow.

Thorax greenish, its dorsal stripes fused, enveloping the carina and forking with it above to meet the humeral. Antehumeral stripe isolated above, sometimes meeting the humeral near its upper end, but well separated through most of its length by a narrow greenish line. A partial brown line on the 1st lateral suture and a narrow complete one on the 2nd.

Legs black, front femora paler below.

Wings hyaline, often flavescent at base, costa black ; stigma cinereous.

Abdomen cylindric, a little narrower in its middle two-thirds, superiorly blackish with a maculose yellowish middorsal line of lanceolate spots on segments 3 to 7, of quadrangular basal spots on 8 and 9. Inferiorly, whitish with fuscous apical spots on most of the segments. Ten yellow ; fuscous at both ends.

Male appendages : superiors, longer than the 10th segment, cylindric ; seen from above, with acute apices divergent ; seen from side, fusiform, with truncate apices, denticulate beneath for one-third their length. Inferior appendage (see plate, fig. 7) bifid by a rounded notch, each branch somewhat flattened with four distal angles (as shown in the figure) or sometimes with only two (merely obliquely truncate); always with an upturned tooth at the outermost angle, sometimes with another at the innermost.

The genital hamules are shown at fig. 6 in the plate. These appear to be quite constant in form.

Female appendages fuscous, longer than 10 ; anal segment as long as the 10th. Vulvar lamina about as long as the 9th segment ; bifid except basal fourth, the branches enclosing an oval notch beyond which their incurved apices meet and then abruptly separate in short, oval, divergent points.

Described from more than seventy bred specimens (some of which will find their way into the collections of all my correspondents), from a single ♀ in the Cornell University collection, and from five specimens captured in May by Mr. Chester Young and Mr. J. O. Martin. I collected nymphs in October which emerged on my table in March. It was easy to collect the nymphs by hundreds in April, and in May the banks of the waters they frequented were fairly covered with exuviæ. Yet, outside of my breeding cages I saw but one live imago, notwithstanding I was doing much collecting at all times and in all places considered favourable. Where were they?

I have recently bred *A. villosipes*, Selys, by scores, and I find its exuviæ sticking to every bank about Ithaca, yet I have not seen a single imago at large. The imagoes, where are they?

Arigomphus australis, n. sp.   Male.—Gotha, Fla.

Length, 52 mm.; abdomen, 39; hind wing, 27.

Black and olive.

Face yellow with dense black pubescence.

A black stripe across base of labrum and another across the anterior margin of the frons. Rear of frons and whole of vertex black. Occiput yellow, convex, ciliate with black. Rear of eyes black above, yellow below.

Prothorax black with a median twin spot and a larger spot each side yellow.

Thorax olivaceous, striped with brown as follows: Dorsal stripes fused to form a cuneiform dorsal spot, not reaching the base, and narrowly divided with yellow along the extreme summit of the carina. Its narrow upper end is met by the strongly incurved antehumeral stripes, which are well separated from the narrower humeral stripes. Narrow but distinct stripes on both lateral sutures.

Legs black. Front femora pale within.

Wings hyaline, costa yellow, stigma brown. Veins black. Hind wing chalky near anal margin.

Abdomen long, slender. Segments 3 to 6 cylindric, narrower than terminal segments, entirely black. Remaining segments black, marked with yellow as follows: Sides of 1 and 2; dorsal lanceolate spots on 7 and 8; sides of 7 apically, and sides of 8 to 10 entirely yellow, 8 one-half longer than 9. Superior appendages about equalling 10, pale brown, divaricate at a right angle. Seen from above the inner margin is straight, the outer margin ends in a stout tooth, beyond which it is cut to a long acute point. Seen from the side each is gradually narrowed to a pointed apex, with a large acute tooth directly under the basal fourth, not visible at all from above. Inferior appendage with branches more divaricate, shorter, very little upcurved, ending under the apex of the lateral tooth.

One finely coloured ♂ taken by Mr. Adolph Hempel, in Orange Co., Fla., on the 21st of April, 1897.

At the same time Mr. Hempel took a *Progomphus obscurus*, Ramb., with its skin, in transformation. While the nymph was known by fair supposition, it appears not to have been reared before.

Mr. Hempel sent me also a nymph of the extraordinary type referred by Hagen (Trans. Amer. Ent. Soc., XII., 277, 1885) to *Aphylla producta*, Selys. It is time for someone to find the imago in Florida.

Gomphus umbratus, n. sp.   Male and female.—Ithaca, N. Y.

Length, 50-54 mm.; abdomen, 35-39; hind wing, 30-32.

Brown and olive, variable.

Face yellow, washed with brown in indistinct lines across the base of the labium and close under the frontal prominence. Rear of frons above and whole of vertex brown. Antennæ black. Occiput yellow, its hind margin convex (male and female), ciliated with black.

Prothorax variable, but always showing a median twin spot of yellow.

Thorax *brown* with a pair of nearly parallel dorsal stripes of yellowish-green, each sending at its lower end a spur against the carina, and at its upper end another spur around the isolated upper end of the antehumeral stripe of brown. Humeral and ante-humeral stripes of brown fused at lower end and near the upper end, and sometimes all the way between. Brown stripes of the lateral sutures overspreading the area between them, or sometimes the sides of the thorax wholly brown.

Femora brown, with numerous long spines in females. Tibiæ black, with a yellowish external line on each. Tarsi black.

Wings hyaline; their basal articulation and stigma rich brown when fully coloured. Costa yellow externally, veins black.

Abdomen cylindric in the female, slightly narrowed between the ends in the male, fuscous; basal fourth of middle segments paler and including a yellowish spot inferiorly. Middorsal stripe of yellow continuous at the base, reduced to lanceolate spots on segments 4 to 8, on 8 very short, on 9 wanting, 10 with a yellow spot in the female, uniform olive-brown in the male.

Male superior appendages flattened, a little arched. Seen from above the inner margin is nearly straight; at two-thirds their length they are cut obliquely to form a long point with an obtuse angle on the external margin. Seen from the side a low obtuse lobe appears on the interno-inferior carina just beyond the external angle. Inferior appendages a little shorter, more divergent and strongly upcurved at apices.

Female vulvar lamina transverse, one-third as long as wide, notched in the middle.

Described from seventeen specimens (14 males and three females), several of them bred, all obtained at Ithaca, N. Y., in May. A common species; next to *G. descriptus*, Banks, perhaps the commonest of the season; more variable in coloration than any other Gomphine I have seen.

Stylurus segregans, n. sp. Male—Havana, Ill.

Length, 61 mm.; abdomen, 44; hind wing, 35.

Face yellowish. Frons yellow, infuscated superiorly. A narrow fuscous stripe in front of ocelli. Frons and the ridge-like elevations behind each lateral ocellus pilose with soft black hairs. Occiput yellow, its border straight, ciliated with stiff black hairs.

Thorax fuscous; dorsum with two isolated lateral yellow stripes, divergent anteriorly. A narrow antehumeral line and a broad stripe down the middle of each of the lateral sclerites, yellow.

Legs brownish, paler internally, with black spinules. Claws pale, with apex and inferior tooth black.

Wings hyaline.

Abdomen fuscous, marked with yellow as follows: Dorsum of segment 1, a line on 2, basal middorsal spots on 3 to 8, extreme apex of 8, sides of 1 and 2, basal lateral spot on 3 to 7, sides of 7 and 8 except extreme lateral margin and apex, and all of 10.

Male superior appendages yellowish-brown, much longer than 10, divaricate at almost a right angle, slightly incurved toward the tip and cut obliquely to form an obtuse external angle at two-thirds their length, and a supero-internal point. The bevelled portion is minutely denticulate opposite the apices of the inferior appendage. No teeth or spines. The inferior appendage is bifid half its length with branches strongly divergent and strongly upcurved, their apices resting outside the bevelled portion of the superiors. Posterior genital hamule simple; pointed, directed forward at an angle of 45 degrees with the axis of the abdomen.

Name refers to its extremely local occurrence.

The single imago was obtained by Mr. C. A. Hart and myself, by rearing a nymph which we found crawling from the water upon bur-rush leaves, 23rd June, 1896, in the mouth of Quiver Creek. I obtained several exuviæ there, and several others later at McHairy's mill-dam some miles further up.

The nymphs of this and of the preceding species will be described in a forthcoming bulletin of the Illinois State Laboratory of Natural History.

Since this paper was written, I have obtained at Ithaca, N. Y., nymphs which can be none other than *Dromogomphus spinosus*, Selys. They fall in the same section of the table with *Arigomphus*, *Stylurus* and *Gomphus*, from all which they are distinguished by a sharp middorsal longitudinal ridge, ending in a straight apical spine on the 9th abdominal segment.

Explanation of Plate 7.

Figs. 1, 2, 3 and 4.—The occiput of the female of *Ophiogomphus carolus*, seen from the front, showing variations in occipital spines.

Fig. 5.—Genital hamules of *Ophiogomphus johannus* from the left side, inverted.

Fig 6.—Do. of *Ophiogomphus carolus.*

Fig. 7.—Inferior abdominal appendage of *O. carolus* seen from below.

Fig. 8.—Head of nymph of *Lanthus parvulus*, seen from above and in front.

Fig. 9 —Mentum of labium of do. from above.
Fig. 10 —End of abdomen of do.
Fig. 11.—End of abdomen of *Gomphus fraternus*, nymph.
Fig. 12.—Part of labium of do.
Fig. 13.—End of abdomen of *Arigomphus pallidus*, nymph.
Fig. 14.—Part of labium of do.
Fig. 15.—End of abdomen of *Stylurus segregans*, nymph.
Fig. 16.—Part of labium of do.

THE COLEOPTERA OF CANADA.

BY H. F. WICKHAM, IOWA CITY, IOWA.

XXVI. The Cerambycidæ of Ontario and Quebec.—(*Continued.*)

BELLAMIRA, Lec.

With this genus begins a series of beetles in which the neck is longer than in *Encyclops,* owing to the constriction of the head being near the eyes. *B. scalaris,* Say, is a very fine insect, varying in length from .75 to 1.20 inch. The form is slender, the elytra tapering greatly to and rounded at tip, deeply sinuate at sides, the tip of the abdomen uncovered. The prothorax is bell-shaped, with prominent, rather flattened hind angles. Colour brownish ; most of the head, the greater portion of the fore and middle legs, the bases of the hind femora and the bases and tips of the ventral segments inclining to reddish or even yellowish. Antennæ rufous. Elytra brownish, with a large common, lighter (golden-sericeous), wedge-shaped mark (wavy on the edges and sometimes interruped at about one-third its length by a transverse brownish band) which extends about two-thirds to tip. The body is finely and densely, in most places rugosely, punctured, clothed with fine golden pubescence, which is much denser on certain parts, notably the abdomen. Recorded as breeding in birch, and has been seen ovipositing in maple stumps.

STRANGALIA, Serv.

Includes two extremely elongate slender species, having the general form of *Bellamira,* but much smaller. *S. bicolor,* Swed., is entirely rufous except the eyes, the tips of the mandibles, the incisures of some of the tarsal joints and the elytra, which are black. Length .48–.56 inch. *S. luteicornis,* Fabr., is rufo-testaceous or yellowish ; the eyes, some markings on the under side, a ring at the tip of the hind femora, two dorso-lateral stripes on the prothorax, a narrow basal and three other transverse bands on the elytra, black. Length .36–.52 inch.

TYPOCERUS, Lec.

The impressed poriferous spaces on the antennæ, which separate this genus from *Leptura,* are to be looked for near the bases of the sixth and following joints, appearing as elliptical smoother spots. A good plan is to take the common *T. velutinus* as a type for examination, since in this species they are very distinct, and having once seen them their detection is easy in the . remaining species. The four Canadian forms may be separated by their colour, but ·it is to be remembered that the elytral pattern is subject to variation. None of them have the pro-thorax strongly rounded on the sides, but the form of this segment is campanulate. Excluding the extra-limital forms, those belonging to our fauna may be thus known :

A. Body above and beneath, legs and antennæ entirely black, ex-cept occasionally a rufescent spot in humeral region. .36-.44 in.. . ...:..*lugubris*, Say.

AA. Body beneath variable, antennæ blackish, legs black or rufous, elytra never entirely black, usually banded.

    b. Legs black. Elytra black with three transverse bands and basal spot yellow, the anterior two bands sometimes united at suture. .36-.40 in............*sparsus*, Lec.

    bb. Legs ferruginous.

        Prothorax very coarsely sparsely punctured. Elytra black, with basal spot and three transverse bands (the anterior two frequently united at suture) yel-low. .40-.52 inch.... ....*zebratus*, Fabr.

        Prothorax more finely and densely punctured. Elytra brownish or reddish, with yellow markings much like those of the preceding species, but these may be incomplete or even wanting. .40-.56 inch........................*velutinus*, Oliv.

T. zebratus (fig. 28) is said to mine in white oak. It bears considerable resemblance to *Leptura nitens*, from which it, however, easily separates by the generic character. *T. sparsus* is unknown to me, and the description is taken from Mr. Leng's table ; *velutinus* is often very abundant on flowers in the summer months.

Fig. 28.

LEPTURA, Serv.

This genus is of very large size, and many of the species are quite abundant. There is no uniformity of facies to give a ready clue to its separation from allied groups, some of the species resembling *Strangalia* in the shape of the prothorax, while others are very different.

The succeeding synopsis follows those of Dr. Leconte and Mr. Chas. W. Leng, with but few changes, chiefly such as are made neces-sary by later corrections of synonymy. ·

Prothorax more or less triangular, or campanulate, widest at base . . 2.

Prothorax nearly quadrate, or else more or less rounded or subcampanulate, usually constricted in front and behind, hind angles not prolonged . . . . 23.

2. Hind angles of prothorax prolonged . . . . 3.

Hind angles not prolonged . . . . 16.

3. Very large species (1.20 in.), prothorax strongly narrowed from the base, which is broadly but deeply bisinuate, posterior transverse impression distinct. Elytra widest at base, gradually narrowed behind, truncate and emarginate at tip, which is not margined. Black with velvety pubescence, elytra red, apex black, antennæ feebly serrate, elytra not sulcate . . . . *emarginata*, Fabr.

Small or moderate sized species . . . . 4.

4. Prothorax without distinct transverse basal impression. Small species, elytra sub-parallel at sides, not spotted nor banded, but uniformly dark. Prothorax often red, hind angles usually small. 10.

Prothorax with transverse basal impression often deep. Moderate sized species, elytra usually narrowed behind, often very much so, and frequently spotted, striped or banded . . . . 5.

5. Prothorax convex, with the sides much rounded in front of the middle, a transverse depression at base, hind angles small. Elytra black and yellow . . . . 14.

Prothorax with sides not much rounded in front of the middle . . . . . 6.

6. Prothorax strongly narrowed from the base, usually regularly so . . . 7.

Prothorax subcampanulate, transverse basal impression deep, hind angles broad, laminate, fourth joint of antennæ shorter than usual, elytra not banded nor spotted . . . . 15.

7. Elytra black or testaceous with black tip. Abdomen with the third, fourth and base of fifth ventrals red. Prothorax finely punctured. .48–.52 in . . . . *plebeja*, Rand.

Elytra with black and red or yellow markings . . . . 8.

8. Antennæ annulate . . . . 9.

Antennæ not annulate. Black, elytra very dehiscent, and not narrowed behind, coarsely punctured, sides of elytra, metathorax and abdomen red, thighs red with black tips. .36 in . . *cruentata*, Hald.

9. Female reddish-yellow, varied with black beneath, legs more or less black ; above with top of head, a discal thoracic stripe or spot, scutellum, sutural and side margins and transverse sub-median elytral band, black. Male black, base of legs and discal elytral

vitta (usually broken), as well as a small spot under the humerus, yellow. Antennæ annulate in both sexes. .48–.60 in...... ....*subhamata*, Rand.

Blackish, region of the mouth often yellowish. Legs and elytra testaceous or yellowish, the latter with sutural discal and lateral marginal vittæ black. .32–.52 in............ ......*lineola*, Say.

10. Elytra margined and usually rounded at tip...................11.

Elytra not or scarcely margined at tip. Blackish, pubescence white, head, legs and first antennal joint sometimes reddish or partly so. .24–.30 in......................*subargentata*, Kirby.

11. Black, elytra blue, polished coarsely and sparsely punctured, antennæ and legs either black or yellow. .24 in.........*chalybea*, Hald.

Black or piceous, head and prothorax often reddish, legs and antennæ frequently in part yellow................................12.

12. Elytra shining, very coarsely punctured, tip subtruncate. Colour black, legs black, head and prothorax reddish. .26–.36 in .... *capitata*, Newm.

Elytra more finely punctured, pubescence fine, white, prothorax and head rarely (never?) at once red or yellow, though often separately so..13.

13. Antennæ piceous ; anterior femora and base of middle ones yellowish. Upper surface piceous or (in the var. *hæmatites*, Newm.) the prothorax may be reddish. Terminal ventral segment of female simple. .16–.24 in......................... ....*nana*, Newm.

Antennæ piceous, basal joint yellow. Anterior femora and the bases of middle and hind ones yellow. Terminal segment of female with a slight tuberosity near apical margin. Colour piceous or blackish, thorax usually with yellow margin. .22–.28 in.....*exigua*, Newm.

14. Black, antennæ brownish, legs and tips of abdominal segments ferruginous, pubescence golden, so dense as to conceal most of the surface colour except on the legs, antennæ, tips of abdominal segments, middle of prothorax, five elytral bands and the sutural margin. .40–.52 in ... .*nitens*, Forst.

Black, antennæ and tibiæ often reddish, pubescence cinereous, not concealing the colour. Elytra yellowish, base, tip and two intermediate (usually interrupted) bands black. .31 – .38 in................................*sexmaculata*, Linn.

15. Black, elytra sometimes rufous or testaceous, prothorax very densely coarsely punctured, elytral punctuation less dense. Antennæ not

annulate, elytra sharply obliquely truncate at tip. .40-.60
in.......... .. ..*nigrella*, Say.
16. Antennæ annulated (except in ♂ of *canadensis*)..............17.
Antennæ not annulated....... ....19.
17. Elytra parallel, elongate, truncate at tip, front of head with transverse
impression. Colour black, punctuation fine and dense. Legs
reddish or brownish. .40 in...................*pedalis*, Lec.
Elytra narrowed from the base ....18.
18. Tips of elytra deeply truncato-emarginate, antennæ serrate in the ♂.
Punctuation very coarse and close, sub-confluent. Black, elytra
usually with large red basal spot, which may extend (in the var.
erythroptera) over the entire surface. .48-.76 in.. *canadensis*, Oliv.
Tips of elytra truncate or feebly emarginate, body of ordinary form,
not very stout ; punctuation of elytra finer, well separated. Black,
elytra reddish, abdomen red ♂ or black ♀ ....*rubrica*, Say.
Tips of elytra nearly rounded, very dehiscent. Form very short and
stout, head broad, elytra coarsely punctured. Black, elytra often
with reddish or yellowish submarginal stripe or entirely testaceous.
.36-.48 in.....................................*vagans*, Oliv.
19. Body densely golden pubescent. Blackish, elytra testaceous, often
darker at sides. .48-.56 in.................*chrysocoma*, Kby.
Body only moderately or sparsely pubescent.................20.
20. Black, elytra reddish or testaceous, wholly or for the greater part. 21.
Black, elytra black, each with four yellowish spots, thighs pale at
base. .40-.48 in.......................*octonotata*, Say.
21. Elytral margin very deeply sinuate (on viewing the insect from the
side). Prothorax with a tolerably well-marked median channel, at
bottom of which is an abbreviated raised line. Black, elytra
reddish except at tip, which is rather broadly obliquely marked
with a black blotch and truncate. .52-.75 in......*proxima*, Say.
Elytral margin not deeply sinuate.........................22.
22. Larger, prothorax with very distinct median channel which is wider
behind. Brownish red, elytra paler, with a submarginal dark spot
near the middle, tip obliquely truncate. .48-.52 in.. *biforis*, Newm.
Smaller, prothorax without median channel. Black, elytra reddish to testaceous, tip blackish, squarely truncate. .40-.48
in.......... ....*sanguinea*, Lec.
23. Prothorax hardly narrowed anteriorly and not constricted behind.
Seventh and following antennal joints with a raised line beneath.

Black, elytra sometimes (in var. *luridipennis*, Hald.) testaceous or with the tip alone dark. .32–.52 in . . . . *mutabilis*, Newm.

Prothorax much, often suddenly, narrowed anteriorly, with or without distinct constrictions . . . . 24.

24. Basal prothoracic constriction very deep, sides strongly rounded. . 26.

Basal prothoracic constriction feeble or absent . . . . 25.

25. Prothorax densely punctured, with median smooth line. Neck very close to eyes. Black, without markings. .36–.40 in . . *pubera*, Say.

Prothorax sparsely punctured, head longer behind the eyes. Usually black, elytra with or without a narrow discal yellow vitta. Varies to entirely testaceous. .40–.52 in . . . . *vittata*, Oliv.

26. Black ; legs more or less yellow ; prothorax (in var. *ruficollis*, Say) sometimes red, nearly smooth, except at base. Antennæ with tendency to become reddish at tips of joints. .28–.32 in . . . . *sphæricollis*, Say.

Black, legs almost entirely yellow in most specimens, prothorax occasionally red, finely and sparsely punctured, except at base, where it becomes more pronounced. Antennæ with the tips of the joints more evidently reddish, elytra with side margin and long discal vitta yellow. .24–.40 in . . . . *vibex*, Newm.

Probably the only serious difficulty to confront beginners in the use of the above table will arise in making the choice between the first two divisions; *i. e.*, 2 and 23. Should doubt arise here the assumption may be made that it belongs in the latter, when reference to other thoracic characters or to those of colour will soon show if the student is on the wrong track. The measurements here, as elsewhere, are in the main those of Mr. Leng, though I have frequently extended them, as shown by specimens in my own collections.

With regard to food habits very little can be said, so few of the *Lepturæ* having been bred ; while the perfect insects are commonly found on flowers, these give little or no clue to the feeding habits of the larvæ. Mr. Harrington has taken *L. subhamata* (fig. 29) on oak and also in a beech log, while the pupa of *L. canadensis* has been found in a hemlock stump. *L. nitens* bores, as a larva, in black oak, *L. vagans* in the yellow birch and pignut hickory, *L. proxima* has been reared from maple.

FIG. 29.

It will probably be noted that the authorities cited for certain of the species are not the same as those in the Check List. The reasons for these changes will be found in Mr. Leng's paper on the genus. Both *L. nana* and *L. exigua* are included in the table, although I am not sure that the latter occurs within our limits ; the former has been recorded by Dr. Hamilton (CAN. ENT., XXI., pp. 33 and 108). The name *zebra* is replaced by *nitens* on the ground of priority ; *sphæricollis* has been preferred as the specific and *ruficollis* as the varietal name, following Mr. Leng. In all probability *L. lacustris*, Casey, described from Michigan, will be found in Ontario. It differs by description from *sanguinea* in the much stouter male antennæ, and by the apices of the elytra being narrowly and obliquely truncate, the truncation sinuate, the angles, especially the exterior, very acute and prominent.

DESCRIPTIONS OF NEW SPIDERS.

BY NATHAN BANKS, WASHINGTON, D. C.

Teminius affinis, n. sp.

Length ♀ 13 mm.; ceph. 5 mm. long, 3.5 mm. wide ; patella plus tibia IV. 6 mm. long. Cephalothorax red-brown, darkest around head ; mandibles dark red-brown ; legs and palpi yellow-brown, lighter at tips ; sternum dark red-brown ; abdomen nearly black above, with faint indications of a light median streak, in the base of which is a black spear-mark ; venter dark gray ; spinnerets yellow. Posterior row of eyes straight, broader than anterior row ; P. M. E. round, separated by their diameter, nearer to each other than to the larger P. S. E.; A. M. E. about half their diameter apart, and slightly nearer to the A. S. E. than to each other. Legs quite long, no spines above or below on tibia I., and none above on tibia IV.; thick scopulas to all tarsi and metatarsi (except IV.). Sternum broad ; the abdomen long and narrow ; the upper spinnerets distinctly two-jointed and much longer than the lower pair, the second joint more than twice as long as wide. The epigynum shows a rounded cavity, slightly longer than broad, broader behind than in front, the anterior portion paler than the rest ; there is a median septum which in the fore part is narrow, but quite suddenly broadens at the middle and then tapers to the broadly rounded tip.

One specimen, Brazos Co., Texas. It differs from *T. continentalis*, Keys, in the larger size, position of eyes, spines on legs, and shape of the epigynum.

Thargalia canadensis, n. sp.

Length ♀ 7 mm.; ceph. 2.8 mm. long, 2 mm. broad ; patella plus tibia IV. 2.8 mm. Cephalothorax reddish yellow-brown, pars cephalica black ; mandibles dark red-brown ; anterior pairs of legs yellowish, hind pairs reddish, all femora with a black stripe each side, those on the fore pairs are much broader at base, the under side of tibia and metatarsus IV. infuscated ; maxillæ dark brown, pale on margin ; sternum reddish ; coxæ yellowish ; abdomen black above, paler below, reddish around the epigynum, above with a narrow white band near base, and another just before the middle, the latter rather indented on the median line. Posterior eye-row procurved, P. M. E. round, over one and one-half their diameter apart, closer to the equal P. S. E. Anterior eye-row procurved, shorter than the posterior, A. M. E. about as large as P. M. E., about once their diameter apart, very much closer to the equal A. S. E., which latter are well separated from the P. S. E. Two pairs of spines under tibiæ I. and II. Sternum one and one-fourth longer than broad, nearly as broad in front as at second coxæ, rounded to the pointed tip. The abdomen has a horny basal shield which extends but a short distance on the dorsum. The epigynum shows two oval openings marked in front by a continuous sinuous ridge.

One specimen from Ottawa, Canada.   (W. H. Harrington.)

Anyphæna fragilis, n. sp.

Length ♀ 5 mm.; ceph. 2 mm. long, 1.3 mm. broad ; patella plus tibia IV. 1.8 mm. Cephalothorax pale yellowish brown, black around eyes, a black line reaching from between the P. M. E. to the indistinct dorsal groove. Sometimes the sides are rather more infuscated. Mandibles dark brown, with a pale spot at base; maxillæ and lip pale, fringed with black hair; legs pale whitish, with blackish rings at base, middle, and tip of tibia, base and tip of metatarsus and tip of tarsus ; the bristles are arranged in lines so as to leave smooth spaces. Sternum pale, infuscated, darker on the sides. Abdomen pale, above with two rows of black spots, and some on each side ; venter pale, spinnerets infuscated. Cephalothorax not much narrowed in front, radial furrows obscure, P. M. E. about twice their diameter apart, scarcely closer to the equal P. S. E.   A. M. E. smaller than P. M. E., about their diameter apart, and nearly as far from the larger A. S. E.   Mandibles rather large and stout, vertical.   Legs short, two pairs of spines under tibiæ and metatarsi I. and II., the second pair at about middle of length ; hind legs more numerously spined. Sternum one and one-third longer than broad, broadest near middle, sides rounded. Abdomen slender, fully twice as long as broad ; ventral furrow nearer to epigynum than to the spinnerets. The epigynum shows a transversely rounded area, trilobate behind, the median lobe smaller and pointed, in each side a curved reddish opening.

Jacksonville, Florida; April. Collected by Messrs. Laurent and Castle.

Theridium dorsatum, n. sp.

Length ♀ 4 mm.; femur I. 2.1 mm., femur III. 1.2 mm. Cephalothorax dark yellow-brown, brown on the edges, eye region blackish, and behind is a triangular brown spot with its apex on the dorsal groove. Abdomen grayish, with a pale central mark bordered by black, from the projections faint marks run to the sides ; sides pale ; venter black, with a large central triangular silvery spot, spinnerets surrounded with black; a curved black line reaches from the anterior portion of the abdomen across the sides to the middle of the venter, where it joins the dark ventral area ; sternum brown ; legs pale yellowish, with brownish bands at the middle and ends of the joints, those on middle of femora I. and II. are narrow and oblique. P. M. E. are about their diameter apart, A. M. E. much more than their diameter apart ; sternum triangular, a little longer than broad in front ; legs moderately long and slender, metatarsus I. about equal to tibia I.; abdomen a little longer than broad and not very high. The epigynum shows a rounded semi-triangular lobe projecting behind.

Olympia, Washington. (Trevor Kincaid). Readily known by the large silvery spot on venter.

Theridium elevatum, n. sp.

Length ♀ 4 mm.; femur I. 2 mm. Cephalothorax yellow, with a black stripe each side and one on the middle, the latter with a short lateral spur each side at the dorsal groove and growing narrower behind ; mandibles with brown lines. Abdomen gray, mottled with white and brown ; the white is in the form of curved lines ; venter dark, with two white spots in front of the spinnerets ; sternum yellow, with some short black lines reaching from the sides ; legs pale, banded and thickly spotted with dark brown, bands at ends of joints, base and middle spotted. P. M. E. hardly their diameter apart ; A. M. E. equal to P. M. E., more than their diameter apart ; mandibles slender ; sternum triangular, barely longer than broad in front ; legs short and stout, femur

I. not quite twice as long as femur III., metatarsus I. barely longer than tibia I.; abdomen higher than long, globose ; region of epigynum swollen ; there is a small median triangular black projection or finger.

Brazos Co., Texas ; Sept.

Plæsiocrærius lobiceps, n. sp.

Length 1.5 mm. Cephalothorax yellowish with a black margin, each eye with a black ring, a black line on each side of the lobe ; mandibles yellowish, legs and palpi yellowish, sternum red, black on margins ; abdomen black, spinnerets pale. Head of male moderately elevated into a large lobe, bearing the P. M. E., which are large and scarcely twice their diameter apart ; a hole on each side just behind the S. E.; the mandibles show a series of transverse lines on the outer side ; legs moderately long, first pair longest, no spines above on the tibiæ ; sternum broad, triangular, bluntly pointed behind. Male palpi quite long ; the tibia with a broad extension above and a hook on the inner side ; the tarsus short, truncate at tip ; the bulb, in side view, is constricted near the middle, the upper part crossed by two transverse dark lines, the black style coiled around the tip once, a small triangular hook near base of bulb. In the female the head is scarcely elevated ; the epigynum shows a semicircular area limited by a concave ridge in front, from which there extends behind a gradually broadening furrow with its margins at tip, curved outward and backward.

One from Chicago, Ill,, under leaves in October ; others from Salineville, Ohio. (A. D. MacGillivray.)

Icius canadensis, n. sp.

Length ♀ 5 mm.; ceph. 2.4 mm. long, 1.9 mm. broad ; tibia plus patella IV. 2 mm. The male but little smaller. Cephalothorax red-brown, black in eye-region ; mandibles reddish ; leg I. reddish except the yellowish tarsi, other legs wholly pale yellowish. Sternum infuscated ; abdomen brownish with a narrow white line around base, and pale chevrons toward tip, venter pale gray, with a straight jet black stripe each side, and a narrow basal median spear-mark ; a black spot each side at base of spinnerets ; in ♂ more white hair around the A. M. E. Eye-region one and a fourth broader than long, broader behind than in front, first eye-row curved ; eyes of second row half way between dorsal and lateral eyes ; cephalothorax moderately high ; mandibles vertical, with one stout tooth on inner edge of fang-groove. Legs moderately long, IV. pair longest, I. pair very stout, three pairs of spines on the tibia and two on metatarsus I., metatarsus IV. spined only at tip, anterior coxæ separated by nearly width of labium. Sternum once and a third longer than broad, broadest between coxæ I. and II. Abdomen once and a half longer than broad, rounded at base, pointed behind, moderately high. The epigynum shows two oval cavities, more than their diameter apart, some distance in front of a posterior median indentation. The male palpus is short ; the tibia has a short, sharp projection on the outside ; the bulb projects beyond the base, and the upper part is much smaller than the lower, showing a curved tube on the outside, and terminating in a stout, straight, black stylus.

A few specimens from Ottawa, Canada ; collected by Mr. W. H. Harrington.

DIPTERA FROM YUCATAN AND CAMPECHE.—I.

BY C. H. TYLER TOWNSEND, FRONTERA, MEXICO.

A few specimens of Diptera were taken in the Yucatecan region, in April and May, 1896, by the writer. The present paper describes the new species. More material from that interesting fauna will doubtless be secured in time, and will form the subject of future papers of this series. For an account of the peculiar bio-geographical aspects of the *Yucatecan* fauna and flora, the reader is referred to the writer's second paper on the Bio-geography of the Southwestern U. S. and Mexico (Trans. Texas Acad. Sci., 1897).

TABANIDÆ.

1. *Tabanus campechianus*, n. sp.

One ♀. April 25th. Taken near Campeche, between that place and Esperanza (State of Campeche). Seems to approach *T. nigrovittatus*, McQ., according to Osten-Sacken's description.

Length, 8½ mm. Palpi almost white, with some white as well as black hairs. Face brownish, covered with a white bloom. Front brown, yellowish-gray dusted ; frontal callosity nearly square, rounded on upper corners ; a smaller longitudinal callosity above it twice as long as wide, and with a tendency to a linear elongation posteriorly. Callosities brown. Front parallel, about one-sixth width of head, parallel portion only a little more than twice as long as wide. First two joints of antennæ pale yellowish, second joint ending above in a sharp spur ; third joint reddish-yellowish, annulate portion black, process of base angular, but not enough developed to form a right angle, greatest width of third joint about twice the extreme basal width. Annulate portion of third joint hardly as long as the basal portion, about four times as long as wide. Thorax cinereous dusted, with a sparse short white pubescence, with four somewhat indistinct wide brownish vittæ. Pleura whitish pollinose. Scutellum cinereous, with a yellowish tinge on margin. Abdomen brownish-yellow, a well-defined, moderately broad median yellowish-white pollinose vitta of even width, becoming indistinct on sixth segment. A brown vitta on each side of and limiting the median vitta, forming a triangle on each side on third and a subarcuate marking on each side on second segment; but these brown vittæ are faintly represented in full width on second and third segments by a shading of brown supplementing the triangular and arcuate markings. On the outside of the brown vittæ on each side there is a lateral yellowish-white pollinose vitta like the median one but not so distinct; while still outside of this is another lateral brown vitta limiting the lateral white one on the inside and parallel with the edge of the abdomen on the outside. The fourth segment has the brownish-yellow considerably more tinged with brownish, and the fifth, sixth, and seventh are quite brownish. Pubescence very scanty, hairs of white portions whitish, of brown portions in main blackish, except on hind margins of posterior segments. Legs brownish-yellow, tips of tibiæ and bases of femora slightly brownish, but front tibiæ brownish on distal half; tarsi brownish, especially front tarsi, while the hind tibiæ and metatarsi are but little tinged with this colour. Wings fuscous-hyaline, costal cells and stigma distinctly yellow. Posterior cells all wide open, no stump nor even angle at the base of anterior branch of third vein. Eyes bare, no ocelli.

2. *Tabanus yucatanus*, n. sp.

Three ♀ s. May 10th. Taken from horses, at the *cenote* of Xcolak, about ten miles southeast of Izamal, Yucatan. This is the first record of a Tabanid of any genus or species, so far as I can find, from Yucatan. Nor can I find any recorded from Campeche. I have searched through all the multitude of existing descriptions of *Tabanus* from North and South America, including Walker's and Bigot's numerous species, and have been unable to identify this and the preceding species with any of them.

Length, 10 to 11 mm. Differs from *campechianus* as follows: Palpi pale watery-yellowish. Gray bloom of face slightly tinged with brownish. Front much narrower, about one-twelfth width of head, parallel portion fully five times as long as wide, just perceptibly narrowed anteriorly, with a callus swollen-conical or rounded posteriorly, prolonged into a second elongate spindle-shaped callus. Third antennal joint clearer reddish, annulate portion not so black ; process more developed, ending in a sharp-pointed angle, basal part of joint rather widened and shortened ; annulate portion short and comparatively stout, pointed elongate conical, hardly three times as long as basal width in two of the specimens, slightly longer and comparatively less stout in the other. Thorax saturate yellowish-brown, with four indistinct whitish lines, the middle ones sometimes obsolete. Scutellum concolorous with thorax. Median whitish vitta of abdomen formed of whitish pubescence in triangles, under which the ground colour is seen to be paler than the brownish-yellow of rest of abdomen. Pale brownish vitta on each side of median one is composed of coalescent oblique markings, like a vitta broken at the incisures, the marking on each segment directed posteriorly outward. A nearly similar, hardly less broken lateral whitish vitta outside of this on each side ; the last is bounded by a broken brown vitta on edge of abdomen, serrate on inner edge. Fourth to seventh, especially fifth to seventh segments, more deeply tinged with brown, or quite dark brown in ground colour. White incisures on sides of abdomen. White vittæ and incisures white-hairy, brownish vittæ black-hairy. Front femora quite brownish, hind metatarsi well tinged with brown, front tarsi almost black. Wings uniformly clear, except the pale yellowish oblique elongate stigma. Otherwise as in *campechianus*, including the venation, bare eyes, and absence of ocelli.

A NEW METHOD OF STUDYING NEURATION.

BY HENRY SKINNER, PROF. ENT. ACAD. NAT. SCI., PHILADELPHIA, PA.

The opprobrium cast on the lepidopterist has been that he did not study the anatomy of his specimens, but depended too much on maculation and colour. There has been much truth in the reproach, as there are few of us who would destroy a rare or unique specimen to examine the neuration. Fortunately the time has arrived when the neuration can be studied with the greatest ease and accuracy, and permanently recorded in a photograph, or, more strictly speaking, a radiograph. The anatomy of a living chrysalis may be studied without removing the cocoon, and also the internal anatomy of the thorax and abdomen can be fairly well seen, and in time the process may be improved for this work. With the aid of the Röentgen or X rays and the photographic plate one could make a picture of the neuration of the beautiful, rare and curiously shaped *Ornithoptera paradiseæ* and not disturb a scale on its superb wings. With the fluoroscope one could doubtless see all the neuration without even going to the trouble of making a picture. This is indeed a wonderful age, and in the future no entomologist will have any excuse for not studying the neuration of the lepidoptera, as he cannot say that he must denude the wings of his specimens, bleach them and mount in balsam as of old and thus destroy them.

BOOK NOTICES.

GUIDE TO THE GENERA AND CLASSIFICATION OF THE NORTH AMERICAN ORTHOPTERA. By S. H. Scudder: 8 vo., pp. 89. W. H. Wheeler; Cambridge, 1897. (Price $1.00.)

The above work, like all of Dr. Scudder's books, is exactly what the title states. It is simply a guide for the use of students of the Orthoptera, by means of which they may determine the genera of their specimens. It consists of excellent and most carefully prepared tables of the seven families into which the Orthoptera of North America are divided. These are followed by most valuable bibliographical notes, in which the student is referred under the head of each family of insects to all the works which refer to it. Then follows a full list of all the works which refer to North American Orthoptera, arranged alphabetically by authors and a complete index. All who have attempted to study Orthoptera know how badly such a book was wanted, and it is well for the science of entomology that the work was done by such a careful and experienced hand. J. F.

THE GENERA OF NORTH AMERICAN MELANOPLI. By S. H. Scudder. (Proc. Am. Acad. of A. and S. V. 32, pp. 195–206. Jan., 1897)

Almost simultaneously with Dr. Scudder's "Guide to the Genera of Orthoptera" two other important and extremely useful papers appeared, one on "*The Genera of North American Melanopli*" and the other on "*The Species of the Genus Melanoplus.*" These are both really advance issues of chapters in Dr. Scudder's great work on the *Melanopli*, which is to be published by the U. S. National Museum. The *Melanopli* are divided into 30 genera, 17 of which are new and 4 have been previously published by the author. The genus *Melanoplus* is characteristically American and is widely disseminated. There are 131 species recognized, grouped under 28 series. The name *furcula* is given to the processes of the last dorsal segment of the male abdomen. J. F.

The Canadian Entomologist.

Vol. XXIX.   LONDON, SEPTEMBER, 1897.   No. 9.

THE COLEOPTERA OF CANADA.

BY H. F. WICKHAM, IOWA CITY, IOWA.

XXVII. The Cerambycidæ of Ontario and Quebec.—*(Continued.)*

With this paper we begin the consideration of the Lamiinæ, the third great subfamily of Longhorns. They have recently been worked up by Mr. Leng and Dr. Hamilton in a joint publication* which has been largely used and followed in the preparation of the succeeding pages. The essential characters are to be found in the oblique sulcation of the outer side of the front tibiæ, the lack of prothoracic margin and the cylindrical pointed terminal joint of the palpi. None of the Canadian forms offer exceptions to the above rule. It will also be noticed that the front of the head is usually vertical instead of being oblique or nearly horizontal. Compare a *Prionus, Romaleum* and *Saperda* and this point will be made clear.

In order to construct a dichotomous table of the Canadian genera it has been necessary to disturb the sequence somewhat. The student will understand, however, that no implication of relationship is meant to be expressed in the succession as adopted in this paper, but convenience of identification has been given the most prominence. Probably the only characters that will be found difficult to a beginner are those relating to the claws (which, however, are sufficiently commented upon in the table), the antennal cicatrix and the front coxæ. The cicatrix is a sort of scar which is to be easily seen in *Monohammus* near the tip of the first antennal joint ; it is, in the above genus, limited by a distinct raised line. The angulation of the front coxal cavities is readily noticeable in the same insect, especially if the leg be removed, when it is seen that the cavity, instead of being circular in outline, has a V-shaped nick in the outer margin.

It is, perhaps, hardly necessary to state so self-evident a fact as that the "Classification" of Drs. Leconte and Horn has furnished the chief

*The Lamiinæ of North America. Trans. Am. Ent. Soc., XXIII.

material for the table, which is in the main only a slight rearrangement of the numerous short ones of their own.

  Humeral angles not prominent, wings wanting. Form very convex, prothorax rounded, unarmed. Elytra with bands of pubescence . . . . *Ipochus.*

  Humeral angles usually distinct, wings and elytra fully developed, not abbreviated . . . . 2.

2. Usually large or moderate-sized species; elytra not spinose at base..4.

  Small or minute species. Elytra with a spine or gibbosity near the scutellum . . . . 3.

3. Humeri rounded, elytra very convex and with large spine near scutellum . . . . *Crytinus.*

  Humeri distinct, elytra less convex, with oval gibbosity near scutellum . . . . *Psenocerus.*

4. Scape of antennæ with apical cicatrix. Nearly all large species, antennæ sometimes greatly elongate in the males. Prothorax with lateral spine present, often very large . . . . 5.

  Scape of antennæ without apical cicatrix . . . . 6.

5. Legs long, anterior pair elongate in the males . . . . *Monohammus.*

  Legs equal, not elongate . . . . *Goes.*

6. Front coxal cavities rounded. Body usually broad. Elytra attenuate behind. Antennæ usually very long in the males . . . . 7.

  Front coxal cavities angulate . . . . 14.

7. Scape of antennæ club-shaped. Prothorax with dorsal tubercles and large, acute, nearly median lateral spine . . . . *Acanthoderes.*

  Scape of antennæ nearly cylindrical. Lateral spine or tubercle, if present, behind the middle . . . . 8.

8. Female without elongated ovipositor . . . . 9.

  Female with elongated ovipositor . . . . 12.

9. Prothorax fully tuberculate or angulate. Mesosternum broad . . . . *Leptostylus.*

  Prothorax distinctly angulate, or more frequently with a short spine or acute tubercle behind the middle. Mesosternum narrow . . . 10.

10. Antennæ without traces of ciliæ beneath, first joint of hind tarsus as long as the next two. Prosternum narrow, body without erect hairs . . . . *Liopus.*

  Antennæ distinctly ciliate beneath, first joint of hind tarsi as long as next three . . . . 11.

11. Elytra without lateral carina, usually with transverse angulated markings...................................*Lepturges.*
    Elytra with lateral carina and marked with numerous small black spots ....*Hyperplatys.*
12. Body above pubescent, without intermixed erect hairs ; antennæ with at least joints 3–4 densely fringed with hairs beneath..*Acanthocinus.*
    Body above with erect hairs mixed with the pubescence ....13.
13. Mesosternum broad, antennæ not much longer than the body and not ciliate beneath except feebly on the scape...... *Graphisurus.*
    Mesosternum narrow, antennæ of male twice as long as the body, ciliate beneath .... *Ceratographis.*
14. Antennæ very elongate, prothorax cylindrical, slightly tubularly narrowed behind (in our species) without lateral armature or dorsal tubercles. Colour black.....................*Dorcaschema.*
    Antennæ not more than moderately elongate.. ....15.
15. Claws (at least on front tarsi) divaricate ; *i. e.*, extending in a plane at right angles to the length of last tarsal joint...............17.
    Claws divergent ; *i. e.*, not in plane as described above, but forming an angle ....16.
16. Rather large species, prothorax sinuate or feebly tuberculate on sides, front of head large, flat. Shape *Saperda*-like. Claws simple.. *Oncideres.*
    Rather small species. Black, front of head in part and sides of prothorax yellow, claws cleft ....*Amphionycha.*
17. Claws simple (except outer one of front and middle tarsi in some male *Saperda*) ....18.
    Claws cleft or appendiculate.... ....22.
18. Smaller species, prothorax spinose or tuberculate on sides..... .19.
    Larger species, prothorax never armed nor tuberculate ....*Saperda.*
19. Thighs clavate, vertex concave, antennal tubercles prominent....20.
    Thighs not clavate, vertex flat or convex, antennal tubercles not prominent. Eyes coarsely granulated, lower lobe as wide as long, body with flying hairs, antennæ pilose, joints 5–10 shorter, equal ....*Eupogonius.*
20. Lower lobe of eyes elongate. Lateral spines of prothorax large, median. Pubescence mottled, gray and black, mixed with short, scattered hairs on elytra..............................*Hoplosia.*
    Lower lobe of eyes subquadrate or subtriangular ....21.

21. Prothorax with lateral spine, flying hairs long........*Pogonocherus.*
　　Prothorax with feebly rounded sides, pubescence short......*Ecyrus.*
22. Eyes not divided, prothorax not distinctly tuberculate, form
　　slender.............................. ....*Oberea.*
　　Eyes completely divided, the upper and lower portions widely sepa-
　　rated, prothorax with large lateral tubercle, form stout. Colour
　　red with black spots .... *Tetraopes.*

Ipochus, Lec.

A record of the Californian species *I. fasciatus*, Lec., is existent
upon the Society's list, but I am unaware of the original authority. It is
a convex, heavily-built beetle, blackish, pubescence long, erect. Prothorax with large punctures, and bearing a transverse row of four spots
of white pubesence. Elytra with irregular transverse bands of whitish
pubescence, varying in width. Length, .18–.30 inch.

Cyrtinus, Lec.

Represented by one extremely small, somewhat antlike species, *C.
pygmæus*, Hald., easily recognized by the convex elytra with rounded
humeri and large juxta-scutellar spine. Colour nearly black, elytra with
a whitish pubescent spot before the middle, antennæ annulate. Length,
.08–.12 inch. Said to occur on dead oak branches.

Psenocerus, Lec.

P. supernotatus, Say (fig. 30), is recorded as boring during larval
life in the stems of grape, currant, gooseberry, and apple. I have
frequently beaten it from crab-apple trees. It is a
small beetle of somewhat elongate form, reddish to
nearly black, the elytra with a darker blotch behind
the middle which is bordered anteriorly and posteriorly by a band of whitish pubescence, the anterior
band usually much the narrower and interrupted

Fig. 30.

near the suture. Antennæ shorter than body in both sexes. Small
specimens are often almost entirely black, and may lack the elevation at
the base of the elytra. Length, .12–.24 inch.

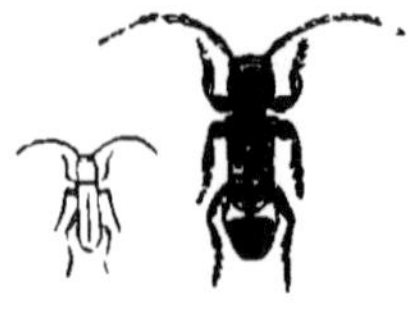

Monohammus, Serv.

Includes several very large species with long legs and antennæ,
especially in the males. Some or all of them are injurious to pine
lumber, and *scutellatus* and *confusor* are usually abundant in the eastern

coniferous forests. *M. maculosus* is more essentially western, but often common; while *marmorator* is very rare. Dr. Horn separates the species thus :

A. Tips of elytra rounded, sutural angle acute or spiniform, especially in the male. Piceous or black, more or less bronzed, elytra irregularly mottled with patches of brownish and grayish or whitish pubescence. Punctuation very coarse and close. .66–1.06 inch . . . . *maculosus*, Hald.

AA. Tips of elytra rounded, sutural angle not prolonged, usually very obtuse.

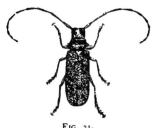

FIG. 31.

b. Black, distinctly bronzed. Elytral patches of pubescence few or wanting ; female antennæ annulate. Scutellum densely clothed with white pubescence. .64–1.24 inch. (fig.31). *scutellatus*, Say.

bb. Brown, elytra sparsely mottled with patches of gray and brown pubescence. Female antennæ not annulate. 1.10–1.24 inch. (fig. 32) . . *confusor*, Kirby.

AAA. Tips of elytra obliquely prolonged and acute. Elytra brownish, surface feebly punctured, clothed with ochreous white and brown patches intermixed. 1.00 inch . . . . *marmorator*, Kirby.

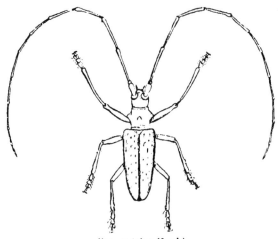

FIG. 32 (after Harris).

GOES, Lec.

Contains several species, mostly of rather or quite large size, resembling *Saperda* somewhat in form, but with a strong lateral thoracic spine. All have the upper surface mottled with pubescence, sometimes arranged in tolerably distinct transverse bands. Since only one of the North American species is lacking from Canada, we reproduce Dr. Horn's synopsis as far as it concerns us :

A. Surface colour of body brownish ; antennæ of male at most one and one-quarter times the length of the body.

    b. Elytra with conspicuous denuded fascia one-third from apex. Pubescence of surface white.   1.00 in........*tigrina*, DeG. Pubescence ochreous or luteous, basal region of elytra darker, less pubescent.   .92 in...........*pulchra*, Hald. Pubescence marmorate, whitish and ochreous, the apical region darker ochreous.   .44–.52 in........*debilis*, Lec.

    bb. Elytra without conspicuous denuded fascia, pubescence cinereous or almost white, uniform, sometimes with faint trace of denuded fascia.   .80–.88 in........*pulverulenta*, Hald.

AA. Surface colour black, shining, pubescence whitish, a small conspicuous black spot on each elytron, one-third from apex. Antennæ of male twice as long as the body.   .40–.44 in......*oculata*, Lec.

A few notes on food-habits have been published, from which it appears that *debilis* has been found on hickory and white oak, *tigrina* on hickory (as an adult) and in oak (as larva); *pulchra* and *oculata* are found in the mature stage on hickory, while the larva of *pulverulenta* is said to bore in wild cherry and in living beech trees.

ACANTHODERES, Serv.

The three species belonging here may be separated from those immediately following by their antennæ, in which the first joint, or scape, is strongly clavate. They are brownish insects, maculate above with whitish or ashy pubescence, of rather robust form, the upper surface rough, the femora much swollen. Prothorax with strong, sharp lateral spine. The differentials are given by Dr. Horn, thus :

A. Sutural region of elytra vaguely grooved, the groove limited on each elytron by a feeble costa. Elytra with a moderately broad transverse band of white in front of middle, broadly interrupted at suture.   .40 in ....*quadrigibbus*, Say.

AA. Sutural region not grooved, elytra with a more or less distinct M-shaped black mark behind the middle of each.

    Base of elytra irregular, an oblong obtuse umbone at middle of base. .42.–48 in...................*decipiens*, Hald.

    Base of elytra regularly convex without umbone. .80 in................................*Morrisii*, Uhler.

The recorded food-plants of *A. quadrigibbus* are oak, hickory, beech, and hackberry. I have found *A. decipiens* on oak logs, but am unaware of the larval habits.

LEPTOSTYLUS, LEC.

Numerous species are known from Canada, and are arranged mainly on the plan offered by Dr. Horn. The name *commixtus* is replaced by *sexguttatus*. The lateral tubercle of the prothorax is always blunt, sometimes obsolete.

A. Elytra without asperities and scalelike hairs.

    Prosternum between the coxæ as wide as a coxal cavity, elytra slightly truncate at tip. Robust, convex, prothorax with whitish pubescence forming a broad lateral stripe each side, bounded internally by a black stripe which is formed of a row of denuded tubercles, the discal area brownish. Elytra with a broad irregular transverse post-median area of whitish pubescence. .16–.36 in............................*macula*, Say.

    Prosternum narrower than a coxal cavity, tips of elytra separately rounded, thorax moderately densely punctured on flanks as well as disk. Yellowish or reddish brown, elytral surface uneven, basal angles black, polished; a dark brown irregular band often extends backward from these angles to the middle, thence narrows towards suture, followed posteriorly by one or two black spots, but these markings may be absent. Three large basal tubercles on prothorax, and three smaller, sometimes obsolete, subapical ones. .40 in.............*collaris*, Hald.

AA. Elytra with asperities or tubercles, bearing at their summits short black scalelike hairs.

    b. Thorax densely punctured, elytra with densely placed coarse deep punctures. Colour variable, elytra mottled with grayish pubescence. .28–.40 in................*sexguttatus*, Say.

    bb. Thorax not densely punctured, elytral punctures not closely placed, often inconspicuous or concealed. Legs not hairy.

c. Antennæ longer than the body in both sexes, the third
joint only slightly longer than the fourth. Punctuation
of thorax regular.

> Elytra very indistinctly punctured, especially at
> apex, the disk with angulate fascia behind the
> middle, tips feebly obliquely truncate. .16–.24
> in........ ....*biustus*, Lec.
>
> Elytra distinctly punctured over entire surface,
> disk with acutely angulated fascia, apex slightly
> prolonged, not obliquely truncate. .16 – .24
> in............................*parvus*, Lec.

cc. Antennæ scarcely longer than the body even in the male,
the third joint conspicuously longer than the fourth.
Robust, brownish, surface of prothorax roughly tuberculate, pubescent, punctures sparse, irregularly
placed. Elytra with raised tubercles or
ridges, and with grayish and whitish pubescence which tends to form a post-median
transverse band broadest at the suture, the
pubescence becoming darker anteriorly. Tip dark.
.32–.40 in. (fig. 33)..................*aculifer*, Say.

Fig. 33.

The food plants of several of the above are recorded. *L. macula*
is known to breed in beech, hickory, walnut, butternut, and chestnut ;
sexguttatus in pine ; and *aculifer* in oak, apple, sycamore, and osage
orange.

RARE BUTTERFLIES.—On the 8th day of May, Mr. James Walker
captured, in a cedar swamp, near Orillia, Ont., a specimen of *Thecla
tæta*, Edw. This butterfly has hitherto only been recorded in Canada,
from London and York Mills in this Province, and from a few localities
in the Province of Quebec.

Mr. C. E. Grant, of Orillia, has recently taken a specimen of the
melanic form of *Colias philodice*, the yellow on the wings being replaced
by dark scales. It is apparently somewhat similar to the aberration
recorded by Mr. Dwight Brainerd (C. E., XXVIII., p. 305), which he
took at Edgartown, Mass., last year. Mr. Grant has also taken at
Orillia, for the first time, *Papilio troilus* and *Lycæna comyntas*, making
the total number of butterflies from that locality sixty-two.

Papilio Ajax (a perfect specimen) has again been seen at Port
Hope on the 24th of July.

A GENERIC REVISION OF THE HIPOCRITIDÆ (ARCTIIDÆ).

BY HARRISON G. DYAR, PH. D., NEW YORK.

The earliest use of the term Arctiidæ is referred by Dr. Packard to Leach (1815). This is antedated by Hübner's Tentamen terms, Hipocritæ and Hypercompæ. The latter is unavailable, as Hypercompa becomes a synonym. I do not find any plural terms for the family before Hübner.

The faunas of Europe and America are here united. I have included the Indian genera as far as possible, but could not do so completely, as Hampson's work is much less available here than usual. Hampson does not recognize the Lithosiidæ on the character of the absence of ocelli, but unites under the term Arctiidæ all the species here grouped as Hipocritidæ with Lithosiidæ, Nyctemeridæ, Pseudoipsidæ and Nolidæ. His subdivisions of this aggregation are based on other characters, so that some of the genera that I have not seen can not be placed in the table from his figures and descriptions. Especially Castalba, Tatargena, Sidyma may be Hipocritidæ, though placed in Hampson's Lithosiinæ, while Rhodogastria, Pangora, Nicæa and Leucopardus I can not place from lack of the type species. I do not think that this affects the present revision, as these genera seem to be distinct from any of those included. As far as the American genera are concerned, I exclude Cydosia and Cerathosia, as they are probably Noctuid. Euverna is transposed to the Arctiinæ and becomes synonymous with Ectypia, a result due to the study of additional material, which I owe to Prof. Smith. Cycnia divides into three genera on venational characters, one of the sections supplanting Pareuchaetes ; Halisidota divides into two genera. The names Elpis and Neoarctia fall before European terms and a new genus is required for the species *virginalis*, Boisd. Pygoctnucha is transferred from the Euchromiidæ on account of the presence of vein 8 of secondaries. Three genera, Eucereon, Bertholdia and Euerythra, lack vein 8 and would seem strictly to be Euchromiidæ, but I hesitate to transfer them, as the habitus is Arctian, the larvæ are unknown and the condition of vein 8 is distinctly led up to in Eupseudosoma, which has a short spurlike vein 8 in the male and none in the female. The Phaegopterinæ may be further modified when the large South American fauna is worked up. In the meantime I dedicate to Mr. Schaus the new section of Halisidota, which he has shown to be of generic value (Journ. N. Y. Ent. Soc., IV., 138) in recognition of his work on this group as well as on the allied Euchromiidæ and in the anticipation of still further and more comprehensive labours.

The following table is based on the work of Prof. J. B. Smith, which
appeared in Can. Ent. some years ago, and was worked over in the revision of Bombyces by Mr. Neumœgen and myself. Following the table
is a list of genera and species; italicized names are North American.
Bibliographical references are omitted, and they can readily be found in
Kirby's catalogue if wanted. The types of genera are recognized as determined by Kirby.

KEY TO THE GENERA.

1. Head prominent, tongue moderate or strong..................2.
   Head more or less retracted, tongue weak or small........... 13.
2. Secondaries large and ample, habitus lithosiiform .... 3.
   Secondaries trigonate, often small, primaries pointed at apex ....44.
3. Vein 5 of secondaries faint or absent.......................4.
   Vein 5 distinct ....5.
4. Primaries long and narrow........................... Coscinia.
   Primaries broad, trigonate........................... Eubaphe.
5. Primaries broad, trigonate.............................:..............6.
   Primaries narrow, apices rounded.................... Utetheisa.
6. Vein 5 of secondaries arising close to vein 4...................7.
   Vein 5 of secondaries from near the middle of the cell........Doa.
7. No accessory cell; veins 7 to 10 of primaries stalked ....10.
   No accessory cell; vein 10 free, from the discal cell .....Axiopœna.
   Accessory cell present, vein 10 arising from it . ...8.
8. Anal angle of secondaries rounded in the male, spurs of tibiæ long..9.
   Anal angle produced to a point; spurs very short ....Argina.
9. Vein 6 arising beyond the angle of the discal cell.....Macrobrochis.
   Vein 6 arising from the discal cell................. Callimorpha.
10. Vein 11 free from vein 10 ....11.
    Vein 11 almost or quite touching vein 10....................12.
11. Secondaries over three-fourths the length of primaries ......Haploa.
    Secondaries less than three-fourths the length of primaries....Areas.
12. Secondaries with veins 6 and 7 from the cell............. Sebastia.
    Secondaries with veins 6 and 7 stalked .... Calpenia.
13. Vein 8 of secondaries wanting...................... Euerythra.
    Vein 8 present..14.
14. Veins 7 to 10 of primaries stalked from apex of cell........... 15.
    Vein 10 arising from the discal cell........................38.
15. Vein 11 free .... 16.
    Vein 11 joined to vein 10 to form an accessory cell.......Hipocrita.

16. Median spurs of hind tibiæ wanting ....17.
    Median spurs of hind tibiæ present.....................26.
17. Anterior tibiæ unarmed ....18.
    Anterior tibiæ armed at tip ....20.
18. Antennæ of male simple............................19.
    Antennæ of male pectinated......................*Leptarctia*.
19. Palpi exceeding the front.........................*Ecpantheria*.
    Palpi not reaching the front......................*Creatonotus*.
20. Inner prong of tibial armour-plate produced into a spine..... ..21.
    Inner prong not much produced, spine short.................24.
21. Wings of male with the outer margin upright, of female
    aborted...*Pachylischia*.
    Wings narrow, outer margin somewhat oblique................22.
22. Costa of primaries convex.........................*Seirarctia*.
    Costa of primaries straight ....23.
23. Robust, with hairy vestiture, blackish ....*Alexicles*.
    Slenderer, the vestiture smooth, white ....*Aloa*.
24. Male and female antennæ simple ....*Phissama*.
    Male antennæ pectinated, female simple.... ....25.
    Male and female antennæ pectinated.................*Estigmene*.
25. Wings with short erect scales, slightly transparent ....*Diaphora*.
    Wings with appressed scales, not transparent.........*Hyphantria*.
26. Antennæ of male simple..............................27.
    Antennæ of male pectinated...........................30.
27. Vestiture of thorax scaly, appressed.........................28.
    Vestiture hairy, smooth ....*Pericallia*.
    Vestiture hairy, short, erect ....29.
28. Wings broadly trigonate ....*Camptoloma*.
    Wings elongate, more rounded......................*Arachnis*.
29. Apex of primaries acuminate......................*Pyrrharctia*.
    Apex of primaries square ....*Phragmatobia*.
30. Ocelli close to margin of eye (about the diameter of the ocellus)..31.
    Ocelli distant from the margin of the eye..................36.
31. Primaries broad, outer margin somewhat erect ....32.
    Primaries narrower, outer margin somewhat oblique ....*Alphæa*.
32. Body slender, secondaries ample.....................*Diacrisia*.
    Body more robust, secondaries moderate...................33.
33. Female wingless ....*Ocnogyna*.
    Female with fully-developed wings ....34.

34. Costa of primaries not depressed before apex .... .35.
    Costa of primaries depressed before apex.............. *Rhyparia.*
35. Wings opaque........................ .... *Spilosoma.*
    Wings somewhat translucent .... *Thygorina.*
36. Front narrowed above and below .... *Arctinia.*
    Front square, not narrowed .. ..37.
37. Rough hairy, wings subdiaphanous.................. *Eucharia.*
    Somewhat smooth, wings opaque .... *Hyphoraia.*
38. Accessory cell wanting ....39.
    Accessory cell present..... . ...40.
39. Wings broad, size large, vestiture smooth and short.... *Platyprepia.*
    Wings moderate, size smaller, vestiture rough .... *Euprepia.*
    Wings elongate, size very small................... *Kodiosoma.*
40. Spurs of posterior tibiæ long or moderate....................41.
    Spurs of posterior tibiæ short .... *Ectypia.*
    Median spurs of posterior tibiæ wanting .... *Ammobiota.*
41. Size small, body rather slender ....42.
    Size large, body more robust ....43.
42. Wings short and broad...... .... *Parasemia.*
    Wings long and narrow .... *Pygoctnucha.*
43. Wings broad ....*Arctia.*
    Wings narrow..................................*Antarctia.*
44. Vein 8 of secondaries absent........................52.
    Vein 8 present ....45.
45. Veins 7 to 10 of primaries stalked..................... *Cycnia.*
    Vein 10 from the discal cell...................... ....46.
46. Accessory cell present ....47.
    Accessory cell absent ....48.
47. Antennæ long, secondaries proportionately small.......... *Theages.*
    Antennæ shorter, secondaries larger............... *Pygarctia.*
48. Primaries broad, secondaries proportionate .... *Euchœtes.*
    Primaries narrow, produced at apex, secondaries smaller........49.
49. Male antennæ simple .... *Pelochyta.*
    Male antennæ pectinate ....50.
50. Vein 8 of secondaries double .... *Schausia.*
    Vein 8 long, single ....,...... *Halisidota.*
    Vein 8 very short, spurlike...........................51
51. Vein 5 of secondaries present......................... *Aemilia.*
    Vein 5 of secondaries wanting.................. *Eupseudosoma.*

52. Vein 10 of primaries from the discal cell..... .... *Euccreon.*

Veins 7 to 10 of primaries stalked ....'...... *Bertholdia.*

LIST OF GENERA AND SPECIES.

Coscinia, Hübn. (= Eulepia,Curt. =
    Emydia, Boisd.)
    striata, Linn.
    cribraria, Linn.
Eubaphe, Hübn.(= Crocota, Hübn.
    = Holomelina, H. S.)
    laeta, Guér.
    intermedia, Graef.
    ostenta, H. Edw.
    costata, Str.
    opella, Grt.
    immaculata, Reak.
    aurantiaca, Hübn.
Utetheisa, Hübn. (= Deiopeia,
    Steph.)
    bella, Linn.
    venusta, Dalm.
    ornatrix, Linn.
    pulchella, Linn.
    formosa, Boisd.
Doa, Neum. & Dyar.
    ampla, Grt.
    dora, N. & D.
Axiopoetra, Ménét.
    maura, Eichw.
Macrobrochis, H.-S.
    gigas, Walk.
Callimorpha, Latr. (Euplagia, Hb.
    = Tripura, Moore.)
    dominula, Linn.
    quadripunctaria, Poda.
    prasena, Moore.
    pallens, Hamps.
    principalis, Koll.
    similis, Moore.

plagiata, Walk.
equitalis, Koll.
nyctemerata, Moore.
Argina, Hübn.
    argus, Koll.
    syringa, Cram.
    cribraris, Clerck.
Haploa, Hübn
    clymene, Brown.
    colona, Hübn.
    Lecontei, Guer.
    contigua, Walk.
    confusa, Lyman.
Areas, Walk. (= Melanareas, Butl.)
    galactina, Van d. Hoev.
    imperalis, Koll.
Sebastia,Kirby(= Moorea,Hamps.)
    argus, Walk.
Calpenia, Moore.
    khasiara, Moore.
    Saundersi, Moore.
Euerythra, Harvey.
    phasma, Harv.
    trimaculata, Smith.
Hipocrita, Hübn. (− Euchelia,
    Boisd.)
    jacobææ, Linn.
Creatonotus, Hübn.
    interruptus, Gemel.
Ecpantheria, Hübn.
    garzoni, Oberth.
    ocularia, Fab.
    permaculata, Pack.
Leptarctia, Stretch.
    californiæ, Walk.

Pachylischia, Ramb. (= Artimelia, Ramb.)
  corsica, Ramb.
  Latreillei, Godt.
Seirarctia, Packard.
  echo, Sm. & Abb.
Alexicles, Grote.
  aspersa, Grt.
Aloa, Walk. (= Bucæa, Walk.)
  emittens, Walk.
  simplex, Walk.
  fumipennis, Hamps.
Phissama, Moore (= Amphissa, Walk.)
  transiens, Walk.
Estigmene, Hübn. (= Leucarctia, Pack.)
  acræa, Drury.
  Rickseckeri, Behr.
  albida, Stretch.
Diaphora, Stephens.
  mendica, Clerck.
Hyphantria, Harris.
  cunea, Dru.
Camptoloma, Felder.
  interioratum, Walk.
  binotatum, Butler.
Arachnis, Geyer.
  aulea, Geyer.
  picta, Pack.
  maia, Ottolengui.
  citra, N. & D.
  zuni, Neum.
Pericallia, Hübn.
  matronula, Linn.
Pyrrharctia, Packard.
  isabella, Sm. & Abb.

Phragmatobia, Steph.
  fuliginosa, Linn.
  assimilans, Walk.
Rhyparia, Hübn.
  purpurata, Linn.
Diacrisia, Hübn. (= Euthemona, Steph.)
  sannio, Linn.
Ocnogyna, Lederer (=Cletis, Ramb. =Somatrichia, Kirb.).
  zoraida, Grasl.
  maculosa, Herm.
  parasita, Hübn.
Spilosoma, Steph. (=Spilarctia, Butl.)
  urticæ, Esp.
  lubricipeda, Linn.
  punctarium, Stoll.
  lutea, Hufn.
  virginica, Fab.
  prima, Slosson.
  antigone, Strecker.
  latipennis, Stretch.
  vestalis, Pack.
  multiguttum, Walk.
  sangaicum, Walk.
  subfascia, Walk.
  dalbergiæ, Moore.
  punctatum, Moore.
  dentilinea, Moore.
  stigmata, Moore.
  mona, Swinhoe.
  gopara, Moore.
  ummera, Swinhoe.
  bimaculatum, Moore.
  jucundum, Butler.
  flavale, Moore.
  todarum, Moore.
  montanum, Guer.

strigulatum, Walk.
castaneum, Hamps.
rubilinea, Moore.
erythrophelps, Hamps.
brunneum, Moore.
casignetum, Koll.
bifasciatum, Hamps.
comma, Walk.
lacteatum, Butl.
melanopsis, Walk.
rubitinctum, Moore.
erythrozona, Koll.
fuscipenne, Hamps.
Thygorina, Walker.*
indica, Guer.
multivittata, Moore.
nigrifrons, Walk.
unifascia, Walk.
discalis, Moore.
obliquivitta, Moore.
venosa, Moore.
flavens, Moore.
biseriata, Moore.
sordida, Moore.
sikkimensis, Moore.
eximia, Swinhoe.
rhodophila, Walk.
melanosoma, Hamps.
Alphœa, Walker.*
fulvohirta, Walk.
florescens, Moore.
imbuta, Walk.
quadriramosa, Koll.
tigrina, Moore.
leopardina, Moore.
vittata, Moore.
biguttata, Walk.

nigricans, Moore.
dentata, Walk.
pannosa, Moore.
siaphi, Moore.
Arctinia, Eichw. (= Elpis, Dyar. =
   Eupatolinis, Butl.)
cæsarea, Gœze.
rubra, Neumœgen.
vagans, Boisd.
Eucharia, Hübn. (= Neoarctia, N.
•   & D.)
casta, Esper.
Brucei, H. Edw.
Beanii, Neum.
Hyphoraia, Hübn. (= Platarctia,
   Pack.)
aulica, Linn.
hyperborea, Curt.
Yarrowi, Stretch.
Platyprepia, Dyar.
virginalis, Boisd.
Euprepia, Ochsenheimer*
pudica, Esp.
fasciata, Esp.
intercalaris, Evers.
virgo, Linn.
virguncula, Kirby.
michabo, Grt.
intermedia, Stretch.
parthenice, Kirby.
rectilinea, French.
anna, Grote.
ornata, Pack.
arge, Dru.
Quenselii, Paykull.
obliterata, Stretch.
proxima, Guer.

*See Hampson for the generic synonymy.

cervinoides, Streck.
Bolanderi, Stretch.
Blakei, Grote.
superba, Stretch.
favorita, Neum.
Williamsii, Dodge.
phyllira, Dru.
figurata, Dru.
placentia, Sm. & Abb.
nais, Dru.
phalerata, Harris.
vittata, Fab.

Kodiosoma, Stretch.
fulvum, Stretch.

Ectypia, Clemens (= Euverna, N. & D.).
bivittata, Clemens.
clio, Packard.

Ammobiota, Wallengren.
festiva, Hufn.

Parasemia, Stephens.†
plantaginis, Linn.
petrosa, Walk.

Pygoctnucha, Grote.
Harrisii, Boisd.
terminalis, Walk.
Robinsonii, Boisd.
funerea, Grote.

Arctia, Schrank (=Epicallia, Hbn. = Hypercompa, Hbn. = Zoote, Hübn.)
villica, Linn.
caja, Linn.
opulenta, H. Edw.

Antarctia, Hübner.
vulpina, Hübn.

Cycnia, Hübn. (=Tadana, Walk.= Pareuchætes, Grt.)
tenera, Hübn.
sciurus, Boisd.
insulata, Walk.

Pygarctia, Grote.
abdominalis, Grote.
vivida, Grote.
murina, Stretch.
Bolteri, H. Edw.
elegans, Stretch.
scepsiformis, Graef.
albicosta, Walk.

Euchætes, Harris.
egle, Dru.
eglenensis, Clemens.
oregonensis, Stretch.
perlevis, Grote.
Spraguei, Grote.
zonalis, Grote.

Pelochyta, Hübn.(=Amerila, Walk.)
astræa, Dru.

Halisidota, Hübn.(=Lophocampa, H. =Euhalisidota, Grt.)
tessellaris, Sm. & Abb.
Harrisii, Walsh.
cinctipes, Grote.
Edwardsii, Pack.
labecula, Grote.
maculata, Harris.
alni, H. Edw.
Agassizii, Pack.
minima, Neum.
caryæ, Harris.
pura, Neum.
longa, Grt.
propinqua, H. Edw.

†See Neumœgen and Dyar for the generic synonymy.

bicolor, Walk.
Courregesi, Dognin.
atra, Druce.
daruba, Druce.
ergana, Druce.
aconia, H.-S.
thalassina, H.-S.
Schausia, Dyar.
argentata, Pack.
subalpina, French.
sobrina, Stretch.
mixta, Neum.||
ingens, Hy. Edw.
ambigua, Strecker.

albigutta, Boisd.
lugens, Hy. Edw.
Aemilia, Kirby.
roseata, Walk.
occidentalis, French.
Eupseudosoma, Grote.
floridum, Grote.
Eucereon, Hübn.
carolinum, Hy. Edw.
Theages, Walker.°
strigosa, Walk.
Bertholdia, Schaus.
specularis, H. S.
trigona, Grote.

LARVA OF TITANIO HELIANTHIALES, Murtfeldt.

BY HARRISON G. DYAR, PH. D., NEW YORK.

Miss Murtfeldt's interesting discovery of this leaf-mining Pyralid suggested to me the inquiry as to how far the setæ of the larva had been affected by this unusual habit. The leaf-mining Tineids have tubercles iv. and v. remote, while all the Pyralids that I have seen have these tubercles united. I was interested to learn how far fixed this Pyralid character is, especially as the setæ have been studied in but a few microlepidoptera.

Miss Murtfeldt very kindly sent me her alcoholic specimens. The larva has the flattened retracted head and large cervical shield of a leaf-miner, but the body is not flattened and the slender legs are normal. The setæ are perfectly normal for the Pyralidæ, iv. and v. closely united. There is also the little additional tubercle before and above the spiracle, which is present in other Pyralids and also in the Cossidæ. In fact, the larva strongly suggests a little Cossid, except that the feet are longer and the circle of crotchets is broken on the outside. The pupa tells a different story. It might belong to the Pyraloid Obtectæ, which Dr. Chapman says have obtect characters in practically all respects except the possession

||In the male type vein 10 of primaries arises from the apex of discal cell on one wing, distinctly stalked on the other wing, but with a basally directed spur, indicating an accessory cell. On secondaries the supplementary vein preceding vein 8 is very short.

°Type of Theages not examined. The characters in the table are those of our species.

of traces of maxillary palpi ; but I can only find with difficulty a slight trace of the maxillary palpi. This would make it almost a true obtect pupa, which is far removed from the Cossidæ.

The following descriptions contain some details not specially mentioned in Miss Murtfeldt's article :

Larva.— Head rounded, flattened, small, partly retracted ; clypeal sutures depressed, upper segment of labium forming a ridge ; dark brown, blackish on the narrow lateral angle ; width, 1.3 mm. Body segments distinct, creased several times in the incisures but not distinctly annulated, joint 13 divided. Cervical shield large, bisected, irregularly marked in black. Setæ distinct, from rather large, flat dark tubercles ; i. and ii. in trapezoidal form, iii. lateral, iv. and v. from a single substigmatal tubercle, vi. posteriorly, vii. above the base of the leg with three setæ, viii. single ; a small secondary tubercle with one little seta before the upper part of the spiracle. On the thorax normal, the setæ of i. and ii. united in pairs, iv. and v. united, vi. with one seta on joints 3 and 4. Thoracic feet well developed, armed with setæ and claw. Abdominal feet distinct, rather slender ; crotchets in a narrow ellipse, broken on the outer side, a single row, but doubly clawed, a slight hook on the outside as well as the more distinct one on the inside, both small.

Pupa.— Smooth, obtected, thickest through the second abdominal segment, slightly tapering each way, rounded, the head a little projecting. Anal end rounded, cremaster without projection, but with four rather long, stout, recurved hooks. Fifth and sixth abdominal segments moveable. Cases reaching to the end of the fourth segment ; eye covered by a single piece, separated below by the small, lanceolate labium ; maxillæ reaching about one-third the length of the cases, a small piece indistinctly segmented off at the base next to the labium ; first leg reaching two-thirds the length of the cases, enclosing a small elliptical piece of its basal part next to the maxilla ; second leg reaching to the tip of the cases, apparently touching the eye, but on careful focusing a small piece seems to be cut off at the base, which I take to represent the maxillary palpus ; antennæ not attaining the extremity of either the second legs or the wing cases ; third legs concealed. The spiracle on the first segment is concealed by a projection of the hind wing case which extends to segment 3. Light yellowish-brown, all the sutures narrowly and distinctly marked in dark brown. Smooth, shining, no distinct punctures or wrinkles of any kind. Length, 6.5 mm.; width, 2.5 mm.

GRAPTA INTERROGATIONIS, ETC.

This insect is not by any means abundant in my neighbourhood, and for several years I only captured one or two of the pale variety *Fabricii*. About four or five years ago I saw a worn female of that variety depositing eggs upon a wild hop I had trained over the front of my house. I did not subsequently see any other female near the plant. I left the larvæ upon it until they were nearly full grown, when I collected about a dozen. I think they all hatched out safely, and the result was about one-third of the dark form *Umbrosa* to about two-thirds of the pale. The larvæ were all of a size, and pupated within a day or two of each other, so I think it reasonable to suppose they were all from the eggs I saw being deposited, and from one and the same mother. Never having before seen or taken the dark form, and not then having any book upon Canadian butterflies, I was rather surprised at the result. On looking over my notes for last year I do not see anything of special interest, except that I took a specimen of *Chionobas varuna* on 21st June, and the only one that I saw. The occurrence here of *Colias cæsonia* has already been noted.

Owing to a conversation I had some time ago with Dr. Fletcher, I paid particular attention to *Colias eurytheme* and its varieties. I did not detect a single instance of *Eriphyle* "in coitu," or even flirting with other than its own female, though there were many flying about of the early small yellow form of *Eurytheme* and also of *Keewaydin*, nor vice versa.

Neither did I notice any intercouse between *Eriphyle* and the large orange form *Amphidusa*, Scudder, but the males of each variety seemed to single out the corresponding females of that variety. I am aware I am venturing upon dangerous ground, but so far as I am able to judge from observation, I should certainly say that *Eriphyle* was a species distinct from *Eurytheme*. Unfortunately, I am unable to give the time required to the rearing of the large number of larvæ necessary to the determination of this question. What I want particularly to convey is that I have never noticed promiscuous intercourse between the different broods, if such they are, though they overlap each other, and are flying at the same time.　　　　　E. Firmstone Heath,

Cartwright, Manitoba.　　　　　　　　　　　　"The Hermitage."

A RARE CATOCALA.

BY ARTHUR J. SNYDER, EVANSTON, ILL.

Early last July, while examining the collection of Prof. G. H. French at Carbondale, Ill., I saw for the first time a specimen of *Catocala Sappho*. Being especially interested in this genus of the Noctuids, I was somewhat surprised to see for the first time so striking a species, and felt sure that I would have no difficulty in recognizing the species should I ever see another example.

On July 6th, near Makanda, Ill., I began a search for Catocala. From the first hickory I " whipped," a *C. Sappho* started and lighted upon a white oak near by, but about fifteen feet from the ground. Through the aid of a fence rail placed against the tree, and by using the net, I easily captured my first *C. Sappho*, a perfect specimen, with the exception that a few scales were removed from the thorax. July 14th I was collecting four miles south of Makanda and captured two more *C. Sappho*, one in fair condition and one a badly worn example. Another in very poor condition was taken on July 13th. Two others were seen and captured, but allowed to escape through sheer anxiety not to injure them. It may be interesting to collectors to know that this rare moth is one of the slowest flyers in the genus, and is easily captured. It usually lights low, and is not easily frightened. On account of its light colour it is quite conspicuous. In all seven *C. Sappho* were seen in the vicinity of Makanda, Ill., in four days' collecting. It has been my pleasure to examine 78 or more of the species and varieties of North American Catocalæ, but I have seen nothing which approaches *C. Sappho* closely enough to be confusing even to an amateur.

THE NEW MEXICO SPECIES OF ANTHIDIUM.

BY T. D. A. COCKERELL, MESILLA, N. M.

The bee-genus *Anthidium* is not very well represented in New Mexico, the following being all yet observed.

(1.) *Anthidium larreæ*, n. sp.— ♀ . Length about 12½ mm., fairly stout, but the abdomen not subglobose ; black, with yellow markings, those of the thorax recalling *Steniolia duplicata*. Head large, face nearly square, moderately shining, closely punctured, sides of vertex with punctures of unequal size ; end of mandibles not developed into distinct teeth. Antennæ short, black. Clypeus, broad triangle above, and lateral face marks, bright yellow ; the last occupy all the space between clypeus and eyes, narrowing obliquely upwards so as to form nearly a right-angled triangle, continuing narrowly a little way along the orbital margin, then enlarging near the top of the eyes to a mark which points inwards towards the ocelli. Cheeks yellow, the yellow continuing across vertex as a narrow line. Mandibles yellow except ends. White pubescence rather sparse on face and cheeks ; also on thorax, becoming dense on lower part of pleura. Tubercles, sides of thorax except a black patch on lower part of pleura, tegulæ except a pair of fuscous spots (one much larger than the other), sides of mesothorax broadly, extending along the front some distance to an oblique truncation, two longitudinal stripes on mesothorax, and scutellum except median base, all bright yellow. Mesothorax and scutellum granular from a very close punctuation. Tubercles with a prominent keel. Hind margin of scutellum rounded, with a wide median emargination. Tegulæ punctured. Wings subhyaline, strongly smoky in upper part of marginal cell, nervures black, second recurrent, going beyond tip of second submarginal cell. Posterior truncation of thorax shining black, with a pair of broad hammer-shaped yellow marks. Legs yellow ; some black on anterior coxæ above, and at base of anterior femora, also at base of middle tibiæ and on basal two-thirds of hind tibiæ ; inner sides of all the legs largely ferruginous. Middle and hind tibiæ, and basal joint of hind tarsi, all greatly broadened. Abdomen shining, microscopically tessellate, with large sparse punctures. Entire apical yellow bands on segments 1–5, broadest at the sides ; apex yellow. Ventral scopa dense, white.

♂.—About the same size, abdomen more slender. Antennæ longer, scape yellow in front. Yellow spot near tip of eyes much reduced, line on vertex broken and nearly obsolete. Stripes on dorsulum wanting. Tegulæ with one large dark spot. Posterior truncation all black ; upper part of pleura largely black. No spine on posterior coxa. First three bands of abdomen emarginate at sides. Rounded median hind border of sixth segment projecting. Apex rounded, broadly emarginate.

Hab.—Mesilla Valley, N.M., close to Agricultural College ; a ♀ at flowers of *Larrea* (Creosote bush), May 6 [Ckll.]; also a ♀ taken May 18 by Mr. F. Garcia, and a ♂ taken some years ago by Prof. Townsend, both in the Mesilla Valley. Unfortunately the ♂ is reddened by cyanide. Mr. Fox kindly compared this species with Cresson's collection, and returned it marked " near *occidentale* and *zebratum.*" It can be distinguished from these by the colour of the legs and the sides of the thorax.

(2.) *Anthidium occidentale*, Cress. — Described from specimens taken in New Mexico by Dr. Samuel Lewis is 1867. Not observed by me.

(3.) *Anthidium gilense*, n. sp. — ♀. Length hardly 10 mm.; robust, with long wings ; black, with lemon-yellow markings. Head, mesothorax and scutellum with close, extremely large punctures, closest on front, largest on scutellum. Edge of mandibles with small, short, but quite distinct, teeth. Tubercles forming an oblong, sharp-edged lobe. Hind edge of scutellum straight, sharp, overshadowing metathorax. Second recurrent nervure going considerably beyond end of second submarginal cell. Abdomen of the subglobose type, shining, with large punctures, close enough to produce a subcancellate effect. Small spot on each side of clypeus ; broad lateral face marks, extending only as far as level of antennæ, where abruptly truncate ; continuous line on vertex, lateral thirds of front margin of mosothorax broadly, ends of tubercles, four spots on scutellum (the middle ones large and elongate), all yellow. Cheeks, pleura and shining posterior truncation, black. Tegulæ rufous, with an elongate yellow mark. Wings fuliginous, with a hyaline spot just beyond and partly in the third discoidal cell, and a much smaller one just beyond apex of second submarginal. Base sub-hyaline. Legs ferruginous, anterior femora blackened, a yellow stripe on anterior and middle tibiæ, a yellow spot at extreme base of hind tibiæ. First abdominal segment with an oblong yellow spot on each side. Second with a band, narrowly interrupted in middle, and produced into a short tooth on each side behind. Third to fifth segments with a pair of large quadrate yellow marks, and a small spot on each extreme side. Apical segment black. Ventral scopa white. Pubescence of legs, thorax and head white, but very little of it ; a small but conspicuous patch behind the wings.

Hab.—West Fork of Gila River, N. M., July 17, one specimen [C. H. T. Townsend]. Of the N. M. species it most resembles *pudicum*, but it is quite distinct.

(4.) *Anthidium pudicum*, Cress.—Five at Santa Fé, N. M.: two on flowers of *Grindelia squarrosa*, Aug. 2 and 3, in company with *Heriades*, *Melissodes*, *Megachile* and *Podalirius ;* two resting in hole in adobe wall, Aug. 2. A ♀ was submitted to Mr. Fox, and returned marked *pudicum ;*

the N. M. form is perhaps a distinct race, as all have the markings yellow, whereas the typical form from Nevada has them white.

(5.) *Anthidium emarginatum*, Say.—Taken in 1867 by Dr. Lewis, and described by Cresson as *atrifrons*.

(6.) *Anthidium interruptum*, Say.—Las Cruces, N. M., and Chaves, N. M.: four, all taken by Prof. Townsend. Determined by Mr. Fox.

(7.) *Anthidium maculifrons*, Smith.—Taken in 1867 by Dr. Lewis. One taken by Prof. Townsend in Soledad Canon, Organ Mts., Aug. 15, 1896, on plant No. 40.

(8.) *Anthidium maculosum*, Cress.—Tuerto Mtn., near Santa Fé, 8,025 feet, Aug. 7, on flowers of *Senecio*. Besides the difference in the markings, this differs from the last in the abdominal punctation.

There is in New Mexico another bee which might easily be taken for a small *Anthidium*, namely *Stelis costalis*, Cresson. This is a very variable species, both as to size and colour. It was taken by Prof. Townsend on the West Fork of the Gila R., July 16, and by me at Santa Fé, on flowers of *Rudbeckia laciniata*, July 19. It is the only *Stelis* yet observed in New Mexico.

A NEW ATTID SPIDER.

BY T. D. A. COCKERELL, MESILLA, N. M.

Icius Peckhamæ, n. sp.

Length not quite 5 mm. Cephalothorax above brilliant peacock green, slightly intermixed with brassy in front; white hairs above the row of eyes forming a weak band, also narrowly encircling the eyes; an irregular patch of white hairs beneath the hindmost eyes; lateral (inferior) margins of cephalothorax with a broad, well-defined white band. Legs black with white hairs, the hairs so arranged as to divide the legs into alternate sections of black and white; the tibiæ black at base and middle, the tarsi narrowly black at base. Palpi covered with white hairs. Mandibles black. Abdomen above brilliant metallic magenta, with the base yellowish green; the sides and the under surface white, minutely speckled with black.

Legs approximately 4 (31) 2. Quadrangle of eyes occupying less than half of cephalothorax. First row of eyes a little curved; middle eyes almost touching, lateral hardly half their diameter, and separated from them by a very short interval. Posterior eyes of the same size as anterior lateral, further from each other than from the lateral borders of the cephalothorax. Sternum with white hairs.

In alcohol the abdomen is not so brilliant, and most of those parts of the legs covered by white hairs appear brown. The legs have a little metallic colour.

First legs 2¾ mm. long, second 2½, third 3, fourth 4. Width of abdomen, 1⅓ mm. Length of cephalothorax, 2 mm.

Hab.—In the course of some investigations of the ,codling moth, this beautiful little spider was found not rarely hibernating under the bark of apple trees in Mesilla, N. M. Mr. G. W. Peckham, to whom specimens were sent, confirms it as new. *I. Peckhamæ* is respectfully dedicated to Mrs. Elizabeth G. Peckham, who, in conjunction with her husband, has done such admirable work on the Attid spiders. The present description will serve to fix the name ; Mr. and Mrs. Peckham will no doubt figure the palpus, etc., when they come to revise the group.

SPHINX LUSCITIOSA, Clem.

On the morning of the 9th of June, 1897, Mr. Bice took from an electric-light pole in London a fine male specimen of that rare moth, *Sphinx luscitiosa*, Clem.

All the writers upon the Sphingidæ that I have consulted are agreed in pronouncing it rare. Mr. Grote says : " This is probably our rarest hawk moth of these kinds, proper to the Middle States." Dr. J. B. Smith states that " the species is very rare." This is the first report of its being taken in this section of the Province that I am aware of.

Prof. Fernald, upon information received from the Rev. G. D. Hulst, says that it had been bred near Newark, N. J., on willow. Dr. Smith says : " The species has been frequently raised in the vicinity of New York on willow." But whether willow is its natural food plant, or that the larvæ merely feed upon willow in preference to other plants offered to them, is not stated. If willow proves to be its natural food plant, it does seem decidedly strange that, with willow everywhere so plentiful, *luscitiosa* should yet remain so very rare, and would lead one to surmise that there must be some special influence at work that is the cause of it. Up to the time of Dr. Smith's writing (1888) no description of the larvæ was obtainable.   J. ALSTON MOFFAT.

London, Ont.

In my last communication *Agrotis catherina* is printed as a separate species, whereas it ought to have appeared as a synonym of *Semiophora tenebrifera*, Walk.   J. A. M.

Mailed September 2nd, 1897.

FIG. 1.—MAP SHOWING DISTRIBUTION OF BROOD XV., CICADA
SEPTENDECIM, IN OHIO, IN 1897.

FIG 2.

The Canadian Entomologist.

VOL. XXIX.   LONDON, OCTOBER, 1897.   NO. 10.

BROOD XV. OF CICADA SEPTENDECIM IN OHIO.*

BY F. M. WEBSTER, WOOSTER, OHIO.

Having had the opportunity of working out the distribution of broods V., VIII. and XXII. in Indiana, brood XV. in Ohio possessed a peculiar interest for me, as in studying it I was able to profit considerably by my acquaintance with the others.   I perhaps ought to say a word in regard to the three other broods mentioned, as one of them (XXII.) is treated of at considerable length in the Report of the Entomologist of the United States Department of Agriculture for the year 1885, and it was while connected with the Department as one of its special agents that these three broods were studied.   Brood XXII. covered the entire State of Indiana, except a narrow strip of land around the lower end of Lake Michigan, from ten or fifteen to twenty miles wide, which area was exactly covered by brood V. in 1888.   The coloured map which accompanies the report mentioned is defective in that the two points extending southward, not indicated as being covered by this brood, were, as was afterwards learned, within the area covered by brood XXII. and not covered by brood V., the line of separation being about ten miles east of the lake on the line between Michigan and Indiana, and running nearly south-west to the east line of Porter county, the course then trending slightly more to the westward to the Illinois line ; in no case, I believe, extending to the Kankakee River, thus making the line of separation much more uniform than the one indicated on the map cited above, on which the dividing line is quite irregular.

Brood VIII. occurred in southern Indiana, becoming excessively abundant only in Harrison county, but covering the area south of a line drawn from Vincennes to Greencastle, Franklin, and eastward to northern Dearborn county.   Singularly enough, a single female was brought me at Lafayette, fully 60 miles north of Greencastle, which probably marked the northernmost point where the species could be said to occur in any numbers.

* Read before Section " F," Zoology, of A. A. A. S., at Detroit, Michigan, August 10th, 1897.

One of the most striking peculiarities of brood XV. in Ohio was its exceedingly uneven distribution within the boundaries of its range. On driving over the country during the midst of the season of greatest activity, one would suddenly find himself in the midst of a din that was almost deafening, and the woods would be browned with discoloured twigs, while within a mile he would find himself in the midst of a silence that from contrast was almost oppressive, while there was not a discoloured twig to be observed. This lack of uniformity in distribution rendered the work of locating the exact boundaries of the brood quite difficult in some cases, as one must often go miles beyond it in order to be quite sure that he had found the last outlying colony. But in other cases the effect was the reverse. In going southward from Painesville, over the P. & W. Ry., which cuts through what is locally known as "Johnnycake Ridge," not a note was to be heard, and not a discoloured twig was to be seen on tree or shrub, but on leaving the cut, which is by no means a long one, the combined notes of the thousands of Cicadas were clearly heard above the noise of the train, while scarcely a tree or bush escaped the attack of the females, and some of them would not have been more thoroughly browned if a fire had broken out among them. In the city of Lancaster they were reported as abundant in the east part of the town, while there were scarcely any in the western portion, and it turns out that the dividing line between this brood and brood XXII. is practically indicated, as nearly the same conditions were observed to occur seventeen years ago.

The brood is certainly becoming weakened each time it reappears, and the boundaries of its occurrence did not in many cases extend as far as they did when it last appeared, sometimes the difference amounting to several miles. Near Painesville it occurred some three miles nearer to the lake shore in 1846 and in 1863 than it did in 1880 or the present year. It was at Bellevue in 1880, but did not extend so far west this year, and the same is true of its occurrence northward toward the lake. Where it was quite abundant in 1880 it did not appear at all this year. It was reported by two observing correspondents as having been present in limited numbers in Ashtabula county in 1863 and again in 1880, but no trace of it could be found this year. In short, it seems to be slowly but surely dying out, and will in time be known only in history. Brood XXII. is a much stronger one—at least it was in Indiana—but I question if in time the Periodical Cicada is not wholly exterminated in Ohio, and there seems no reason why this should not be true of many other States. The gradual extinction of the native forests will have much to do with this, but their natural enemies, especially the English sparrow, are having a much more fatal effect.

In 1885, in Indiana, I first saw the English sparrow come in contact with the Periodical Cicada. In the city of Lafayette the insect appeared in considerable abundance, and for a few days there was no lack of the well-known notes of the male, but suddenly there was a decided falling off, and by listening carefully one would occasionally detect a note suddenly cut short at its very height, and close watching revealed the fact that the sparrows had come to recognize the note as well as the form of the musician, and as a result, within a few days, though there were myriads in the woods, not a single one could be found in the city, the abundance of wings upon the pavements showing too well the tragedies that had been enacted there.

With these observations in mind, I watched for the coming of brood XV. in Ohio with considerable interest. On the morning of May 28th a full complement of wings was found on the pavement under a shade tree, and during the following days these detached wings became more numerous, but not a Cicada note was heard. Going out into the residential portion of the town at dusk, I would observe pupæ emerging from the lawns and making their way to the shade trees across the pavement bordering the street, but not one could be found the next morning, though the pavement was littered with detached wings. While back in the woods a half mile away there were great numbers of them, creating almost a continual din during the day; in town during the whole season I only saw a single living adult and heard not a single note.

In southern Ohio I one day watched the Cicadas attempting to make their way across a clearing, from a bit of woods to an orchard situated some distance away and below the woods, which was on a bluff. The afternoon sun shone directly across the clearing, thus enabling me to witness every attempt of the insects to fly from woods to orchard. The sparrows were in the latter, and the moment a Cicada appeared its silvery wings would glisten in the sunlight for a few moments, when a sparrow or sometimes two of them would make a dash for it, and if the prey was missed, as was sometimes the case, the bird would turn suddenly and try again, generally with better success. I watched the actions of birds and insects for a couple of hours, but did not see a single Cicada cross the clearing. Though there were numbers of *Pieris rapæ* and some other butterflies winging their way about over the clearing, I did not see a single mistake made on the part of the sparrows. They had become adept enough in two or three weeks to be able to distinguish a Cicada with an unerringness that was simply surprising, when we come to consider that none of their immediate progenitors could have seen or tasted a Cicada.

Other bird enemies appear to be very few, and these not over-voracious. Mr. J. J. Harrison, of Painesville, Ohio, saw the crow black-bird feeding upon them in 1846, while the labourers on the Station Farm at Wooster claim to have observed the robin to attack them. A species of Tachina fly seemed to play havoc with the latter portion of the brood, and either owing to this or some other reason, they suddenly disappeared between June 24th and June 28th. On the former date, in the Experiment Station orchard, they were excessively abundant, while on the latter there was not a living Cicada to be found there, while the stench arising from the dead bodies was quite apparent to one walking through the orchard.

As usual, the injury inflicted was slight, except in cases of very young orchards, and I saw in one case a, to me at least, unique form of attack. This is shown in the plate (fig. 3), and instead of the regular, quite con-spicuous punctures (fig. 2) made by the female for a nidus, she appeared to have simply thrust her ovipositor into the wood, and with no further external wound deposited her ova.

The distribution of the brood in Ohio is illustrated in the accom-panying map, plate 8, fig. 1.

In its distribution, rivers do not appear to have had much influence, as it will be noticed that in southern Meigs county a small area outlined by a bend in the Ohio River is only partly covered ; in one township, Letart, the Cicada not being found at all ; while a corresponding point of West Virginia comes within the range of distribution, even though lying across the river. From this point the dividing line trends slightly to the south-west, passing north of Gallipolis, and extending to the Scioto River, at a point a few miles above its mouth, but not extending beyond this to the westward. North the line follows the east bank of the river until the bend between Waverly and Chillicothe is reached, when it crosses the river and holds to its nearly northerly course to near Circle-ville. Here the line makes a sharp curve to the north-east to the city of Lancaster, in Fairfield county, but trends north-west to the eastern line of Franklin county, thence almost northward along the east line of

Delaware and Morrow counties to a point in Richland county about a mile west of the village of Ontario, when it changes again to the northwest, crossing the north-east corner of Crawford and the south-east corner of Seneca, then a little east of north to a point near Lake Erie, a mile and a half south-west of the city of Huron, Erie county. This area in Erie county is, however, but little more than a peninsula-like extension, and will probably not appear again. Near the south line of Erie county the line of demarcation makes a broad sweep to the south-east, thus leaving both the north-west and north-east corners of Huron county unoccupied, as well as all of Lorain county, except the southern portion and south-eastern border, and the western end of Cuyahoga county. Just west of Cleveland another peninsular extension occurs lakeward, where the Cicada appeared for a few days at first, but suddenly disappeared before the brood reached its maximum in numbers in the adjacent counties. This also will hardly appear again. From this point to near the eastern end of Lake county the insect keeps well back from the lake, though it formerly occupied ground much nearer to the shore. The eastern terminus also comprises but little more than a promontory, as the course here changes broadly to the south-west and then to the south-east, leaving a considerable portion of eastern Geauga county and the north-east corner of Portage county unoccupied. The dividing line here only touches the south-west corner of Trumbull county and includes the western end of Mahoning and Columbiana counties and the southern border of the latter, the line passing into West Virginia or Pennsylvania, near East Liverpool, Ohio, where this year a very few Cicadas appeared, and where brood XV. overlaps brood XX.

I shall be obliged to confess that when I began to map out the area covered by brood XV. it was with more enthusiasm than I could command when I finished the survey. The map indicates, with a good degree of accuracy, the area over which the brood occurred in 1897, but that it will as accurately show the area covered by the brood in 1914, I have no expectations. The continued destruction of forests and the inroads made upon the brood by its natural enemies will result in great changes, not only in the outline of the area of habitation, but this will be composed of more and more isolated and continually decreasing " Cicada Islands," as I might term them, until the well-known notes of the male will have ceased forever, and the voiceless female will have followed her spouse into the shades of oblivion.

THE NINTH ANNUAL MEETING OF THE ASSOCIATION OF ECONOMIC ENTOMOLOGISTS, DETROIT, MICH., AUGUST 12TH AND 13TH, 1897.

The Association met in Room 212, Central High School Building, immediately following the adjournment of Section F. Thirteen active members were present, together with many visitors, prominent among the latter being Dr. C. A. Dohrn, Prof. E. B. Poulton, Dr. C. P. Hart, Dr. C. S. Minot, and Dr. C. W. Stiles. The Association was called to order by the President, and in the absence of Secretary Marlatt a secretary *pro tem.* was chosen. The address of the retiring president, Prof. F. M. Webster, treated of "The Present and Future of Applied Economic Entomology in the United States," and contained, among other very interesting features, an admirable tribute to the value of the systematist and a somewhat caustic criticism of the "species maker," helpful suggestions for the experiment station worker, and a very frank discussion of the unfortunate results which attend the attempts sometimes made to combine politics and science. The following were elected to active membership :

G. B. King, Lawrence, Mass.

Gerald McCarthy, Raleigh, N. C.

E. P. Felt, Albany, N. Y.

A. F. Burgess, Malden, Mass.

W. B. Barrows, Agricultural College, Mich.

R. H. Pettit, Agricultural College, Mich.

W. S. Blatchley, Indianapolis, Ind.

The following were elected foreign members :

Claude Fuller and Richard Helm, both of Perth, West Australia.

These additions increase the numbers of this Association to ninety-three active and thirty-one foreign members.

Following the election of members, Dr. L. O. Howard presented "Additional Notes on the Parasites of *Orgyia leucostigma.*" This paper gave the results of the rearing of a large number of primary and secondary parasites, and contained a general discussion of the different phases of insect parasitisms.

"Temperature Effects as Affecting Received Ideas Concerning the Hibernation of Insects," by the same author, showed that a sudden alternation between low and high temperatures was remarkably fatal to the larvæ of clothes moths, Buffalo carpet beetles, and other insects of allied habits.

An abstract of " Notes on Certain Species of Coleoptera that Attack Useful Plants," by F. H. Chittenden, was read by the secretary *pro tem.* These notes treated chiefly of the food plants and habits of certain Chrysomelids.

A letter from Miss E. A. Ormerod called particular attention to the fact that the house sparrow had been very abundant and very obnoxious in certain parts of England, and it seemed probable that some legislation or public measures would need to be adopted to control this bird. The arrival from Tripoli of a cargo of wheat badly infested by the Angoumois moth was recorded and reference made to the occurrence in injurious numbers of *Xyleborus dispar* at Teddington.

Prof. P. H. Rolfs presented notes on " A Fungus Disease of the San José Scale." This disease seems to be confined to the southern part of the United States, but is very helpful to fruit growers there. The scale has been almost eradicated from several orchards by this disease. Laboratory and field experiments now in progress promise helpful results, but it does not seem probable that this disease will be of value in the northern part of the United States, since warmth and moisture are necessary for its development.

Mr. Barrows made a brief statement concerning the distribution of the San José scale in Michigan. The insect had been found scattered throughout the southern counties of the State, where it had probably existed for eight years. In discussing this paper, Mr. Craig spoke of the occurrence of the scale in southern Ontario, where there were at least seven infested localities.

A paper from Prof. C. P. Gillette, on " Insects Taken at Light and Sugar," evoked considerable discussion, and was followed by " A Study of the Possible Origin and Distribution of the Chinch Bug," by Prof. F. M. Webster. The author advanced the idea that this insect had originated in the southern part of the United States, and spread by two diverging streams up the Mississippi Valley and along the eastern Atlantic coast. In the former region the long-winged form predominated, while the coast form was short-winged. In the discussion following this paper the general opinion seemed to be that the length of the wings depended upon environment rather than heredity. Mr. C. W. Mally recorded the capture at Ohio of a specimen having one long and one short wing, thus throwing additional light upon the relationship between the two forms.

" Notes on the Common House Fly," by Mr. Howard, gave the

negative results of a series of experiments with lime, land plaster, etc., used to destroy the larvæ of the house fly. He emphasized the necessity of greater cleanliness in the management of horse stables.

A paper from Mr. Gillette, on "Vernacular Names of Insects," was read and referred to a committee consisting of Messrs. Howard, Fernald and Lintner. A communication from C. P. Lounsbury, giving very interesting notes on "Cape of Good Hope Insects," particularly the locusts of that region, was then read.

Mr. H. G. Hubbard presented an account of the "Insect Fauna of the Giant Cactus," recording the capture of a large number of insects on this plant and giving notes on their habits.

Mr. Howard described "A Valuable Coccid" lately discovered in Arizona and New Mexico, from which, by suitable treatment, a good grade of white wax could be obtained. The refuse from this operation is of the nature and consistence of India-rubber, and may be of commercial value.

"Notes on Insects of the Year," by Messrs. Webster and Mally, recording interesting experiences with several of the common insect pests. The negative results of a series of experiments with kainit, against the insects attacking the roots of the grape, caused considerable discussion, and the need for further experimentation along this line was pointed out.

A paper by A. H. Kirkland, on "Preparation and Use of Arsenate of Lead," detailed a method of preparing this insecticide at a cost of about seven cents per pound. Work against the Gypsy moth was mentioned, and the condition of the infested region was reported as generally better than that of last year. This undertaking, however, is still handicapped by insufficient financial support.

"A Malodorous Carabid," by Mr. Barrows, gave extensive notes on the annoyance and discomfort caused by the almost unbearable odour of this insect, *Nomius pygmæus.*

At the final adjournment of the session it was voted to hold the next meeting at Boston, Mass., August 19th and 20th.

Several resolutions were passed, among which were (1) a resolution requesting the publication of the proceedings as a bulletin of the Division of Entomology, U. S. Dept. of Agriculture, and (2) expressing familiarity with the efforts of the State of Massachusetts to exterminate the Gypsy moth, and commending the results already accomplished.

The election of officers resulted as follows: President, Herbert Osborn, Ames, Iowa; 1st Vice-President, Lawrence Bruner, Lincoln, Neb.; 2nd Vice-President, C. P. Gillette, Ft. Collins, Colo.; Secretary and Treasurer, C. L. Marlatt, Washington, D. C.

A. H. KIRKLAND, Secretary *pro tem.*

A LIST OF THE COLEOPTERA OF THE SOUTHERN CALIFORNIA ISLANDS, WITH NOTES AND DESCRIPTIONS OF NEW SPECIES.

BY H. C. FALL, PASADENA, CAL.

Early in May of the present year (1897) the Pasadena Science Club sent three of its members on a month's general collecting trip to certain of the Santa Barbara islands lying off the coast of Southern California. While none of the members of the expedition were, strictly speaking, entomologists, a considerable experience in collecting, combined with some preliminary instruction, enabled them to devote intelligently a portion of their time to the collection of insects, more especially of the Coleoptera.

The islands visited were in the order named, Santa Barbara, San Nicolas, and San Clemente, distant respectively forty, sixty, and fifty miles from the nearest point on the mainland. Inasmuch as the entire material in Coleoptera, consisting of forty-six species and upward of one thousand specimens, has been submitted to me for study, it seems a fitting occasion for presenting as complete a list as possible of the coleopterological fauna of the entire group of islands, from Santa Cruz to Guadalupe.

To Eastern collectors it may seen a matter for wonderment that so interesting a field should so long remain, entomologically speaking, practically unexplored; yet it must be remembered that entomologists are here exceedingly few and far between, and the islands are, with the exception, for the past few years, of Catalina, nearly or quite uninhabited and not conveniently accessible. Every now and then, to be sure, an Eastern man appears with bottles and net, but to him the whole vast region is a terra incognita. Mountain and desert and valley offer opportunities without number; he takes the goods the gods provide and troubles not himself about possibilities in lands hull down in the Pacific. And so it happens that the few beetles recorded from the islands we owe to the kindness of one or another of the botanists or ornithologists who have at long intervals found their way there.

It is believed by those best competent to judge that these islands are the summits of a submerged mountain range forming a part of the mainland, or at least connected with it as a peninsula, until after the beginning of the Quaternary Period, when it was separated and broken up into islands by subsidence. The close similarity between the flora and fauna of Guadalupe and California has several times been cited in support of this view and seems in itself almost conclusive.

As far as I can learn from the literature at hand, the following fifteen species of Coleoptera are all that were described or reported from the islands up to 1875 :

From Santa Cruz—*Asaphes tumescens, Malthodes laticollis, Phobetus comatus, Ernobius debilis, Helops Bachei;* from Santa Catalina—*Pristoscelis punctipennis* and *P. pedalis* ; from San Clemente — *Coniontis lata, Eusattus robustus, Eulabis grossa, Amara insularis* ; from Santa Barbara — *Eleodes scabripennis, Cibdelis Bachei* and *Meloe barbara.* Pristoscelis ænescens is said to be "from the islands off Santa Barbara," and it is more than probable that the same reading should be applied to the three preceding species. Nearly if not all the above named species were taken either by C. M. Bache or Dr. J. G. Cooper and given to Dr. Leconte, by whom they were described (1861–1866), with the exception of *Amara insularis,* which was described by Dr. Horn in 1875.

In 1875, Dr. Edward Palmer brought from the Guadalupe Islands the following twenty-three species, which were enumerated by Dr. Horn, Trans. Am. Ent. Soc., V., 1876 :

Calosoma semilæve, Lec.
    "   Palmeri, Horn.
Bembidium striola, Lec.
Amara insignis, Dej.
    "   californica, Dej.
Platynus maculicollis, Dej.
Calathus obscurus, Lec.
Tachycellus nebulosus, Lec.
Anisodactylus piceus, Mén.
Anisotarsus flebilis, Lec.
Necrophorus nigrita, Mann.
Dermestes vulpinus, Fab.

Trogosita virescens, Fab.
Saprinus lugens, Er.
Cardiophorus luridipes, Cand.
Pristoscelis pedalis, Lec.
Necrobia rufipes, Fab.
Cœnonycha socialis, Horn.
Atimia dorsalis, Lec.
Cœlotaxis muricata, Horn.
    "   punctata, Horn.
Conibius seriatus, Lec.
Helops Bachei, Lec., var.

In an appendix to the annual report of the Chief of Engineers for 1876, Dr. Leconte gives a list of species taken in So. California the previous year by the expedition for geographical survey under Lieut. Geo. M. Wheeler, among which are the following seventeen species from the island of Santa Cruz :

Omophron dentatum, Lec.
Bembidium transversale, Dej.
Calathus ruficollis, Dej., var.
Platynus brunneomarginatus,Mann.
Pterostichus lætulus, Dej.
Amara californica, Dej.
Anisodactylus consobrinus, Lec.
Hippodamia vittigera, Mann.
Dermestes talpinus, Mann.

Tropisternus californicus, Lec.
Hydrocharis glaucus, Lec.
Carpophilus pallipennis, Say.
Polycaon Stoutii, Lec.
Phlœodes diabolicus, Lec.
Coniontis viatica, Esch.
        subpubescens, Lec.
Cratidus osculans, Lec.

In 1892 — Zoe, Vol. III., p. 262 — Mr. F. A. Seavey gives a short list of insects taken by him on Santa Catalina in August of that year. This list includes fourteen species of Coleoptera, of which three — *Balaninus obtusus, Pristoscelis quadricollis* and *Anthonomus canus* — are quite surely incorrectly determined and will not be included in the following list.

During parts of July and August, 1892 and 1894, about four weeks in the aggregate was spent by the writer on Catalina. The island was then very dry and collecting was rather unremunerative. Nevertheless upward of one hundred species were added to previous records.

To these must be added more than thirty species out of seventy-five taken by Dr. Gustav Eisen on Santa Rosa during May of the present year; a few taken on Catalina by Dr. A. Fenyes, of Pasadena, at about the same time; and finally, about half of the forty-six species collected by the expedition alluded to at the beginning of this article. The material collected by them is of especial interest inasmuch as it is probable that no insects have before been brought from either Santa Barbara or San Nicolas—the most remote of all the islands of the Santa Barbara group—and but four beetles from San Clemente. The catch of Dr. Eisen on Santa Rosa is of nearly equal interest for similar reasons. The following abbreviations are used in the subjoined list :

B. Santa Barbara.
Ca. Santa Catalina.
Cl. San Clemente.
Cz. Santa Cruz.

G. Guadalupe.
N. San Nicolas.
R. Santa Rosa.

 * Species hitherto recorded from same island.

 † Species not known to occur on mainland.

Cicindela oregona, Lec. R.
Omophron dentatum, Lec. Cz.* R.
Cychrus mimus, Horn. Ca.
Calosoma semilæve, Lec. G.* R.
†   "   Palmeri, Horn. G.*
Dyschirius gibbipennis, Lec. R.
Schizogenius depressus, Lec. Ca.
Bembidium transversale, Dej. Cz.
   * R.
Bembidium striola, Lec. Ca. G.*
   "   platynoides, Haywd. R.
   "   indistinctum, Dej. R.
   "   ephippiger, Lec. R.
   "   iridescens, Lec. Ca.
Tachys vittiger, Lec. Ca.
   "   sp. nov.? Cl.
Pterostichus Isabellæ, Lec. Ca. Cl.
   "   Menetriesii, Mots. R.
   "   lætulus, Lec. Cz.* R.
   "   sp. indet. R.
Amara insignis, Dej. G.* Ca.* R.
† " insularis, Horn. Cl.* N. B.
   "   californica, Dej. G.* Cz.* R.
Calathus ruficollis, Dej. Ca. Cz.*
   "   obscurus, Lec. G.* R.?
Platynus brunneomarginatus, Mann.
   Ca.* Cz.* R.
Platynus funebris, Lec. Ca.
   "   maculicollis, Dej. G.* R.
   "   variolatus, Lec. Ca.
Brachynus carinulatus, Mots.? Ca.
Chlænius obsoletus, Lec. Ca.
Agonoderus lineola, Fab. R.
Stenolophus limbalis, Lec. Ca.
Bradycellus rupestris, Say. Ca.
   "   californicus, Lec. R.
Tachycellus nebulosus, Lec. G.*
   "   nitidus, Dej. Ca. R.

Anisodactylus dilatatus, Dej. R.
   "   piceus, Mén. G.* R.
Anisodactylus consobrinus, Lec.
   Ca. Cz.* R.
Anisodactylus californicus, Dej.
   Ca. R.
Anisotarsus flebilis, Lec. G.*
Deronectes striatellus, Lec. R.
Hydroporus vilis, Lec. Ca. R.
Agabinus glabrellus, Mots. Ca.
Agabus lugens, Lec. R.
Ochthebius discretus, Lec. Ca.
Tropisternus ellipticus, Lec. Ca.
Tropisternus californicus, Lec. Ca.
   * Cz.*
Hydrocharis glaucus, Lec. Cz.* R.
Chætarthria nigrella, Lec. Ca.
Laccophilus ellipticus, Lec. Ca. R.
Cymbiodyta dorsalis, Mots. Ca. R.
Cercyon luniger, Mann. Ca.
Necrophorus guttula, Mots. Cl.
Necrophorus nigritus, Mann. G. *
   Cl. R.
Silpha ramosa, Say. R.
   "   lapponica, Hbst. R.
Aleochara bimaculata, Grav. Ca. Cl.
   "   sulcicollis, Mann. R.
Polistoma arenaria, Csy. Ca. B.
Heterothops californicus, Lec. Ca.
Creophilus villosus, Grav. Ca. Cl.
Hadrotes crassus, Mann. R.
Philonthus longicornis, Steph. Ca.
   "   nigritulus, Grav. Ca.
   "   Lecontei, Horn. R.
Actobius puncticeps, Horn. Ca.
Cafius canescens, Mann. N.
   "   lithocharinus, Lec. R.
   "   luteipennis, Horn. R.

Cafius sulcicollis, Lec. R.

" opacus, Lec. Ca.

Lathrobium jacobinum, Lec. R.

Caloderma reductum, Csy. Ca.

" mobile, Csy. Ca.

" sp. Ca.

Tachyporus californicus, Horn. Ca. R.

Pseudopsis, sp. Ca.

Haploderus flavipennis, Csy. Ca.

Apocellus analis, Lec. Ca.

Hippodamia vittigera, Mann. Cz.*

Hippodamia ambigua, Lec. Ca. * Cl. R.

Hippodamia convergens, Guér. Ca.*

Coccinella, v. californica, Mann. Ca. Cl. N. R.

Cycloneda oculata, Fab. Ca.

" sanguinea, Linn. Ca. *

Psyllobora, v. tædata, Lec. Ca.*

Chilocorus bivulnerus, Muls. Ca.*

Cryptognatha catalinæ, Horn. Ca. Cl.

Hyperaspis lateralis, Muls. Ca.*

Scymnus guttulatus, Lec. Ca.

" nebulosus, Lec. Ca.

" cervicalis, Muls. Ca.

" marginicollis, Mann. Ca.

" ardelio, Horn. Ca. Cl.

Cephaloscymnus occidentalis, Horn. Ca.

Cephaloscymnus ornatus, Horn. Ca.

Rhizobius lophanthæ, Blaisd. Cl.

Aphorista morosa, Lec. R.

Cryptophagus, sp. Ca. Cl.

Atomaria, sp. R.

Dermestes marmoratus, Say. Ca. Cl. N. R.

Dermestes Mannerheimii, Lec. Cl. N. B. R.

Dermestes talpinus, Mann. Cz.*

" tristis, n., sp. R.

" vulpinus, Fab. G.*

Trogoderma sternale, Jayne. Ca.

Hololepta vicina, Lec. Ca.

Saprinus interstitialis, Lec. Ca.

" lugens, Er. Cl. N.B.G.* R.

" fimbriatus, Lec. Ca.

" vitiosus, Lec. Ca.*

" lubricus, Lec. Ca. Cl. R.

" sp. near laridus. Ca.

Cercus sericans, Lec. Ca.

Carpophilus pallipennis, Say. Ca. * Cz. * Cl.

Coninomus fulvipennis, Mann. Ca.

Corticaria distinguenda, Com. Ca. Cl.

Corticaria, sp. Ca.

Trogosita virescens, Fab. G. *

Dryops productus, Lec. Ca.

Cardiophorus luridipes, Cand. G.*

Melanotus variolatus, Lec. Ca.

Asaphes tumescens, Lec. Cz.*

Acmæodera connexa, Lec. R.

Telephorus notatus, Mann., var. Ca.

Malthodes laticollis, Lec. Cz.*

Collops cribrosus, Lec. R.

Endeodes abdominalis, Lec. Ca.

† " sp. Ca.

" sp. R.

Malachius, sp. nov.? R.

† Attalus subfasciatus, n. sp. Cl.

Pristoscelis ænescens, Lec. B.* R.

" punctipennis, Lec. Ca.*

Pristoscelis pedalis, Lec. Ca.* G. * Cl.

Listrus, sp. Cl. R.

† Dasytes, sp. nov. Ca.

†　„　sp. nov. Ci.

Eschatocrepis constrictus, Lec. Ca.

Cymatodera ovipennis, Lec. Ca.

　　„　angustata, Spin. R.

Necrobia rufipes, Fab. G. * Ca. Cl. R.

Necrobia ruficollis, Fab. Cl.

† Ernobius debilis, Lec. Cz. *

† Oligomerus? n. sp. Ca.

Trypopitys tenuilineata, Horn. Ca.

Hemiptychus obsoletus, Lec.? Ca.

Euceratocerus Hornii, Lec. Ca.

Sinoxylon declive, Lec. N.

Polycaon Stoutii, Lec. Cz. *

Cis, sp. Ca.

Cœnonycha rotundata, Lec. Ca.

†　„　socialis, Horn. G. *

Phobetus comatus,Lec. Ca.Cz.* R.

Cyclocephala villosa, Burm. Ca.

Phymatodes juglandis, Leng? R.

Oeme gracilis, Lec. Ca.

Romaleum simplicicolle, Hald. Ca.

Megobrium Edwardsii, Lec. R.*

Xylotrechus obliteratus, Lec. Ca.

Atimia dorsalis, Lec. G.*

Ipochus fasciatus, Lec. Ca.

Pachybrachys, sp. Ca.

　　„　sp. Ca.

Diachus auratus, Fab. Ca. Cl. R.

†Colaspidea subvittata, n.sp. Ca.Cl.

Diabrotica soror, Lec. Ca. *

Monoxia puncticollis, Say. R.

Phyllotreta pusilla, Horn. Ca.

Bruchus pauperculus, Lec. Ca.

Eurymetopon convexicolle, Lec. Ca.

Phlœodes diabolicus, Lec. Cz. *

Nyctoporis carinata, Lec. Ca. R.

Coniontis elliptica, Csy. Ca. R.

†　„　lata, Lec. Cl. * B. R.

†Coniontis lata, var. insularis, Csy. Cz. * R.

Coniontis viatica, Lec. Cz. *

Coniontis subpubescens, Lec. Ca. Cz. *

† Cœlotaxis punctulata, Horn. G.*

†　„　muricata, Horn. G. *

†　„　angustula, Csy. G. *

† Cœlus pacificus, n. sp. N. R.

†　„　remotus, n. sp. Cl.

† Eusattus robustus, Lec. Cl. *

　　„　politus, Horn. R.

Eleodes quadricollis, Esch. Ca.

　　„　dentipes, Esch. Cl. N. R.

　　„　scabripennis, Lec. B.* R.

† Eulabis grossa, Lec. Cl. * N. B.

　　„　pubescens, Lec. Ca.

　　„　obscura, Lec. R.

Amphidora littoralis, Esch. Ca. R.

Cratidus osculans, Lec. Cz.* R.

Cibdelis Bachei, Lec. B. *

Blapstinus rufipes, Csy. Ca.

　　„　brevicollis, Lec.? R.

Conibius seriatus, Lec. G. *

Notibius sulcatus, Lec. Cl.

† Helops Bachei, Lec. R.

†　„　„　var. G. *

　　„　sp. Ca.

Hymenorus infuscatus, Csy. Ca.

Isomira variabilis, Horn. Cl.

Pentaria nubila, Lec. Ca.

Anaspis collaris, Lec. Ca.

Mordellistena, sp. Ca.

　　„　sp. Ca.

Notoxus constrictus, Csy. Ca,

Anthicus californicus, Laf. Ca. Cl.
" sp. Ca. Cl.
† Meloe barbara, Lec. B. *
" sp. Ca.
Rhynchites aureus, Lec. Cl.
" sp. nov.? Ca.
Trigonoscuta pilosa, Mots. Cl. R.
Sciopithes setosus, Csy. var. Cl.

Apion antennatum, Sm. Ca.
" œdorhynchum, Lec. Ca.
Cleonus basalis, n. sp. Cl.
Smicronyx, sp. R.
Anthonomus pauperculus, Lec. Ca.
Tychius, n. sp. N.
Balaninus occidentis, Csy. Ca.
Sphenophorus vomerinus, Lec. R.

Concerning the value of certain names upon which there is a disagreement among authorities I am unable to offer any very well founded opinion. I am, however, inclined to doubt the validity of *Cryptognatha catalinæ*, Horn, and *Cælotaxis angustula*, Casey; and on the other hand it seems probable that *Conibius guadalupensis*, Casey, is a good species and not a form of *seriatus* as recorded by Dr. Horn.

Tachys, sp.—Two specimens from Clemente are closely allied to *corax*, Lec., but seem distinct by the obviously less transverse thorax.

Amara insularis, Horn.—Very abundant on all the islands visited by the Pasadena party. I saw no signs of it on Catalina in midsummer, though the dried remains of *insignis* were common enough.

Agabinus glabrellus, Mots.—Not rare on Catalina. Very scarce on the mainland in the streams in the mountain canons.

Cercyon luniger, Mann.— A small number found in decaying sea-weed on Catalina ; *fimbriatum*, which may be found by thousands along the opposite coast, has not yet been detected.

Hippodamia ambigua, Lec.—Specimens from Santa Rosa might with equal propriety be placed with *convergens*. Unless some other character than thoracic markings can be discovered to separate these two so-called species they cannot be held as distinct. They constantly occur together everywhere in South California, and intermediate forms are frequent.

Rhizobius lophanthæ.—It is certain that this beetle is an importation from Australia, but it seems very probable that the pioneers were not introduced as advertised. How it first got here is a mystery, but it is surely here to stay, and is now quite as much at home as any of our native Scymni. Although already widely distributed in California, its occurrence on an island so distant and so rarely visited as San Clemente was, to say the least, unexpected.

Aphorista morosa.—According to Mr. Ricksecker this and *læta* are sexes of the same species, the latter being the female. *Morosa* is in my experience much more commonly met with in So. California than *læta*.

Endeodes.—The species of this genus are among the most curious of the Coleoptera inhabiting the California sea beaches. They are to be found most frequently in April and May running about on the sand, or concealed under rubbish or driftwood so common in such situations. The two undetermined species are represented by one specimen each. That from Catalina was taken by me in July, and is quite surely nondescript, differing from our described species by the very minute elytra, as well as in coloration. The Santa Rosa example is almost entirely black, and is possibly a colour variety of *collaris*.

Phobetus comatus.—There is a very confusing amount of variation exhibited by specimens of this species from various localities in the State. Specimens taken by myself on Catalina, of small size, subimpunctate thorax with hind angles entirely wanting, seem quite distinct when compared with a series from Fresno county, of larger size, different colour, rather closely punctate thorax with distinct hind angles. I have, however, seen intermediate forms, and it would be unsafe to make a division without a large series from diverse localities. The name *testaceus* was originally given by Leconte to specimens from Santa Cruz Island, and it may possibly have to be revived.

Xylotrechus obliteratus.—A fine series of this beautiful longicorn was taken by Dr. Fenyes on Catalina. All the specimens found were males, the females being indeed very rarely taken. The species occurs on willows.

Ipochus fasciatus.—This occurs rather plentifully on Catalina under the bark and on the branches of dead *Rhus laurina* (or *R. integrifolia*). The form, size, sculpture and markings vary greatly, often in a series taken from the same tree.

Balaninus occidentis, Casey. — This species has heretofore been confounded with *uniformis*, but is abundantly distinct. It is common enough on Catalina, but much less so on the mainland, frequenting several species of oaks.

A certain small species taken on Santa Catalina by myself in 1894, and again found on Clemente this year, has not been included in the list for the simple reason that its affinities are not yet sufficiently clear to admit of placing it even in a family sense. Two of our specialists to whom specimens have been sent have ventured opinions that are quite at variance ; the case is therefore postponed for further hearing.

It is not unlikely that a few species have been overlooked in the preparation of the preceding list, but it is hoped that any such omission may not seriously impair its usefulness as a foundation on which to base any future reports on the Coleoptera of these islands.

It need scarcely be said that the 226 species enumerated here can represent but a fraction of the entire coleopterological fauna.

The following species, it is believed, are now made known for the first time. There are surely a number of other undescribed species, but their description would involve far more study than can now be devoted to them.

Cœlus pacificus, n sp.—Broadly oblong, elliptical, moderately convex, piceous black, surface polished. Epistoma broadly sinuate, antennæ with three-jointed club. Prothorax equal in width to the elytra, a little more than twice as wide as the length at the middle, widest immediately before the base, sides rather feebly arcuate and strongly convergent, moderately densely evenly punctate throughout. Elytra twice as long as the thorax along the median line, not longer than wide, equally densely but more finely punctate than the thorax, the punctures not in the least asperate on the disk, and only very feebly so on the declivity and along the margin. Process of first tarsal joint extending under the next three.

Length, 7 mm.; width, 5 mm.

Very distinct from any of our described species by the conspicuously long prothorax, and from all but the next in the almost entire lack of elytral asperities. The marginal fringe of hairs on the prothorax is noticeably shorter and finer than in any of our mainland species. Described from a single example of unknown sex taken on San Nicolas, May 24. Since the above description was written I have seen numerous examples in the material collected by Dr. Eisen on Santa Rosa. With the exception of some variation in size these differ in no noteworthy respect from the San Nicolas type.

Cœlus remotus, n sp.—Very convex, piceous black, legs and elytra brown. Epistoma broadly sinuate, antennal club four-jointed. Prothorax similar in outline to *pacificus*, but shorter; surface subopaque, densely coarsely punctate. Elytra shining, densely finely punctate, without trace of asperities. Process of first tarsal joint extending beneath the next two.

Length, 6.5–7 mm.; width, 4–4.5 mm.

The above brief description is sufficient to characterize this somewhat remarkable species, of which a single pair was taken (June 3) on San Clemente.

The marginal fringe is longer than in *pacificus*, but thinner than usual.

Both the above described species were found under rubbish at a distance from the shore, and have probably the habits of Coniontis and Cœlotaxis rather than those of the other members of the genus. This might indeed be safely inferred from the less developed marginal hairs and lack of elytral asperities, which have an undoubted connection with the habits possessed by the mainland species of burrowing in loose sand. Whether we have here a change from the ancestral mode of life, due to a change of environment, or whether, as seems to me more likely, the burrowing habit is of recent development and the island species are the surviving representatives of an earlier type, is an interesting question.

Cleonus basalis, n. sp. — Moderately stout, integuments black, polished. Beak three-fourths as long as the prothorax, not dilated at tip, rather thinly clothed above and beneath with short cinereous hairs, sides glabrous, above subcarinate in basal two-thirds, rather coarsely punctate throughout. Prothorax as long as wide, sides very slightly convergent, apex feebly constricted, basal lobe angulate, surface very closely densely punctate, feebly carinate in apical half, deeply excavate behind. Vestiture condensed in four narrow vittæ; the two dorsal approximate in front, posteriorly divergent and incomplete; the lateral vittæ dislocated at the apical constriction. Elytra barely twice as long as wide, humeri rounded, tips separately rounded and scarcely acuminate; striæ composed of large, closely-placed punctures; intervals scarcely wider than the punctures, especially on the disk; base strongly impressed each side, leaving the base of the third and to a less degree that of the sixth interval strongly tumid. The third, fourth, fifth and outer three intervals are so thinly clothed as to appear glabrous; the first is, how-ever, very finely pubescent throughout, as is the seventh behind the middle. The dark areas contain a few small spots of condensed hairs, and there is a larger conspicuous spot at the base of the second interval. Lower surface and legs as usual. The third joint of hind tarsi is small, but obviously wider than the second.

Length, 10 mm.; width, 3.5 mm.

Hab.—San Clemente.

The single male above described must evidently be placed near *quadrilineatus* by Casey's table — Coleop. Not., III., p. 186 — but the deep basal excavations of the thorax and elytra, as well as the dense punctuation of the former, clearly separate them. The ocular lobes are moderately well developed in *basalis*, and are said to be wanting in *quadrilineatus*.

Attalus subfasciatus, n. sp.—Very small, narrow, depressed, black, thorax with sides behind and base narrowly testaceous, elytra with a slightly antemedian pale fascia which is interrupted at the suture. Head broad, antennæ slender, not in the least serrate, reaching the middle of the elytra (♀), or as long as the entire body (♂), the four basal joints pale. Thorax narrowed behind, of the same form as in Endeodes. Elytra parallel (♂), or posteriorly dilated (♀). The pubescence consists as usual of very short semi-erect hairs, with longer erect darker hairs sparsely placed.

Length, 1.5–2 mm.

Hab. — San Clemente.

Described from one ♂ and eight ♀ s. A very peculiar little species, differing in antennal structure, form of thorax and style of elytral coloration from all other species in our fauna. It may for the present be placed next to *lobulatus* in which there is a faint indication of the present form of thorax.

Colaspidea subvittata, n. sp.— Piceous, with more or less distinct greenish-bronze lustre ; legs, more especially tibiæ and tarsi, rufescent. Sides of prothorax not strongly rounded, punctuation moderately close, a little coarser on the elytra. Pubescence long (for the genus), recumbent, distinctly subvittate on the elytra in fresh examples. Length, 3.5–4.5 mm. Found abundantly by me on Catalina, also brought from Clemente. There is practically no variation except in size in the large number of specimens examined. The mainland species, on the contrary, exhibit a bewildering amount of variation in size, colour, punctuation, and even in form. One variety of varicolor is of nearly the same colour, but the pubescence is erect and the thorax more strongly rounded at the sides. The pubescence is much more easily removable in *subvittata* than in any of the other species, and the vittate arrangement is scarcely evident except in very fresh examples. In the males the antennæ are somewhat longer and all the tarsi moderately dilated—characters common to all the species of the genus.

Dermestes tristis, n. sp.—Length, .22–.26 inch. Elongate convex, parallel, black, clothed above with black pubescence, with a sprinkling of paler hairs on the prothorax, and rarely mottled with cinereous hairs on the elytra. Scutellum densely clothed with ochreous hairs, usually forming the only relief to the sombre aspect of the upper surface. Thorax not very obtusely rounded in front, anterior portion of lateral margin invisible from above, sides uniformly rounded, slightly sinuate before the front angles, which are distinct and only slightly obtuse. Surface of thorax densely, more coarsely punctate than usual. Beneath clothed as usual with dense white pubescence, with lateral series of black spots ; prevailing colour of last ventral whitish ; legs annulated with white. Males with median pits on third and fourth ventrals, tarsi clothed beneath with spinous hairs. Occurs in various parts of maritime So. California, and on Santa Rosa Island.

One of our smallest species, perhaps most nearly resembling *talpinus.* The latter is, however, more robust, with ochreous and gray mottlings on the elytra, sides of thorax more strongly rounded near the base, and pubescent male tarsi.

It seems not to have been noticed that in a considerable number of our species of Dermestes the front and middle tarsi of the male are rather densely pubescent beneath. The character is an important one and enables us to establish the distinctness of *Mannerheimii,* which has never looked right as a variety of *marmoratus.* These last named species may then be thus compared :

MARMORATUS.—Size large (.40–.45 in.), elytra mottled with ochreous and cinereous hairs, tarsi spinous beneath in both sexes.

MANNERHEIMII.—Size smaller (.24–.32 in.), elytra mottled with cinereous only, front and middle tarsi (♂) pubescent beneath.

HEPIALUS QUADRIGUTTATUS, Grote.—This large salmon-pink variety was taken this year near Metis, P.Q. Messrs. L. Reford and E. Brainerd, of Montreal, chanced one day to pick up the wing of a specimen on a little dry area in a swamp several miles from Metis. They returned to the village for a lantern, and then tramped back again. Their industry was rewarded by the capture of two specimens. They saw five others, and report that the moths appeared about nine o'clock p.m., and flew in a zigzag horizontally, not up and down like *H. thule.* This species has been taken in Ontario by Mr. Elcome, at Peterborough, and at Roach's Point, Lake Simcoe.

LEDRA PERDITA vs. CENTRUCHUS LIEBECKII.

BY F. W. GODING, M. D., PH. D., RUTLAND, ILL.

In the February CANADIAN ENTOMOLOGIST, page 38, Prof. C. F. Baker contributes an article on *Ledra perdita*, A. and S., in which he attempts to identify the insect described by Amyot and Serville under that name with my *Centruchus Liebeckii*. Those authors describe their species from an admittedly inaccurate figure, the original type having been destroyed. They state that their species is from Northern America. [See note.] Van Duzee states (fide Baker) that *perdita* is from Pennsylvania, on what authority I do not know, and Prof. Baker decides that because Van Duzee gives that State as the habitat of the insect, and my species having been described from the same commonwealth, they must be identical. As there is no proof beyond the dictum of Van Duzee that *Ledra perdita* is from Pennsylvania that point may be dropped until we hear further from him. He is too careful a student of our Homoptera to be guilty of confusing a Membracid with a *Ledra*. The facts regarding Fitch's identification are these : While in Washington a few years ago, and working over the Fitch material, I found an example of *Liebeckii* labeled in Fitch's handwriting, " Ledra perdita, A. and S.," and " capra, Mels.," both names being on the label, which I recorded in the CANADIAN ENTOMOLOGIST, Vol. XXV., p. 172. Fitch never published his opinion regarding this species. Prof. Baker (l. c.) says : " So peculiar in form is it that there is not a possibility of confusing it with anything else in our fauna." Since that was written he has come into possession of a copy of Fowler's great work on the Membracidæ of Mexico and Central America, and I do not doubt that since he has examined Fowler's figure of *Centruchoides laticornis* his opinion has undergone a change, for the figure of *perdita* certainly resembles that figure as closely as it would a figure of *Liebeckii*. The same is true with several others of the Centrotinæ, viz., *callicentrus*, etc., etc., from " Northern America."

NOTE.—I do not know where Mr. Van Duzee publishes my reference to this species beyond a note in his catalogue of the Jassidæ, wherein he says: " One American species of Ledra has been described, but I have not yet seen an example." Doubtless he here refers to Amyot and Serville's species.

Now, regarding *Microcentrus caryæ*, Fitch, and *Centruchus Liebeckii* being congeneric, at the time I wrote the description of *Liebeckii* the close relationship between my species and *caryæ* was recognized, but as Stal says that prothoracic horns are absent in Microcentrus, I looked for some other modern genus in which to locate, temporarily, the species. The Old World genus Centruchus seemed to fit it the best, and that generic term was used although there was an extra discoidal cell which must sooner or later place it in a separate genus. This has been done by Fowler, who has described the genus Centruchoides. Of the two species the neuration is identical, and the entire anatomy (other than the presence of lateral horns in *Liebeckii*) is the same. I have before me the example of *Liebeckii* mentioned by Prof. Baker as having aborted horns. In my collection is an example with horns still more aborted, and while in Washington a few weeks ago I found several similar examples in Mr. Heidemann's collection. That gentleman informed me that he had taken both forms together, with their larvæ and pupæ, while collecting. This shows that the horns are variable, and, as I believe, in some cases absent, as is true of *Platycotis sagittata*, Germ., as recorded in my paper " Fitch's types."

Mr. Fowler has re-described the genus Microcentrus as Phaulocentrus, and after stating that *caryæ*, Fb., belongs to his genus, describes and figures four new species, viz.: *pileatus, proximus, sordidus* and *cornutus :* the first three closely related to *caryæ ;* the fourth, I believe, bears the same relation to one of the others that *Liebeckii* bears to *caryæ,* and I should not be surprised if his **Centruchoides laticornis* was still another instance.

In conclusion, I will say that the name *Ledra perdita,* in my opinion, should be forgotten. The type was destroyed ; the description, which might apply to any one of a dozen species of Centrotinæ, drawn up from an unrecognizable figure, and there is no possible way of determining what insect the artist had in hand when he drew the figure from which Amyot and Serville drew up the description of *perdita.* Let the name be buried in oblivion. I believe *caryæ,* Fitch, and *Liebeckii,* Goding, are one and the same species. As Fitch's name has priority, the name *Microcentrus caryæ,* Fb., should stand, while the horned form may be known as *var. Liebeckii,* Godg.

---

* Centruchoides is not a MS. name. It is described in Fowler's work, page 159.

A PRINCIPLE TO OBSERVE IN NAMING GALLS: TWO NEW GALL-MAKING DIPTERA.

BY WM. H. PATTON, HARTFORD, CONN.

(EDASPIS-SOLIDAGO ATRA.

Galls do not differ from those of *Œ. polita*, as described by Osten Sacken (Tr. A. E. Soc. ii., 301 ; 1869):

This is an addition to the list of gall-making Trypetas given by Osten Sacken in *Psyche* for April, 1880. I bred both sexes from Solidago galls, Sept. 8, 1875, in Connecticut.

Flies.—Female agrees perfectly with Loew's description of a specimen from New York. The eyes in the living flies are green, with two longitudinal purple stripes. The shed puparia are left in the galls, and are of a delicate texture and milk-white colour. The New York specimens from which *atra* was described approach *polita* in all their points of difference from the Mexican specimens. Whether the Mexican specimens belong to the same species is a question which does not concern us in determining the synonomy of *atra*. If the pale gray border of the wing cross-bands was darkened and one of the bristles on the lateral border of the front was lost (differences which might well arise with increased maturity of the specimens) we should have nothing to separate the species excepting the slightly greater divergency of the second and third bands, and it is probable that this greater divergency would disappear with the blackening of the gray borders. *Œ. atra* is a later name than *Œ. polita*.

CECIDOMYIA-CELTIS (new genus) DESERTA, new species.

Galls are hollow, elongate swellings of young twigs, from which emerge, about the first of June, single Cecidomyian flies from a perforation near the base. Length of gall one half inch to one inch.

On Hackberry *(Celtis occidentalis)* ; Orange, Connecticut.

The name describes the genus.

This gall I name and describe to illustrate a principle which may be useful in naming galls of which the makers are unknown. It does not seem proper to refer such galls to the genus of plants alone, as was done by the older botanists, nor to the genus of insects alone, as is at present the fashion, but to a combination of the two, thus : *Cynips-quercus, Cecidomyia-quercus, Cecidomyia-salix,* etc. All Cynips are, it is true, confined to *Quercus,* but it is the gall and not the insect for which I propose this nomenclature ; besides, *Quercus* supports other genera of gall-makers. The combined generic name is in the nominative case and will not conflict with the many specific names which have been drawn from the plant and used in the genitive. In many cases the genitive of the plant genus has been used in combination with a specific name not derived from the plant, as *Cynips-quercus-futilis*. The suggestion made by Osten Sacken that in these cases the genitive or its initial (which is often all that is used) should be dropped seems worthy of being carried into effect, as this genitive appears in most cases to have been inserted by accident or error.

This nomenclature also has the advantage of not presenting the appearance of describing what is unknown ; it has no binding force of priority over the specific name of the insect when that is discovered. It has, however, a priority in the description of *galls*, and the specific name should be retained as the name of the *gall*, even though the insect should by chance receive a different name or it should prove not to belong to the genus under which the gall is described. It also has the advantages of simplicity and of conformity with medical usage in naming gall diseases of animals.

To exemplify the principle I name the following galls described in the *5th Rept. U. S. Ent. Comm* :

    p. 612, 30, *C.-c. oviformis.*
    p. 613, 31, *C.-c. semenrumicis.*
    p. 613, 32, *C.-c. pubescens.*
    p. 613, 33, *C.-c. capsularis.*
    p. 614, 34, *C.-c. spiniformis.*

THYREOPUS ADVENUS (Sm.), PACK., A PROTECTOR OF THE ARMY WORM.—This species is an exception among burrowing wasps in being injurious to vegetation, as I have found it killing and carrying to its nest *Sarcophaga*, *Musca domestica*, and that enemy of the Army worm, *Belvosia unifasciata.* The wasp forms its small hillocks under the shelter of shade trees late in August, in Connecticut. In rainy summers its numbers are much reduced. *Miltogramma* pursues the wasp with felonious intent. The wasp may be destroyed by pouring strong alkaline washes into the burrows.

The *B. unifasciata* varies in having a red tail, contrary to the name *flavicauda* by which it was formerly known.   W. H. PATTON.

Mailed October 4th, 1897.

The Canadian Entomologist.

Vol. XXIX.   LONDON, NOVEMBER, 1897.   No. 11.

NOTES ON THE LIFE HISTORY OF COLIAS INTERIOR, Scud.

BY H. H. LYMAN, MONTREAL.

When in New York, towards the end of May, 1894, I paid a visit to Mr. B. Neumœgen, who, though suffering considerably from the fatal disease to which, after a brave fight, he finally succumbed, received me kindly, and after a short conversation sent me upstairs to Mr. Doll to get the names of certain species which I had brought for determination.

In one of the drawers which Mr. Doll showed me I found several specimens of Colias Interior, one being of a very rich shade of colouring, almost orange, in fact. Asking where they came from, I learned that they had been taken the previous season at Camp Lou, on Osgood Pond, in the Adirondacks, and I immediately determined, if possible, to get eggs. Early in July I wrote to Mr. Neumœgen to ascertain the best time to be on the happy hunting-grounds and for any suggestions, and received a post card, dated 9th July, written on a railway train, and the last communication I received from him, telling me that then was the right time. I was unable to go just then, but on the 20th I left by the evening train over the Adirondack and St. Lawrence Railway, and reached Paul Smith's hotel shortly after 9 o'clock. The 21st it rained all day till late in the afternoon, but the 22nd was fine, and I soon had 2 ♀ ♀ of Interior caged for eggs. For the cage I used a tomato can filled with such soil, chiefly sand, as I could find, and in it I placed two species of Vaccinium, two willows, Kalmia Augustifolia, Trifolium Stoleniferum. The following day I took five more females and two males. One ♂ Philodice was taken courting a ♀ Interior, and was confined with the ♀ ♀ to see if it would copulate with one of them, but it did not do so, so far as I observed. Three of the freshest ♀ ♀ were killed for the cabinet, but the remaining four with 1 ♂ Interior and the ♂ Philodice were kept caged. The 24th was again rainy, and in the afternoon I left for home, carrying my menagerie with me. On the 26th July the plants were changed to a flowerpot of larger size than the tomato can, and the following species of plants were added : Melilotus Officinalis and Alba, Amphicarpæa Monoica,

Vicia Cracca, Desmodium Dellenii. One ♀ was found dead, and one very feeble and apparently dying. The living ones were fed with sugar and water, and here I may be permitted to say that the only success I have ever had in feeding butterflies was when I uncoiled their tongues with a pin bent at the point, and then put a camel's-hair pencil dipped in the syrup to the tongue. They will then continue to feed as long as they are hungry ; but holding the brush in front of them and blowing gently towards them, as the authorities tell us to do, I have found a failure, and putting a saturated sponge in a cage utterly useless. No eggs were observed at this time, but one at least must have been laid some time before, as a larva hatched on 30th. On the 29th I was ill in bed all day, but on the 30th I found that from 28 to 30 eggs had been laid, nearly all on the Vaccinium, and that one larva had hatched as above stated. One egg was laid on Amphicarpæa Monoica. One or two eggs were laid after the 30th. I divided about half of the eggs between Messrs. Fletcher and Scudder, sending eight to the former and six to the latter.

Of the eggs that I kept, one hatched on 30th July, four on 4th August, eight on 5th, and two on 6th. The egg period must therefore have been about six or possibly seven days in one or two instances.

The egg and first stage of the larva have been described by Mr. Scudder in his great work on butterflies, but as that work is unfortunately not available to all entomologists, it will do no harm if I give my notes, imperfect as they may be, in full.

Egg.—Length, 1⅓ mm.; diameter, .48 mm. Similar to Philodice in shape. Number of ribs, about 20. At first, white tinged with greenish-yellow, soon turning reddish-orange. Just before hatching turning dark. The larva can then be seen through the shell, standing on its tail, with a clear, vacant space above the black head. The larva emerges a little below the top, just where the head is. One that was watched crawled slowly down the shell on to the leaf, moving its head from side to side on the leaf as though spinning a silken path, and as soon as it was all on the leaf, it turned round, climbed to the top of the shell, and began to devour it, and ate it all up, its meal taking 40 minutes. Most of the larvæ did not eat more than half of the shell, and some did not eat any.

Young larva.—Length, 1.91 mm.; width of head, .366 mm.; head black, the hairs pellucid. Body brownish-green, finely transversely striated, with about five striations to each segment. Skin faintly shagreened with yellowish-brown ; the striations are of same colour ; the raised points are pale in colour, black at the summit. Hairs pellucid, club-shaped, especially on second segment, where they are considerably longer than on the other segments.

Tried the larvæ with Vaccinium, two or three species of birch, two or three species of willow, Amphicarpæa Monoica, Epilobium Augusti-folium, and several other plants at a venture, but in all cases they crawled off the leaves on to the side of the jar. One that was afterwards placed upon a willow leaf just died and dried up where it was put. On 5th August found the leaves of Vaccinium eaten in several small patches, and a sprinkling of tiny frass in the bottom of the tumbler. All the larvæ behaved as though Vaccinium were not their proper food plant, leaving it and wandering around the glass, and only returning to it when they found that they could not get anything more to their taste. I have, however, since then, seen the same thing done by the larvæ of a Noctuid, the eggs of which were found on the leaves of a shrub, and I therefore judge that it is owing either to a desire to explore their surroundings or because they object to the confinement. The frass from the young larvæ must have been ejected with considerable force, as the jar was always sprinkled half way up the side. The mortality, through drying up apparently, was very heavy, and by the 15th of the month only five remained out of fifteen, and in my despair I wrote to Mr. Scudder for suggestions, and on the 17th received an answer from him describing his method of unconfined feeding. I then filled a homœopathic vial with water, bored a small hole in the cork, and inserted a small sprig of Vaccinium. The vial I placed in a wineglass, with earth around it to support it, in order that should the larvæ fall off the leaves they would be caught, and also to decrease the danger of their straying. The earth I watered, so as to render the air about the larvæ slightly moist. I then transferred all that remained alive, viz., three, as two had perished since the 15th, to the sprig. The following day I found that one had not moved from the spot where I placed it, and was apparently dead and drying up, but the two others were healthy, and thenceforward I had no trouble, and carried these two right through to imago. One of these larvæ passed first moult on 18th August, and the second on the 20th. Just before the moult the larva seems quite smooth.

After first moult : Length about 3.9 mm., rather plump, colour dull green, head same colour as body, head and body covered with very short, minute whitish hairs, giving a shagreened appearance ; faint, darker green longitudinal lines are visible under a lens.

On the 20th I left for a short holiday at Murray Bay, carrying my menagerie with me, and my arrival with it caused a certain amount of curious interest among the guests at the hotel.   The larva eats the paren-chyma of the leaf in small round patches ; mine fed on the upper side, and when resting, they rested along the midrib, head sometimes up and sometimes down.   Mr. Scudder tried his larvæ with Vaccinium Corym-bosum and V. Vacillans, and found that the one on the latter ate with twice the zest of that on the Corymbosum, and further, that the one on the latter fed on the upper surface of the leaf, while that on Vacillans fed on the lower surface.

About the end of August or first of September they ceased feeding and became lethargic, lying along the midrib of the leaf, near the petiole, upon a slight carpet of silk, and as they were plainly hibernating, and I feared they might dry up, I removed the leaves from the sprig, cut away the surplus space of the leaves, and secured them to the bottom of a pill-box with a touch of glue.   When the pieces of leaf seemed per-fectly dry, I put the pill-box in a bottle, corked it and placed it in the refrigerator.   Some time afterwards I found that in some way water had got into the bottle, and the card pill-box was wet and mouldy.   I took it out, removed the mould as well as possible with a camel's-hair pencil, and allowed the box to dry.   The larvæ were apparently healthy, and I then put the box out of doors on a gallery, where the occupants would be as cool as possible and protected from sun and rain.

As soon as the snow came I got a small wooden box, cut several small pieces about an inch square out of the upper edge for ventilating purposes, put it on the ground, with a brick on the bottom inside, placed my box with hibernating larvæ on the brick, and covered the box with an inverted tin tray that I had had made, the tray projecting about an inch all around the box, and then covered it with snow.   In the spring, as the snow gradually melted, I had more brought from the shady parts of the garden to pile over the box, and finally had the much-reduced heap covered with ashes to protect what little snow remained from the genial warmth of the end of April.   I wrote to Mr. W. H. Edwards to try to secure some Vaccinium from the South, offering to pay a boy to get it, but Mr. Edwards wrote that he did not know where to get it, and advised me to try willow.   I then appealed to Mr. Jack, at Jamaica Plain, and a few days later to Mr. Fletcher, at Ottawa.   Both kindly responded promptly, and as a result I received a plentiful supply of shoots with the first tiny leaves coming out. The snow and ashes were removed from the top of the box on 1st May, and the box opened. The card pill-boxes were found very damp and mouldy, but the two larvæ were sound and healthy in spite of the mould all about them, but were naturally somewhat shrunken in size from their long fast.

At midday, on 3rd May, as I found that they had moved from their positions, I placed them very carefully upon the open buds of a sprig of Vaccinium, arranged in water as previously. They soon crawled on to the stem and rested, one head down, the other up.

During the 4th they remained lethargic, in the same position, but by the morning of the 5th the one which previously had its head up had turned round and had its head down, and by the evening I found that they had eaten a little. They now eat the entire leaf, which is young and tender.

On 9th May they moulted for second time.

After 2nd moult. — Length, about 7 mm. Head green, slightly roughened with minute brown points. Body green, with many minute yellowish raised spots, each tipped with a minute brown hair or point. Along the spiracular space there is a raised band like a fold, mottled with white, pink and yellowish.

On 12th May one passed 3rd moult about 3.30 p.m. while under observation. When first seen the old face still adhered to the mouth-parts of the new, but the skin had been worked more than half way to the anal extremity. It only took a very few minutes to get clear of the old skin, and then it proceeded to divest itself of the old face, which it soon accomplished.

After 3rd moult.— Length at rest, 8.6 mm. Head bright green, roughened as before. Body darker green, shagreened with yellowish raised spots, with short brown hairs or points. Spiracular fold as before, whitish, with orange and yellowish patches and markings.

The weather turned cold and wet, and the second larva was two or three days later in moulting than the other, but the exact date was not recorded.

The species of Vaccinium that Mr. Fletcher supplied me with was Canadense, but I had also received V. Vacillans from Mr. Jack. On the 17th a careless servant threw away my supply of V. Canadense, so I gave the larvæ the V. Vacillans, but the following day I found they had refused it and had eaten nothing, so I offered them some of the sprigs that I had first received with the opening buds which I had kept in a tin, and they then began to feed eagerly upon these.   The arrival of a fresh supply of V. Canadense from Mr. Fletcher removed all cause for anxiety.

On 25th May both were observed to be apparently fixed for 4th moult. Length as contracted, 14 mm.   The spiracular fold is pink, bordered with white and interrupted by the spiracles, which show as a green oval ring on the white band with a white centre.   There is no trace of any pink or other stripe above or below the spiracular one.   There is a dark green dorsal line, and the space on each side of it has a yellowish-green appearance from the minute yellowish warts, but the subdorsal or lateral region is of a bluish-green shade, and the warts are whitish. The region above the spiracular fold is thus about evenly divided between bluish and yellowish green.

One was found, about 10.45 a. m., on 26th May, to have passed 4th moult, and was described at 1 p m.

After 4th moult.—Length at rest, 16 mm.   Head and second seg-ment bright green, finely sprinkled with black points, from which arise minute hairs, blackish above, whitish below.   Dorsal region green, with a brownish-yellow tinge, as before.   Lateral and sub-spiracular regions bluish or whitish green, the minute hairs being whitish.   Spiracular fold white, overlaid along the middle with coral-red.   On following morning the other larva had passed 4th moult, and on 31st the former was appar-ently mature, as it left its food and crawled up to top of cage.   I described it in the afternoon, but it was very restless, sometimes crawling very fast, and sometimes in a very funny, jerky manner.

Mature larva.—Length, 25 mm.   Rich dark yellowish-green on head and above, with narrow dorsal dark stripe bluish-green on sides, with innum-erable small papillæ and minute hairs of a dark brown or black colour above, partly white on sides, and white below spiracular fold. Spiracular fold white, with bright crimson stripe included.   Head small ; as finely dotted as the body.   Below bright green ; feet and prolegs the same.

Mr. Fletcher kindly gave me the following note on the general habits of the larva :

. "Larva decidedly sluggish for the greater part of the time, but when feeding, which was generally twice a day, very nervously active, biting with great rapidity, and moving slowly with short, jerky steps."

The following particulars are also taken from Mr. Fletcher's notes on the mature larva :

" Length, 1⅜ inches when extended feeding. Head, 2 mm. broad; narrower than segment 2. Segment 3 slightly the widest of all. Body cylindrical from 4 to 10 inclusive, then tapering slightly to end. Head concolorous with body, evenly reticulated all over with dark green, the interspaces yellowish and pubescent, the bristles on apex short and black, those toward the mouth much longer and white. Mandibles darkened at apex, process beneath the neck honey-coloured. Ocelli six, in two lines, the anterior of 4 slightly curved forward and lying on a yellowish-white stripe, the other two lying behind these, one above the other, smaller than those of the anterior row. In the anterior row the 2nd and 3rd are the largest, all blackened at base, vitreous at apex. No. 1 of posterior series is the smallest and least conspicuous."

On 1st June one fixed for pupation, and the other on 2nd. Pupation occurred on 3rd June.

Chrysalis very similar to that of Philodice. Green, vermiculate with yellowish-white markings over upper part of thorax and all the abdomen, giving a pea green effect. A green dorsal line extends the whole length. When first formed, there is a spiracular band, similar to that of the larva, running from the wing-case to the tail, but the crimson soon disappears, and the band becomes yellowish and inconspicuous. Half way between the band and the ventral surface there is a broken reddish-brown stripe on the first three abdominal segments, beyond the wing-covers. The head is marked with darker green, yellowish-white at apex. The girdle is rather long, and the chrysalis hangs loose from its support. Length of chrysalis, 18.7 mm.; greatest thickness, 6 mm.

On 12th June the antennæ cases were crimson, tipped with yellowish-green, and the outer and apical margins of the wing-covers were the same. All the parts between the antennæ cases were brownish-green, the eyes deep green, the ventral half of the abdomen yellowish, the wings greenish-yellow. While I was describing it, it bent itself from side to side, bending the abdominal joints as much as possible.

Both pupæ disclosed the imago on afternoon of 13th ; the chrysalis stage being thus ten days. Both were males. I had intended sending away one larva to have a coloured drawing made of it when mature, and of the chrysalis when formed, but my ones matured so fast that I was too late for this, so appealed to Mr. Fletcher to send his one specimen, which had lagged a little behind mine in development. He very kindly acquiesced, but, unfortunately, the larva pupated in the mail, with fatal result. From the length of the larva, 1⅜ inches, he judged that it was a female. The form of the species which occurs in the Adirondacks is that with yellow female, but what that form should be called is a matter of some doubt. In Mr. Scudder's " Butterflies of New England," page 1107, he suggests that as this form was first described when the species was re-described by him, under the name Eurymus Philodice, var. Laurentina, it should be designated by the trinominal appellation, Eurymus Interior Laurentina, the pallid, or white, female being called Eurymus Interior Interior. But it seems to me that the doctrine of priority of description cannot govern the matter in the case of a variety, else we may have what is the normal form in nature labelled as the abnormal in our cabinets, and the abnormal variety of nature standing as the normal form in our cabinets. Clearly, where there is dimorphism in one sex of a species, the form which predominates in a marked degree must be considered the normal form, and the other the varietal, all original descriptions to the contrary not-withstanding. Priority must rule in regard to the species, but it must give way where it clashes with nature in regard to varieties.

The question, then, to be settled, is what is the predominating form of the female in this species? Possibly at present the material in cabinets may not be sufficient to settle the matter authoritatively, but I believe it will be found that the yellow or syngenic form is the normal form, and that the antigenic or pallid female is only an albinic form, as in Philodice.

Among the types of Interior there was only one female, and this happened to be of the pallid form described by Mr. Scudder as " white, with a very pale yellowish tinge "; but among the large number brought from Cape Breton Island by Mr. Roland Thaxter there were eight pallid females, and ten which Mr. Scudder called gynandromorphic females, by which not very happy term, I suppose he designated the yellow form.

Besides the seven females taken by me in the Adirondacks and the three from the same region that I saw in Mr. Neumœgen's collection, I have one from the *Godbout river, in the Gulf of St. Lawrence, and one from Nepigon, and all these are yellow, and I do not remember having ever seen a white one, though it is possible I may have done so. Dr. Bethune has informed me that he took a good many at Nepigon, and all were yellow. Mr. Fletcher wrote me that he had taken 18 ♀ ♀ at Nepigon, and of these 11 were of what he calls the pallid form, and 3 at

*The man who collected for Mr. Couper at this locality was named Comeau, not Corneau, as printed in the CAN. ENT.

Ottawa, of which one was pale, and that Prof. Macoun took 3 ♀ ♀, all yellow, in Prince Edward Island; but I do not think Mr. Fletcher has ever had a white female Interior. Mr. Fletcher has also one ♀ of the yellow form from British Columbia. If my belief in this matter should prove to be well founded, the species should simply be known as Colias Interior with an albinic variety of the female, and the name Laurentina should simply fall into the synonymy.

Mr. Scudder further says that "the males are very much more numerous than the females," and among his "desiderata" asks why this is so.

On general principles I should think such a condition of things extremely doubtful, and I believe Mr. Scudder's assertion to be founded on insufficient evidence, especially as in the collection which Mr. Roland Thaxter made in Cape Breton, and which furnished Mr. Scudder with his types of C. Laurentina, there were 18 females to 21 males, certainly no great discrepancy.

I have only twice met with this species in numbers, but neither experience would lead me to form such an opinion. The first occasion was on 8th July, 1890, along the line of the C. P. R. west of Sudbury, when travelling to Nepigon in company with Mr. Fletcher. Whenever the train stopped for a minute or two we jumped off with our nets, and I think we took a dozen between us, and I believe all were males, but it was evidently too early for the females, as the males were quite fresh, and the next day when we arrived at Nepigon, where the season is later, we found that the males had not yet appeared. I think it probable that a fortnight later plenty of females would have been flying near Sudbury.

I may say, however, that the evidence of Mr. John D. Evans, of Trenton, who collected for a number of years at Sudbury, is rather on the other side, as out of a series of 31 specimens in his collection only 4 are females. This is probably accounted for by the fact that out of the 31 no less than 19 were taken prior to July 5th, and for 4 others the date of capture is unknown, and I have already pointed out that the females probably appear later. Twenty-one out of the 31 were taken by Mr. Evans in 1886, who found this species comparatively scarce in later years.

In 1894, at Paul Smith's, I took seven females to two males, but, of course, the former are easier of capture. When this matter has been further investigated, I am confident it will be found that no serious discrepancy in numbers exists between the sexes.   In speaking of the probable or possible life history of the species, Mr. Scudder says (page 1110): "Mr. Fletcher obtained them (the eggs) July 16-24, and they hatch in seven days.   This gives ample time for the caterpillars to attain maturity and pass into pupa for the winter; but what the creatures actually do, and how winter is passed, is unknown.   There is, however, certainly but one brood anywhere."   It seems to me, however, that it may be mathematically demonstrated that any species of which there is only one brood in the year and which does not appear on the wing till July or the very end of June must pass the winter in the larval condition not more than half-grown.

THE LIFE HISTORY OF EPIRRANTHIS OBFIRMARIA, Hbn.

BY REV. THOMAS W. FYLES, SOUTH QUEBEC.

Epirranthis obfirmaria is a swamp insect.   I take it in "The Gomin" near Quebec, where, in ordinary seasons, it is on the wing from early in June till the close of that month.   The following is a brief description of this beautiful insect :

♂.   Expanse of wings one inch ; length of body half an inch ; length of antennæ three-tenths of an inch.   Colour a rich, warm brown. The primaries have a broad ochreous band, widest at the costa, outlined with dark brown.   In this band, not far from the costa, is a dark brown spot.   The secondaries have the outer third of the same warm brown as the primaries, with an ochreous patch at the outer angle of it.   The rest of the wing is ochreous, clouded towards the base.   The marginal dark brown line of this ochreous portion is somewhat angulated.   In the part of lighest colour in the wing is a conspicuous dark brown spot.   The antennæ are pectinated.

♀.   Expanse of wings an inch and one-fifth ; length of body nine-twentieths of an inch ; length of antennæ two-fifths of an inch.   The marks in the wings are similar to those in the wings of the male, except that there are no brown dots in the primaries.   The colours are much brighter : the darker portions are of a rich brick-red, and the lighter of a clearer yellow than in the male.   The brown spots in the secondaries are small.   The antennæ are filiform.

Eggs of E obfirmaria.

The eggs of *E. obfirmaria* are laid dispersedly and unattached. They are pale greenish-yellow, small, and bluntly oval in outline.   They have minute granulations on the surface. A batch of the eggs, laid on the 14th of June in the present year, hatched on the 27th of the same month. The larvæ fed on *Vaccinium, Cassandra, etc.*

Newly-hatched larva.

A "looper," one-tenth of an inch long, suspended itself by a line. It was black with white patches on each segment, and presented a strangely checkered appearance. The head was large and black; the mouth-organs white. The feet also were white. The claspers were wide apart—beside them it had but one pair of prolegs. There were a few bristles at the anal extremity, and along the sides of the larva. It moulted July 3rd.

Larva after first moult.

One-fourth of an inch long; brownish-green in colour; had five conspicuous brown warts on each side of the body. The head was light brown, and the legs brownish-green.

[NOTE.—The habit the larva has of eating its *exuviæ* makes it exceedingly difficult to follow its changes. The insect I am telling of, however, certainly moulted on July 16th.]

Larva after moult of July 16th.

Length three-fifths of an inch. Colour brownish-ash above, with fine paler lines. The fourth and terminal segments were somewhat lighter in colour. Underneath the larva was of an Indian yellow shade. The face was flat, outlined with brown, and had two white spots near the upper edge. The spiracles were dark brown and appeared in a line of folds or broken ridges. The larva moulted July 24th. After moulting it ate its old skin all but the mask.

Full-grown larva.

Length four-fifths of an inch. Colour brownish-ochreous. It had a dorsal line faintly outlined with brown, and on either side of this a row of dark brown spots. It had also a row of similar spots just above the spiracular line. This line was pale ochreous and warty. Below it was a row of oblong, dark brown patches. The spiracles were dark brown.

The larva ceased to feed in August, and towards the middle of that month gathered a few leaves together and spun a light cocoon somewhat after the manner of *Caterva catenaria.*

DESCRIPTIONS OF FIVE NEW GENERA IN THE FAMILY CYNIPIDÆ.

BY WILLIAM H. ASHMEAD, ASSISTANT CURATOR, DEPARTMENT OF INSECTS,
U. S. NATIONAL MUSEUM.

Subfamily VIII.—CYNIPINÆ.

XYSTOTERAS, gen. nov.

This new genus somewhat resembles *Philonix*, Fitch (= Acraspis, Mayr), and agrees with it in having 14-jointed antennæ, but otherwise is quite different. The head, thorax and abdomen are highly polished, impunctate, the mesonotum being entirely without any trace of the parapsidal furrows, and in this character differing widely from all other of the agamous genera of the Cynipinæ. The absence of the parapsidal furrows being peculiar only to the sexual genera *Neuroterus* (Ameristus, Förster) and *Dolichostrophus*, Ashmead. The third joint of the antennæ is not quite as long as joints 1 and 2 or 3 and 4 united, joints 10–13 a little longer than thick, while the last joint is fusiform, a little longer than the penultimate. The scutellum has a depression across its base, but is without distinct foveæ, and is also not separated from the base of the mesonotum by a delicate grooved line ; apically it is obtusely rounded, shagreened and somewhat densely pubescent. The mesopleura have a large, rather deep vertical femoral impression. The wings are represented by very short pads which do not extend beyond the apex of the meta-thorax or just reach to base of abdomen. The abdomen is about twice as long as the head and thorax united, polished, bare ; the second segment dorsally occupies about half the whole surface ; the third segment dorsally is not quite as long as segments 4 and 5 united ; the segments 4–7 subequal ; while the hypopygium terminates in a long, pubescent spine. The hind tarsi are as long as their tibiæ, the claws being simple.

Xystoteras volutellæ, sp. n.

Gall.—A conical, bluish-gray gall, from 3 to 3.5 mm. high, by 2.5 mm. in diameter at base ; occurring singly or in great numbers on the under surface of the leaves of *Quercus macrocarpa* in Riley County, Kansas. The top of the gall is truncate and internally it is hollow, with the larval cell or kernel, resembling a minute nipple, situated at its base. The gall is attached to the leaf by a few fibres and may easily be detached. The colour of the gall is produced by a powdery or pruinose bloom which completely covers it when fresh.

Gall - wasp.— ♀ . Length 2 mm. Polished black, very sparsely pubescent. Antennæ 14-jointed, about two-thirds the length of the, body, the first joint or scape obconical, slightly curved, swollen at tip, the second joint about 1½ times as long as thick, both much stouter than the flagellum. Mesopleura smooth, shining, with a deep vertical femoral fovea. Legs, except knees or the extreme apices of the femora, which are dull honey-yellow, entirely black.

Hab.—Manhattan, Riley County, Kansas.

Described from a single specimen received nearly ten years ago from Mr. C. L. Marlatt. The wasp, according to Mr. Marlatt, issues from the gall early in January.

ZOPHEROTERAS, gen. nov.

This genus also comes very close to *Philonix*, Fitch, but is readily separated from it by the shape of the scutellum, by antennal characters, by bareness of abdomen, and by the claws of the hind tarsi being simple, not toothed.

The frons and mesonotum are alutaceous or feebly shagreened, the latter having distinct traces of the parapsidal furrows, or at least these are more or less distinct posteriorly. The scutellum is rounded or semi-circular, rounded off posteriorly and separated from the mesonotum by a delicate grooved line and carina. The antennæ are long, 14-jointed, the third joint as long or nearly as long as joints 4 and 5 united ; joints 6–13 are a little more than twice as long as thick. Claws of hind tarsi simple. The abdomen is longer than the head and thorax united, bare, or at the most with only some sparse pubescence at sides towards the base ; the second segment dorsally occupies fully half the whole surface; segments 3–5 short, subequal ; segments 6 and 7 very short ; while the hypopygium is armed with a hairy spine.

To this genus belongs *Acraspis vaccinii*, Ashm.

XANTHOTERAS, gen. nov.

Head shagreened, the frons without a distinct ridge or carina be-tween the antennæ. Mesonotum polished, with deep, distinct parapsidal furrows. Scutellum with more or less distinct lateral margins or with a frenum, two indistinct shallow foveæ at base and the same separated from the base of the mesonotum by a delicate but distinct transverse grooved line and a carina. Antennæ 14-jointed, the third joint a little longer than the fourth, or the latter about two-thirds the length of the third ; joints

7–14 a little stouter than joints 2 to 6, and much shorter, joints 11–13 being hardly longer than wide. Tarsi shorter than their tibiæ, the claws with a tooth within.

This genus, although closely allied to *Biorrhiza*, Westw., is readily separated by the absence of the middle frontal ridge between the antennæ, by the shape of the scutellum and by the claws having a tooth within.

The type of the genus is *Biorrhiza forticornis*, Walsh.

PARATERAS, gen. nov.

Last joint of labial palpi somewhat enlarged, ovate. Antennæ 14-jointed, the third joint long, but much shorter than joints 4–5 united, joints 11–13 scarcely twice as long as thick, the last joint hardly so long as the two preceding united. Head and thorax alutaceous or finely shagreened, the mesopleura finely delicately sculptured, without a femoral fovea. Mesonotum with two distinct parapsidal furrows which converge and meet at base of the scutellum. Scutellum small, highly convex or elevated, with a distinct tranverse fovea at base (in reality two foveæ united). The hind tarsi are longer than their tibiæ, the claws with a distinct tooth at base beneath. Abdomen polished, bare. This genus comes nearest to *Sphæroteras*, Ashm., but is readily separated by having 14-jointed, not 13-jointed, antennæ, by the scutellum having a fovea or foveæ at base, and by the hind tarsi being longer, not shorter, than their tibiæ.

Parateras Hubbardi, sp. n.

Agamous ♀.— Length 2 mm. Head and thorax reddish-brown, the vertex and scutellum somewhat obfuscated. Abdomen black, piceous towards base. Antennæ with the first two joints ferruginous, dusky above, the flagellum black or brown-black, except first joint basally. Legs, including coxæ, pale ferruginous, with all the tibiæ, or at least outwardly, dark fuscous or blackish, the tarsi more or less fuscous. Abdomen with the second segment not quite occupying half the whole surface, the third segment dorsally not quite as long as four and five united, the fifth about two-thirds the length of the fourth, the following segments retracted.

Hab.--Detroit, Michigan.

Described from two specimens received from Mr. H. G. Hubbard, to whom the species is dedicated.

Asclepiadiphila, gen. nov.

This new genus comes very close to *Antistrophus*, Walsh, and might easily be confused with it, since it agrees with it in all particulars except as follows :

The female antennæ are 13-jointed, not 14-jointed, the third joint being shorter than the fourth ; in the male the antennæ are 14 jointed, not 15-jointed. The second abdominal segment occupies fully two-thirds the whole surface, while in *Antistrophus* the second segment is considerably shorter.

Asclepiadiphila stephanotidis, sp. n.

Gall.—A small, rounded or pea-like gall averaging from 6 to 8 mm. in diameter, growing from the stems of a species of *Stephanotis*. Externally it is opaque and varies from a gray to a brownish colour, while internally it is whitish and composed of a dense pithy substance with a single larval cell in the centre.

Gall-wasp.— ♀. Length 3 mm. Head, thorax and legs reddish-brown, the sutures of the thorax dusky, the mesonotum with a dark streak down the middle, while the middle and hind tarsi are more or less obfuscated. Antennæ 13-jointed, brown. Abdomen black, highly polished. Wings hyaline, the veins pale yellowish ; the first branch of the radius is straight or nearly so ; areolet entirely wanting, the transverse cubitus about two-thirds the length of the first abscissa of the radius, the first branch of the cubitus very delicate, indistinct, and originating from about the basal third of the basal nervure.

♂.—Length 2.6 mm. Black ; tips of femora and more or less of anterior and middle tibiæ basally dark honey-yellow, rest of legs black. Antennæ 14-jointed, the scape and pedicel black, the flagellum brown.

Hab.—Oregon, Missouri.

Types, No. 3737, U. S. N. M.

A NEW FOOD PLANT FOR PAPILIO ASTERIAS.

BY G. H. FRENCH, CARBONDALE, ILL.

A few days ago I received a letter from Mr. A. V. Thomsen, Chicago, giving a new food plant for Papilio Asterias. But I can give it best by quoting part of Mr. Thomsen's letter. He says :

"Having made a very interesting observation in my study of Lepidoptera, I herewith enclose the notes regarding the same. Aug. 26, '97, I received from Mr. Higgins, in charge of Dept. of Hardy Perennials and Wild Flowers, Lincoln Park, four larvæ of *Papilio Asterias*, nearly full-fed. Found on *Ruta Graveolens* (English Rue). These larvæ pupated as follows : One on Sept. 4, three on Sept. 7. On Sept. 13, '97, I received from the same source eight larvæ of *P. Asterias* in third, fourth and full-fed stages. One pupated Sept. 16. Two of these were found on *Ruta Graveolens*, the balance on *Fœniculum officinale* (common Fennel). Now, I have never seen nor heard of any previous records of *P. Asterias* being found on anything else than members of the Umbelliferæ, and I consider it a very strange occurrence that they should choose a family so widely separated from the Umbelliferæ as the Rutaceæ, of which *Ruta Graveolens* is the type.

"If it had been *Papilio Cresphontes* which I had found upon that plant I should not have wondered, as its food plants here are *Xanthoxylum* and *Ptelea*, two of the Rutaceæ, but *P. Asterias !* "

I should like to ask here if any one has found *Papilio Philenor* feeding on anything but *Aristolochia ?* The species of this genus are rare here, but the butterfly is rather common.

ANOTHER NEW SPECIES OF PROTANDRENA, Ckll.
BY S. N. DUNNING, HARTFORD, CONN.

Protandrena Bancrofti, n. sp.

♀.—Length 9-11 mm.; not as stout as *Cockerelli*, Dun.; a few gray hairs on face and cheeks and on under side of thorax, hair bands on seg. 2-3-4, seg. 5-6 covered with hair growing rufous towards tip, legs and venter with sparse rather longer hairs ; black except a T-shaped mark resting against upper edge of clypeus, spot on tegulæ, tubercles, and four anterior knees which are pale yellow. Head subquadrate, broader than high, venter with fairly deep, not close, punctures; clypeal and sub-clypeal punctures larger and shallow, mandibles piceous; antennæ black at base, growing brownish towards tip, 1st jt. flagellum not as long as 2nd and 3rd combined. Mesothorax deeply and closely punctured ; scutellum with large, coarse punctures ; post-scutellum not smooth, shining ; metathorax with a smooth, shining spot on upper lateral angles, closely and finely punctured, a narrow suture extending upwards. Abdomen finely and closely punctured ; venter with large shallow punctures. Wings sub-hyaline, much darkened outwardly (very much more so than in *Cockerelli*); stigma and nervures ferruginous, a light spot before the stigma.

Two specimens (D. 1102, July 6, 1897, on *Solanum rostratum ;* D. 1262, July 11, '97, on *Medicago sativa*, or alfalfa) taken on the Bancroft Farm, near Denver, Colo. One has been deposited with the American Entomological Society.

SOME NEW AND LITTLE-KNOWN COCCIDÆ COLLECTED BY PROF. C. H. T. TOWNSEND IN MEXICO.

BY T. D. A. COCKERELL, N. M. AGR. EXP. STA.

The Coccidæ herein described were collected by Prof. Townsend in 1896, and kindly transmitted to me for study by Dr. L. O. Howard. The collection made by Prof. Townsend will be fully enumerated in a paper to be published by him elsewhere, so the present contribution is confined to descriptions of the new species and descriptive notes on one hitherto imperfectly known. I have also included the description of a new variety of *Comstockiella* from Mexico, not found by Townsend.

(1.) *Aspidiotus reniformis*, n. sp.— ♀ scale circular, diam. 2 mm., flat, pale reddish-brown ; exuviæ concolorous or slightly darker, covered, but both skins very distinctly visible, large, laterad of the middle. First skin when rubbed shining coppery.

♀.—Reniform, yellow with a brown margin ; the posterior portion large, pale yellow, projecting with the outline of a cone, unusually produced and narrow, the sides meeting at less than a right angle. Pygidium (so-called) minutely striate ; anal orifice oval or subtriangular, a long distance from hind end. Four very small low broad inconspicuous lobes, the plates between them scarcely visible ; these lobes are twice as broad as long, the second about or nearly as broad as the first. Immediately cephalad of the second lobe comes a pair of small diverging spinelike plates; then after an interval somewhat greater (sometimes less) than the distance from the hind end to the plates just mentioned, comes a depression in which is a larger, but still small, pair of diverging spinelike plates; beyond this the margin is distinctly but very minutely serrate, with three small pointed prominences at rather long intervals, and a small rounded notch about half way between the first of these and the largest plates.

There are long tubular glands opening at the bases of the lobes, and also at the place of the obsolete third lobe ; these are three on each side, with others, shorter and smaller, between them. Caudolateral grouped glands a long distance cephalad of the anal orifice. Four groups of ventral glands, caudolaterals 4 to 7, cephalolaterals 8. The antennæ are represented by small tubercles, each emitting a bristle. On each side of the mouth, some distance from it, is a small reniform orifice, its convexity directed laterad.

Hab.—Numerous on under sides of entire, lanceolate leaves, about 60 mm. long. Tehuantepec City, Mexico, May 26th (Townsend ; Div.

Ent., No. 7196). This is related to the subgenus *Chrysomphalus*, and comes nearest to *A. perseæ*, Comstock. It resembles *A. mimosæ* in some respects, but the tubular glands are much longer than in that species, or in *smilacis*. The scale might be taken, at a superficial glance, for *amantii, dictyospermi*, or one of the *uvæ* group, all of which are quite different structurally.

(2.) *Aspidiotus (Hemiberlesia) tricolor*, n. sp.— ♀ scales 1⅔ mm. diam., crowded on twig, approximately circular, very little convex, white with a brownish stain ; exuviæ central or sublateral, covered by a film of secretion, appearing as a blackish spot ; first skin in many examples uncovered, black or dark brown ; second skin rarely uncovered, deep orange. Removed from the twigs, the scales leave a whitish film, quite conspicuous.

♀.—Circular, orange-brown. Only a single pair of lobes, these very large, entire, broad and low, much broader than long, gently rounded at ends, shaped like the end of an axe-blade ; separated by a pair of well-developed spinelike plates. On the margin cephalad of the lobes is a group of five more or less serrate spinelike plates ; then come three very short spinelike plates, after which the margin is more or less, irregularly, crenate. Anal orifice large, oval, distant from bases of median lobes less (sometimes a little more) than its own length. No groups of ventral glands. A few oval glands marking the lines of the obsolete segments. Two small saccular incisions with thickened edges on each side, one immediately laterad of the median lobe, the other cephalad (or laterad) of the obsolete second lobe.

Hab.—Salina Cruz, Mexico, May 29th (Townsend : Div. Ent., No. 7193). Distinguished by its very broad entire lobes, and the orange second skin. It will form with *A. rapax*, Comstock, and *A. ulmi*, W. G. Johnson, a little group, to which the name *Aspidites* is applicable, thus :

Subg. *Aspidites*, Berlese and Leonardi, 1896, s. str.—Scale white or whitish, no groups of ventral glands, only one pair of lobes.

Exuviæ black or at least very dark..................*rapax*.

First skin black or very dark, second orange.. ....*tricolor*.

Exuviæ wholly orange-yellow........................*ulmi*.

A. rapax is the type of *Aspidites*. *A. perniciosus, tenebricosus* and *smilacis*, included in it by Berlese and Leonardi, are not closely related to *rapax*, and should be placed elsewhere. [Since writing the above I have found that *Aspidites* was proposed by Waagen in 1895 for a genus of Cephalopoda ; *Aspidites*, Berl. & Leon., may therefore be changed to *Hemiberlesia*.]

(3.) *Diaspis persimilis*, n. sp.— ♀ scale about 1½ mm. diam., snow white, slightly convex ; exuviæ sublateral, brownish-orange, first skin wholly on second. ♂ scale unknown.

♀.—Circular, orange-brown, hind end strongly striate. Five groups of ventral glands, median 25 or more, nearly touching cephalolateral, cephalolateral about 15, caudolateral 7 to 12. Anal orifice small, caudad of caudolateral glands, but some distance from hind end. Only one pair of distinct lobes, these rounded, not particularly large, very slightly inclined to be crenate on edges, nearly touching at base; at outer base of each lobe a spine ; then a spinelike plate, the branching tips of which slightly exceed the lobe ; then a pair of minute tubercles representing the second lobe, then a spine ; then a very large and stout spinelike plate, branched at tip ; then three minute tubercles, then a spine ; then a spine-like plate resembling the second but not quite so stout ; then a slight notch, followed by a minute tubercle, then on the margin at intervals twelve ordinary spinelike plates of moderate size, and a few spines. At the bases of the median lobes are short dark sacs, a pair to each ; and smaller sacs mark the places of the obsolete lobes on the margin. The oval and elongate glands in rows marking the obsolete segments are comparatively few in number.

Hab.—Crowded on fruit of " Chico Sapote," Laguna, Carmen I., Mexico, April 24, 1896. (Townsend : Div. Ent., No. 7184.) Very near to *D. amygdali (lanatus);* it may be recognized by the small number of orifices in the caudolateral groups of glands, the form of the lobes, and other minor details.

(4.) *Comstockiella sabalis*, v. *mexicana*, v. nov.— ♀ oval, orange-yellow. Grouped glands as follows : Caudolaterals 14–17 (6–10 in type); mediolaterals 11–15 (4–7 in type); cephalolaterals 7–10 (4 in type). Scale as in type.

Hab.—On palms which arrived at San Francisco from Mexico ; found by Mr. Craw, who thinks the palms came from near Maratlan, and were growing wild about 75 or 100 miles inland. The genus is new to the Mexican fauna.

(5.) *Lecanium (Eulecanium) perditum*, n. sp.— ♀. Long. 3, lat. 2 to 2¼, alt. 1½ mm., general shape low-conical or hemispherical ; very dark brown, more or less shiny; sides with linear plications. Boiled in caustic soda turns the liquid yellowish-brown. Antennæ pale, well-developed, tapering, ordinary, 7-jointed, formula 32 (17) 5 (46); 3 extremely long, considerably longer than 4 to 7 together; 2 about as long as 4 + 5; a faint false joint marks the basal ¼ of 3; 1, 2 and 3 each with a pair of bristles, on 1 and 2 about the middle, on 3 near the end; 7 with several hairs, an especially long one, longer than itself, springing from its base. Rostral loop short. Anal plates yellowish-brown, the caudolateral margin somewhat shorter than the cephalolateral. Legs well-developed, pale. Digitules filiform, with large knobs. Tarsus hardly half length of tibia. Derm not reticulated, with sparse small round or oval gland orifices; a broad marginal area with very large round or oval gland-pits, the derm between them exhibiting a faint tendency to minute reticulation. These large gland-pits are double or more often complex; they are often nearer together than the diameter of one.

Embryonic larva (after boiling) pale yellowish-brown; rostral filaments in two coils. Caudal tubercles not or little projecting beyond body margin, though well-developed. Anal ring with six hairs, its broad margin conspicuously striate. Claw long; digitules of claw filiform, distinctly knobbed, extending beyond tip of claw; tarsal digitules stouter, with very distinct knobs, not nearly twice as long as claw-digitules, their origin some distance basad of base of claw.

Hab.—Xcolak, near Izamal, Yucatan, Mexico, March 10th, 1896. (Townsend: Div. Ent., No. 5663.) This is a most interesting species; the first *Eulecanium* ever found in the tropics. The antennæ are like those of *L. antennatum*, Signoret. The compound submarginal glands or pits remind one of the large double glands of *L. Fletcheri*. On the other hand, the large pits of the neotropical species *L. baccharidis* (from Brazil) and *L. batatæ* (from Antigua) are at once suggested, and it seems that we have here an indication of the affinities of these two species, which had been heretofore wholly obscure. *L. perditum* presents some superficial resemblance to small examples of *L. depressum* or *begoniæ*, but these belong to a quite different section.

(6.) *Lecanium chilaspidis*, n sp.— ♀ very dark brown, shiny, but largely encrusted (especially at sides) with a dull dark grayish substance; strongly convex, long. 8½, lat. 6, alt. 5 mm. Beneath, at the lateral (spiracular) incisions, are conspicuous patches of white secretion, only visible after detaching the scale. Younger specimens are flatter, long. 6, lat. 4, alt. 2 mm. There is no waxy secretion on the surface.

♀.—Boiled in soda stains the liquid dark Vandyke-brown. No legs or antennæ found ; probably they are rudimentary and easily deciduous. Anal plates small, pinkish-brown, together forming about a square. Derm pale reddish-brown after boiling, not reticulated, remarkable for an immense number of minute gland orifices, among which are interspersed a lesser number of larger, but still small, glands, which are circular and brown in colour. There are also large brownish patches. In places the tubular ducts of the minute glands are darkened, giving the derm a bristly appearance. The derm may be compared to the sky seen through a telescope, the minute glands being the fixed stars, the larger the planets, and the patches the nebulæ, though of course the sky does not exhibit so many planets or nebulæ.

Embryonic larva (after boiling) very pale pink, with very well-developed, stout, cylindrical caudal tubercles, which are the forerunners of the anal plates ; each emits the usual long bristle, but these are easily broken off. Tarsus hardly or not over ?⅗ length of tibia, femur and tibia approximately of equal length. Digitules all filiform, the tarsal ones very long, twice as long as those of claw, and longer than the tarsus itself. Rostral loop extending considerably beyond the hindmost legs. Anal ring with apparently only six bristles. Last joint of antennæ long.

Hab.—On *Chilaspis linearis*, Tehuantepec City, Mexico, May 26th, 1896. (Townsend : Div. Ent., No. 7216.) On the *Chilaspis* at the same time and place were also taken species of *Aspidiotus* and *Mytilaspis*, but the material is inadequate for proper study. *L. chilaspidis* is a very distinct species, but more nearly allied to other neotropical forms than to anything else.

(7.) *Lecaniodiaspis (Prosopophora) radiatus*, n. sp.-- ♀. Long. 3, lat. 2 mm., often rounder, to long. 2⅗, lat. 2¼ mm., more or less shiny, flattish, pale ochreous, with a longitudinal median keel, low but distinct, and well-defined radiating ribs, marking the segments. Removed from the bark, the scale leaves a whitish mark. Boiled in soda, it turns the liquid greenish. Antennæ pale brownish, apparently 8-jointed, but the joints obscure ; 8 short, buttonlike ; 3 longest, then 4, or these two about equal ; 2 broader than long ; 5 and 6 might be taken for one long joint, fully as long as 3 ; 7 very little smaller than 6. Dermis with numerous very small figure-of-8 glands, which under a low power look like simple oval glands. Mouth-parts large, yellowish. Dermis not minutely wrinkled. Antennal formula (34) (12567) 8. 8 with some bristles, one longer than itself.

Hab.—On bark of branch of some woody plant, Salina Cruz, Mexico, May 29, 1896. (Townsend : Div. Ent., No. 7194.) *L. radiatus* is much more depressed than *quercus*, not marked like *dendrobii*, rounder than *acaciæ*, differently coloured from *eucalypti*, darker, rounder and smaller than *rufescens*, darker and more distinctly radiately ribbed than *yuccæ*. It seems to be very near to *Lecaniodiaspis atherospermæ* (Maskell), by its small size, 8-jointed antennæ, and very minute figure-of-8 orifices ; yet it differs in some particulars, and is, I believe, not the same. *L. atherospermæ* is from Australia, but it may not be a native of that country. Mr. Maskell himself remarks that it is more like the neotropical *dendrobii* than the other Australian members of the genus.

(8.) *Conchaspis Newsteadi*, n. sp.— ♀ scales crowded on the bark, overlapping ; subcircular to oval, dirty white, low conical, diam. 2½ mm. Apex sublateral, no radiating ridges.

♀ oval, orange-brown, similar to *C. angræci* in most respects. Antennæ 6-jointed, joints subequal, variable. Femur longer than tibio-tarsus, coxa about twice as broad as long. The round gland orifices with crenate edges (so strongly crenate as to appear moniliform) are very distinct ; the hindmost segment that shows them is the fourth from the end, this has a pair, close together, on each side. The next segment has on each side four close together, one a little mesad of these, then two at considerable intervals mesad. The next has on each side five in an irregular row, and two pairs at considerable intervals mesad. The next has five and one mesad. The details of the arrangement will differ on the two sides of the same specimen. Long marginal hairs as usual in the genus. Lobes at end of body indistinct.

♂ scale similar to that of the ♀ in texture, but small and elongate.

♂ Pupa red-brown, antennæ stout, of about 7 joints, reaching beyond base of the large rounded wing-pads ; end of abdomen with a short, stout caudal stylus, blunt at tip ; on each side of the last abdominal segment, by the base of the stilus, are three bristles, two very small, one longer.   •

Hab.—On Zuchil tree (*Plumieria*), Vera Cruz, Mexico, Feb. 26th. (Townsend : Div. Ent., No. 7159.) I take the liberty of connecting with this insect the name of Mr. R. Newstead, who, under the name of

Pseudinglisia, has given us the best account of *Conchaspis* yet published. With Mr. Green's Ceylon *C. socialis*, this will make the third species of the genus so far discovered. The ♂ pupa, now described, is very interesting, as it is just like the pupa of a Diaspid.

(9.) *Llaveia axinus* (de la Llave).—Prof. Townsend found at Salina Cruz, on May 27th, specimens of a large monophlelid, which I believe is identical with the imperfectly described *Ll. axinus*. The specimens are red, with mealy powder, and are sparsely marked with small black spots ; dried specimens appear more grayish, and look something like very large *Coccus cacti*. The legs and antennæ are red-brown, the inner side of tibia and tarsus presents a row of short spines, about 11 on anterior tibia, and six, very small, on anterior tarsus. There are two rows of longer spines on the under side of the femur. Dermis rather thickly beset with short hairs. The largest specimen sent to me is perhaps not adult, and has only nine-jointed antennæ. Its dimensions are, long. 13 mm., lat. 6½, alt. 4½ mm. It appears, however, that adults were certainly found by Townsend, as among the material received at Washington were both eggs and young larvæ. Dr. Howard has kindly lent me a mounted larva, from which I have made the following description :

Larva oval, bright red, beset with short, rather stout spines. Seven very long hairs on each side of hindmost half of body, one to each segment, each accompanied by a much shorter and more slender hair, the smaller hair on the penultimate segment longer than its representatives on those anterior to it, and about half as long as the long hair of the same segment. The long hairs of the caudal segment accompanied by two smaller hairs, of which the innermost are the longest. Legs long, femora moderately stout, those of front legs about as long as tibia, of hind legs shorter than tibia. Tibia and tarsus very slender ; tarsus of front legs equal with tibia, of middle legs a little shorter, of hind legs conspicuously shorter than tibia. Claw long, little curved. Eyes very dark, subconical. Antennæ 6-jointed, last joint or club very large, much swollen, longer than 4 + 5, with three whorls of hairs. Second joint a little longer than third, 3 and 4 equal, 5 shortest. The joints from 1 to 4 might be called subequal, and the formula then written 6(2134)5.

I am inclined to suppose that Llaveia and Ortonia will prove to be the same genus, differing at any rate not more than do species now included under *Icerya*.

EARLY STAGES OF BREPHOS INFANS.

BY DWIGHT BRAINERD, MONTREAL.

Eggs laid April 25th, side by side, packed closely together on the twig at fork of leaf bud. The moth standing head downwards with half opened wings and " see-sawing " out a string of from three to twelve eggs. Between times it runs all over the twig as do the Tineids. Egg oblong, rounded at both ends, length .87 mm., width .46 mm. Slightly roughened and punctured like the skin of an orange. Colour at first a delicate pea-green turning yellowish. The number deposited at the base of each leaf varied considerably. Hatched May 3rd to 5th. At birth larva 1.6 mm., semi-transparent, light sap-green with evanescent purple shades. Body cylindrical, of same approximate size throughout, ending in a strongly bifurcate anal segment. Head light yellow-brown ; 1st and 2nd epicranial and 1st clypeal setæ rudimentary, the remaining eleven primary setæ well-developed blunt bristles. Ocelli prominent, dark brown. Shield concolorous. True legs transparent, with dark claws ; 4th prolegs fleshy, rimmed with brown ; the others not showing.

Segments 3-wrinkled, tubercles uniform on the abdominal joints ; a pair each side of dorsal line, a single one above, a pair below spiracles and one above leg plate. Caterpillar a semi-looper, suspending itself by a thread.

Second stage.—3.75 mm. Colour whitish-green. head yellow. Inter-segmental spaces white and much swollen.

Third stage.—Length 12 mm. Sap-green changing to apple-green. Head and appendages, except claws, transparent. Body marked with a double ad-dorsal and a stigmatal white line.

Fourth stage.—Length 30 mm. Colour on dorsum apple-green to blue-green, according to age. Head appendages and venter much lighter ; almost yellowish. Tubercles simple, white, oval to round ; setæ short and spinulate. Ad-dorsal line wavy, obscure, slightly broken. There is a narrow double white line through abdominal segments on lateral surface enclosing a darker area ; and stigmatal band is broad, white to yellow-white. Spiracles red-brown edged with black, set in indistinct white blotches. Body cylindrical, tapering from 12th segment. Pupated June 12th. Food plant white birch. Pupa green at formation, changing to dark chestnut-brown. 14 x 4 mm., smooth. Extremities short, rounded ; medial portion cylindrical, of equi-width ; the whole cocoon approximately oval. Prothorax strongly incised dorsally and pitted. Frontal headpiece convex, hyaline. Maxillæ reach nearly and antennæ fully to extremity of wing-covers. (4th a. s.) Abdominal segments slightly indented down the back. Cremaster with a single stout hook.

Mr. H. H. Lyman kindly measured the eggs, and I had the advantage of Rev. Mr. Fyles's notes on the caterpillar.

Mailed November 5th, 1897.

The Canadian Entomologist.

Vol. XXIX.   LONDON, DECEMBER, 1897.   No. 12.

NOTES ON GRAPTA INTERROGATIONIS, FABR.

BY H. H. LYMAN AND A. F. WINN, MONTREAL.

This species was unusually abundant in this, as in many other localities, during the season of 1896, and afforded an excellent opportunity for studying it, which we took advantage of by rearing it from the egg. The preparatory stages are well known, and a full account of the life history was given by Mr. W. H. Edwards in Can. Ent. XIV., pp. 201-207. As noted by Mr. Edwards, the larvæ vary greatly, and this is true even in those raised from the same batch of eggs, and these variations seem to be in no way connected with the two forms of the imago.

In Mr. Caulfield's List of Diurnal Lepidoptera of the Island of Montreal, published in the Can. Ent. in 1875, this species is called "rare," and its seasons are stated to be "May (hibernated); July to October."

The question as to the number of broods in the season is an interesting one and requires careful examination, but the majority of the authorities are not very clear upon this subject.

Dr. J. G. Morris made no attempt in his "Synopsis" to deal with seasons or broods.

Dr. Harris is not very clear, as he says that the butterfly "first appears in May and again in August and September," and that "the caterpillars come to their full growth in the latter part of August." From these statements it would seem as if he only recognized one annual brood, the individuals of which hibernated and appeared again in the spring; but he says further that "there is probably an early brood of caterpillars in June or July," though he had not seen any on the hop vines before August, but from his remarks on the duration of the pupa stage, viz., "the chrysalis state usually lasts from eleven to fourteen days, but the later broods are more tardy in their transformations, the butterfly sometimes not appearing in less than 26 days after the change to the chrysalis," would seem to indicate that he recognized more than two broods.

Dr. Packard in his "Guide" says of the butterfly : "It is found in May, August, and Autumn," which would not indicate more than two broods.

Mr. W. H. Edwards, who bred this species repeatedly at Coalburgh, says in the CAN. ENT., X., 71, and XIV., 204, that in West Virginia " there are three broods and a more or less successful effort for a fourth." " In Florida," he says, "there are at least four broods, and probably five," but that " in the Northern States, and probably in Canada, it is two-brooded."

Prof. Fernald in " Butterflies of Maine " says nothing of the number of broods, but mentions the dimorphic forms, so he must have recognized that there were at least two broods.

Mr. Scudder in his " Butterflies of New England " says it is double-brooded, the first brood in descent from the hibernators appearing in July, sometimes during the last days of June, and continuing into August, the second brood beginning to emerge towards the end of August and continuing to do so until at least the middle of October.

In regard to the dates at which the hibernators appear in this latitude, Mr. Winn records in his notes April 25, 1890 ; April 14, 1892 ; April 9, 1894 ; and found it quite common in New Brunswick the first week in May in 1896, the specimens seen there being of the form Fabricii. A few Fabricii were seen around Montreal during the latter half of May, but no particular attention was paid to them ; but on the 6th June our Montreal Branch joined the Natural History Society in its annual field day, but separated from the party at Ste. Adele, at which point a number of Interrogationis were seen, and two were taken by one of our members, but both were of the form Umbrosa, though worn, and either hibernators or, perhaps, colonists from the South.

In this connection reference may be made to the experience of Mr. W. F. Fiske, of Mast Yard, N. H., as written to Mr. Lyman, and since then published in the CAN. ENT., XXIX., 26. In this case no specimens of Interrogationis were seen till the middle of May, when a badly worn Umbrosa was observed, and during the rest of the month this form was common, but no Fabricii were seen, and this certainly suggests the idea that these individuals were colonists from the South.

On 13th June our Branch had a little excursion to the Blue Bonnets Swamp, about half way to Lachine, and several Umbrosa were seen and taken. Most of these were worn, but Mr. Winn saw a fresh specimen, and others were seen and one secured on the 14th.

These were evidently individuals of the first brood in descent from hibernators or colonists, and assuming that the eggs were laid during the first week of May, would allow about six weeks from egg to imago, which corresponds with the experience of Mr. Edwards with the first brood in West Virginia, which took 37 days—28th April to 4th June.

On 14th June Mr. Winn also observed two very much worn Fabricii ovipositing on the young leaves of an elm. This late laying of eggs causes the broods to overlap and makes it almost impossible to tell to what generation any captured specimen belongs.

From the 15th to the end of June Umbrosa was quite common, but no more Fabricii were seen. On 24th a number of larvæ, apparently not more than a day old and quite close to the empty egg-shells, were found, and on 25th about 40 eggs and seven young larvæ were found on a bunch of elm leaves plucked at random. These produced the imagos between 19th and 29th July and were 31 Umbrosa and two Fabricii, and were doubtless part of the second brood of the season.

On 1st July Mr. Lyman took at Lachine a ♀ Umbrosa and confined it over leaves of elm, but no eggs were laid for over a week.

On 12th July the butterfly was found to be dead, but had laid 101 eggs, some almost ready to hatch and some just recently laid.

The eggs began hatching that same evening and others continued to hatch during the 13th and 14th. Some of the earliest to hatch passed first moult on the 15th, the third day from the egg. The first chrysalis was formed on 5th Aug., and the first imago emerged on 13th Aug., giving a pupal period of eight days, a period from hatching of egg to imago of 32 days, and a probable period from oviposition to imago of 35, or, at the outside, 36 days.

Some, of course, took a few days longer than this, but all had emerged by the 21st August. Of nearly 60 butterflies which emerged, not more than five were Fabricii, all the others being Umbrosa.

Now it seems clear that the parent butterfly which was taken on 1st July, but would not lay till 8th or 9th, must have belonged to the first brood in descent from the hibernators or colonists, whichever the early ones were, and that the brood thus reared represented the second brood, and there would be abundance of time after the 21st August for a third brood to mature. That such a third brood must exist is practically proved by the fact that the second brood as raised by us was almost entirely composed of the form Umbrosa, while it is well known that Fabricii largely predominates in the autumn, which would not be the case if there were no third brood.

On 26th July, while Mr. Winn's second brood was emerging, he confined a ♀ Umbrosa on elm and obtained eggs the same day, which hatched on 30th. Others were caged on 28th and five more on 2nd Aug., on hop, and many eggs were obtained. Some were left on the food plant, but the others were taken on a holiday trip to Metis, Q., the last hatching 7th Aug. On Aug. 24th the first chrysalis was formed, and imago emerged 4th Sept. and proved to be Fabricii, but at the same time a number of the larvæ were just past the third moult. While at Metis the larvæ were fed on hop, as elm trees were not found, and when brought back to Montreal were again fed on elm.

Either from this change of diet or from the colder climate of the lower St. Lawrence, the majority of this brood were greatly retarded and emerged at intervals all through September, and one as late as 18th Oct. Of nineteen individuals seventeen were Fabricii and two Umbrosa.

One fresh Umbrosa was also seen on 16th Sept., and Fabricii was common on the fine days of the early part of that month.

This makes the third brood, with a varying preparatory life duration of 40 to 77 days.

With Mr. Edwards the period of the third brood varied from 31 to probably over 50 days.

In nature the oviposition of the various broods would doubtless be extended over a longer time and the emergence of the imago similarly spread out, but when a species can go through all its changes in from 31 to 36 days it stands to reason that there must be at least three broods in the season in this latitude.

The third brood must certainly hibernate, and Mr. Winn found that those flying in September did not seem inclined to lay eggs, and careful search failed to produce a single one.

In Can. Ent., X., p. 72, Mr. Edwards states his belief that the scarcity of hibernators in the spring compared with the abundance of the species in the summer is due to the existence of the species being dependent on the partial fourth brood, which he considers the only one that hibernates, and states that the species does not suffer from parasites to any extent.

This statement, published in April, 1878, is strikingly at variance with his former notes upon this species in part 9 of Butt. N. A., I., issued in January, 1872, pages 117–118 of the volume, where, after recounting the large number of enemies which prey upon it, he says, "It is doubtful if much more than two per cent. of the eggs laid produce butterflies."

Mr. Winn collected early in September from off the fence over which his hop vine grew 32 chrysalids, being the result of the eggs laid 3rd and 4th August, which he had left upon the vine. From these only two butterflies emerged, both on 18th September, and, curiously enough, one was a ♂ Umbrosa and the other a ♀ Fabricii. All the others were attacked by parasites, which Mr. W. H. Harrington determined as Pteromalus puparum, Linn.

The following notes upon the eggs were made by Mr. Lyman :

In regard to the colour, number of ribs, etc., of the eggs, there is considerable divergence among the authorities.

In regard to the colour, Scudder, quoting Riley, says that at first they are dull bluish-green, afterwards becoming grayish-green with silvery reflection. Edwards and Fernald call them "pale green," and this I consider correct, as I could see no trace of blue-green about them. Edwards says that the eggs have eight or nine vertical ribs, and is followed by Fernald. Edwards also says that the eggs laid in strings have always the same number of ribs, and hence Scudder deduces the theory that individual butterflies always lay eggs of the same number of ribs, but the latter author gives the number of ribs as "nine to eleven, commonly ten."

Of the 101 eggs laid by my butterfly in confinement, 24 were laid on the leaves, 3 being above and 21 below, and the rest, except 2, on the gauze.

There were ten strings of two, four strings of three, one pyramid formed of two below and one above, and another formed by one standing upright upon one on its side, and sixty-four singles. Some of the strings were very irregular, and some had apparently been laid at different times.

Of 52 eggs examined, 31 had 9 ribs and 21 had 10. One of 9 ribs, with larva nearly ready to hatch, had a green newly laid egg with 10 ribs on top of it.

In striking contrast to its abundance in 1896, only one specimen of this butterfly was seen during the season of 1897 by Mr. Winn.

NEW SPECIES OF CHIONASPIS.

BY R. A. COOLEY, B. S., AMHERST, MASS.

At the request of Prof. T. D. A. Cockerell, through correspondence with Prof. Fernald, I was induced to take up the study of the genus *Chionaspis,* and Prof. Lull the genus *Pulvinaria.* Prof. Fernald prepared and sent out a circular letter to all entomologists whose addresses could be obtained, in this and other countries, and personal letters were also sent to the leading coccidologists, asking for as many species as possible to aid in the preparation of monographs of these two genera. The result has been most gratifying, for already a very large amount of material has been received.

In the material before me the following new species of *Chionaspis* have been found, and are published now in preference to waiting till the monograph is issued. The studies on these insects are being made in Prof. Fernald's entomological laboratory connected with the Massachusetts Agricultural College, where every possible facility is afforded for breeding and studying insects, together with very complete literature of the subject.

Chionaspis Cockerelli, n. sp.

Scale of female.—The female scale is about 3.2 mm. long, straight or very slightly curved, moderately thick in texture, slightly convex, white, with the exuviæ pale yellowish-brown, the second skin being covered with secretion.

Female.—The pygidium is distinctly notched at the end, the sides of the notch being formed by the divergent median lobes. These lobes are firmly united at the base and have serrate edges. Two distinct parallel spines arising from the bottom of the notch are about as long as the distance between the inner edges of the lobes at the base. Compared with the other lobes of the pygidium the median ones are larger and extend farther into the body. Each lobe of the second pair is composed of two well rounded and distinct lobules, the incision between them extending to the base of the lobe. The inner lobule is larger and extends posteriorly about even with the median lobes. The third pair of lobes may be present or aborted ; when present they are broad and low, with an elongated pore anterior to the base of each. Between the median and second pair is a minute spine, followed by a plate which is about as long as the second pair of lobes, and following these is a conical projection bearing a marginal pore. Outside of the second lobe is a spine, a plate and a marginal pore, this plate being a little larger than the first one. Following the third lobe, when it is present, or a space when it is absent, there are two spines, one above and one below. These are followed by a plate and a distinct marginal pore, and after an interval interrupted by one or two spines, another plate, and following this another interval, terminated by a group of about three plates.

The spinnerets are in five groups : median, 7–9 ; anterior laterals, 17–23 ; posterior laterals, 23–34.

Described from dead and shrunken specimens.

Scale of male.—Length, 1.2 mm. ; feebly carinated, white, with the larval skin almost colourless.

Described from a single imperfect specimen.

Male.—Male insect unknown.

The specimens were taken by Mr. Alexander Craw, on palm imported from China to San Francisco, Cal., July 11, 1897.

I take pleasure in naming this insect after Prof. T. D. A. Cockerell, who has made extensive and valuable contributions to our knowledge of the Coccidæ, and has shown me many kindnesses in my work on this group of insects.

Chionaspis aucubæ, n. sp.

Scale of female.—The female scale somewhat resembles that of *Chionaspis Lintneri* in outline, being strongly broadened posteriorly and abruptly rounded at the extremity. It is moderately convex, about 3 mm. in length and about 2 mm. in width. The exuviæ at the apex of the scale have the first skin very pale yellow, and the second yellowish or brownish. The second skin is covered with a slight secretion. The scale itself is white and very thick and strong. There is a partial ventral scale at the anterior end.

Female.—As I had only dead and dry specimens of this insect, I made no attempt to describe anything but the pygidium of the female. Median lobes moderate in size, divergent, united at the base, with their inner edges distinctly serrate. Each lobe of the second pair is composed of two rounded lobules, the incision between the two reaching nearly or quite to the base of the lobe. The inner lobule is larger and projects farther posteriorly than the outer, sometimes surpassing the median lobes. The third lobe is simple and sometimes rudimentary. Between the bases of the median lobes is a pair of minute convergent spines. On each side between the median and second lobes are a spine, a plate and a marginal pore, and between the second and third lobes two spines, one above and one below, followed first by one or two plates, and then by a conical projection bearing a marginal pore. Outside of the third lobe are a spine and from one to three plates, then a slight notch, immediately followed by a marginal pore and after a space two unequal spines and about three plates. Following these plates are a notch and a marginal pore, then after a space a group of about five plates.

Spinnerets arranged in five groups : median, 8–14 ; anterior laterals, 19–28 ; posterior laterals, 19–33.

Scale of male.—The male scales are much more numerous than those of the female. They are white, delicate in texture, about 1.2 mm. in length, the larval skin at the anterior end being colourless or slightly yellowish. The scale itself may be parallel sided or slightly broadened posteriorly, and is indistinctly carinated.

Male.—Male insect unknown.

On Aucuba from Japan. Discovered by Mr. Craw in the course of his quarantine work at San Francisco. The scales are grouped together on one side of the leaf beneath, and the edge of the leaf is folded under, almost completely hiding them from view.

Chionaspis wistariæ, n. sp.

Scale of female.—The female scale is about 2 mm. in length, though some specimens are slightly longer, moderately broadened, dirty white in colour and delicate in texture, being a close imitation of the epidermis of the bark on which it rests. The scales usually occur in the longitudinal cracks of the bark, and are partially concealed under the epidermis. They are very often pressed out of the normal form. The exuviæ are brownish, and the second skin is covered with secretion.

Female.—The following description of the female was made from dead and shriveled insects. The median pair of lobes is large and conspicuous, the second pair considerably smaller, and the third pair obsolete. The median lobes are darker in colour than any other part of the pygidium, firmly joined at the base, their inner edges parallel and nearly touching each other for about half their length, then diverging at about a right angle, with the exposed edges serrate. The second lobe is composed of two lobules, the inner one being the larger. Within the outer edge of each of the median lobes is a spine, and next to this a short blunt plate, followed by a marginal pore. Between the lobules of the second lobe is a spine, and outside of the second lobe are a plate and two marginal pores, followed first by a spine and then by a plate, which is about as long as the median lobes, and often forked at the tip. Outside of this plate are two marginal pores, followed by a spine and one or two plates, then after another marginal pore a group of about four plates.

There are five groups of spinnerets : median, 8–15 ; anterior laterals, 19–31 ; posterior laterals, 13–23.

Scale of male.—The male scale, as in all other species of this genus, is elongated in form and white in colour. The sides are nearly parallel, and it is distinctly tri-carinated. Length, about 1 mm. The larval skin resembles the anterior or smaller one of the female scale.

Male.—Male insect unknown.

Dicovered by Mr. Craw, July 8, 1897, at San Francisco, on the bark of Wistaria from Japan.

Chionaspis pinifoliæ heterophyllæ, n var.

Scale of female.—The scale of the female is indistinguishable from that of *pinifoliæ*, Fitch, having the same range of form and size, the colour of the scale and exuviæ being the same. The scales vary in size from 2 mm. to 3.4 mm., the average length being about 2.5 mm. The scale is white, strongly convex, with the exuviæ at the anterior extremity yellow, both skins being naked.

Female.—The description of the female is made from dead and shriveled specimens. At the anterior end of the body are two distinct, curved bristles, which may be the rudiments of the antennæ ; these are found also in *pinifoliæ*. The last segment terminates in a median notch, the sides of which are formed by the divergent median lobes. The lobes of the second pair are low and inconspicuous, and each one is composed of two lobules of about equal size. Two minute spines, one above and one below, arise from near each median lobe, though back from the edge of the segment. Contiguous to each median lobe is a simple plate, outside of which is a marginal pore. Between the lobules of the second lobe is a distinct spine, and outside of this lobe is a plate with a spine at its base, followed by a marginal pore. Outside of the rudimentary third lobe is a marginal pore, followed by a spine and a plate with a spine at its base. Then follows a pronounced marginal pore, a short interval, another space and a long interval, interrupted only by a spine, and terminated by the fourth and last plate.

There are five groups of spinnerets : median, 4-8 ; anterior laterals, 12-18 ; posterior laterals, 14-16. The chief characters by which *pinifoliæ* and the variety can be separated are the presence of the median notch in the variety and the larger size and more rounded form of the lobes in *pinifoliæ*.

Scale of male.—The male scale cannot be distinguished from that of *pinifoliæ*. It is slightly more than 1 mm. long and .4 mm. wide at the posterior end, where it attains its greatest width. The scale is white, with a moderately distinct median carina. The larval skin is like the first one of the female.

Male.—Male insect unknown.

On Cuban pine, *Pinus heterophylla*, from Florida. I am indebted to Prof. A. L. Quaintance for a bountiful supply of specimens, as well as to Prof. Cockerell, who first called my attention to this insect and sent me specimens.

The scales are found chiefly at the bases of the very long, slender leaves, and mostly on the inner surface. A few specimens occur also on the stems of the new growth. There were circular openings in a few of the female scales, from which parasites had emerged.

The following original description, which has never been published, was sent to me by Prof. Cockerell to be added to this paper :

Chionaspis latissima, Ckll.

C. latissima, Ckll., Calif. Fruit Grower, June 5, 1897, pp. 4–5. (Descriptive note ; no full description.)

" Female scale circular, 2 mm. diam., white, semitransparent, with the light ocreous exuviæ to one side, first skin half overlapping second, second skin oval. Eggs shining, pale pink.

" ♂ scale linear, white, with a very feeble median keel.

" ♀ when boiled in caustic soda turns yellow, marbled and suffused with bright blue-green ; the mouth-parts remain a warm brown. Under pressure the ♀ becomes greatly elongated. Anal orifice level with the lower (caudad) edge of the cephalolateral glands. Five groups of ventral glands, median of 8, cephalolaterals of 18, caudolaterals of 20. Lateral dorsal rows of elongated pores. General characters of *chinensis*, *nyssæ*, etc. Differs from *chinensis* by the median lobes being not or barely brownish, and being decidedly produced, and the second and third lobes each represented by three distinct lobules. The lobes are much more produced than in *nyssæ*. The spinelike plates are large. The scale is very similar to *vitis*, Green, but is smaller than that or *varicosa*, Green.

" On under sides of leaves of *Distylium racemosum*, from Japan, found by Mr. Alex. Craw, April, 1897, in the course of his quarantine work at San Francisco."

PREPARATORY STAGES OF PYRUS TESSELLATA, Scud.

BY G. H. FRENCH, CARBONDALE, ILL.

Egg.—Diameter, .02 inch. Blunt conical, height about the same as the diameter; ridged with about 30 longitudinal striæ, with shallower cross striæ. Colour pale green. Duration of this period six days.

Young Larvæ.—Length, .08 inch; cylindrical; head somewhat cordate, two-thirds the width of the body; the anterior part of joint 2 about one-half the diameter of the head, the posterior part as wide as joint 3; each joint back of 2 with four low transverse folds besides the very narrow fold at each end of the joint. Colour pale greenish with a white sheen; piliferous spots concolorous; hairs erect, forked to about the middle, the forks curving back towards the body anteriorly and posteriorly. These are the hairs from the piliferous spots. Hairs on the body black, hairs on the head and joint 2 white and not furcate. Head jet black; joint 2 pale yellow-brown with a black transverse bar just back of the middle of the joint; dark along the sides; thoracic feet black. There are eight hairs in pairs on the dorsal bar of joint 2. Duration of this period two or three days.

After first Moult.—Length, .15 inch. Shape not materially changed. Head and joint 2 jet black; hairs all white, shorter than before, more numerous, the end capitate instead of bifid; head and neck corrugated. Duration of this period six days.

After second Moult.—Length, .40 inch. Marked as before; hairs still capitate, white; a dorsal and subdorsal line a little more plainly green; head and joint 2 profusely hairy, but the hairs are all short, surface corrugated. Previous to this moult the larvæ mostly lay coiled on the surface of the food plant, but now they straighten out under a thin silky covering. Duration of this period four days.

After third Moult.—Length, .50 inch. Cylindrical, head about the same width as the body; black, covered with white hairs, each of which has about six short side spurs from about the middle up; joint 2 black, with the dorsal bar red-brown with a whitish margin; hairs on this joint of two kinds, short and long, the long about one-sixth the width of the body in length and very shallowly trifid at the end; body, each joint with five folds, the anterior twice the width of the others; two forms of hairs, one very short and the other long, each long one about the length of those on joint 2 and arising from a white conical base, trifid at the outer end; the short ones arising from a shorter cone and capitate at the end. Colour of the body pale yellowish-green, a more distinct dorsal and subdorsal stripe and the subdorsal space with mottlings of darker green ; stigmata sordid white. Duration of this period four days.

After fourth Moult.—Length, .85 to .95 inch. About the same as during preceding period, but the head hairs have a brown tinge, and joint 2 is brown, with a whitish dorsal line, and sometimes subdorsal also. Duration of this period seven days.

Pupa.—Length, .55 inch ; diameter, .15 inch. Nearly cylindrical ; from the head to the posterior part of the wing-cases .37 inch, these extending nearly to the posterior edge of joint 5 ; body pretty well covered with short, simple, white hairs ; head rounded, eyes rather prominent, a prominent tuft of hairs between them (frontal hairs they might be called), another anterior tuft on the inner edge of margin of eye, more on the outer margin, while the space around the eye between these tufts is without hairs. Colour gray ; head gray with a greenish tinge except on the eye-space ; dorsal part of thorax gray with a slight green tinge, three transverse rows of small black spots, the first, one on each side, subdorsal ; the second row six, one each side of a very slight green dorsal line and one on each side of what would be a subdorsal line if such were present, a little anterior to the others ; third row six, one on each side of the dorsal line and one outside and a little anterior to this and one on the shoulder of hind wings. Spiracle just back of the eye large, elevated, dark or Vandyke brown, the outer portion pale. Wing-cases green, ribbed as usual, mottled slightly in two shades, but not strongly contrasting ; abdomen with each joint gray (the gray of the whole pupa a more sordid white with a gray tint, as there is none of the dark gray about it), slightly green tinted, the incisures more distinctly pale green, each joint with its row of small black spots across the middle, supplemented back of the row with a less perfect row of smaller spots, the first row of six spots, of which the outer spot on each end of the row is the black spiracle ; cremaster brown, elongated hooks at the end that fasten into a thin, loose button of silk. Duration of this period eight days.

The larvæ, when ready to pupate, folded a leaf together and loosely fastened it with silk, but there was no lining of silk except a small, thin button to which the cremaster was attached.

The eggs were sent me by Dr. C. Hoeg, of Decorah, Iowa. At first he sent me two eggs under date of July 31st, 1897, that he had found on

Malva rotundifolia. These did not hatch on account of injury in transit, and under date of August 6th he sent me fifteen more, found on the same plant. These hatched out August 12th. I fed them first on a species of Abutilon, but as they did not take to that readily, though eating it a little, I changed to *Althea rosea.* I think they will eat any of the rough-leaved Malvaceæ readily. On account of being away from home part of the time, the larvæ were somewhat neglected in the last stages, but notwithstanding this two passed through all their stages, producing the first imago September 12, 1897.

THREE INTERESTING STAPHYLINIDÆ FROM QUEEN CHARLOTTE ISLANDS.

BY REV. J. H. KEEN, MASSETT, B. C.

At the request of Dr. James Fletcher, I am writing a few notes to accompany the three figures which have been made at his instance, and kindly presented by him. They represent three *Staphylinidæ* taken by me at Massett, on the Queen Charlotte Islands, and were prepared under the direction of Mons. A. Fauvel, the well-known specialist in that order, to whom also I am indebted for the determination of the beetles themselves.

Haida Keeni, Fauvel. New gen. and new sp. (Fig. 34.)

For this interesting little *Homalien,* M. Fauvel found it necessary to construct a new genus, and his description will be given in full as soon as it

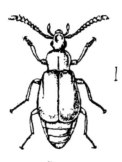

comes to hand. The beetle is of a dark reddish-brown colour, much brighter on the elytra, which have a broad transverse band of black posteriorly. It varies a good deal in size and also in the depth of its coloration. It is found throughout the year, but is most abundant in September, when it frequents rotten leaves on the ground, and seems to have a preference for elder leaves. In winter it occurs in moss about the roots of spruce and other trees. It is somewhat sluggish in its move-ments and feigns death for a minute or more on

FIG. 34.

being disturbed. I have not yet succeeded in taking it on the mainland, though it is fairly common at Massett.

Liparocephalus brevipennis, Mökl. (Fig. 35.)

This submarine species is very abundant on the shores of the Q. C. Islands and not uncommon on the mainland opposite, though until I took specimens of it at Massett it was known only, Dr. Fletcher says, by the type specimen at Washington. The insects are found crawling over barnacle-covered stones and boulders near low-water mark. Occasionally they occur congregated in a mass of several hundreds under a single stone, but for what purpose I have been unable to discover. It is most abundant in autumn.

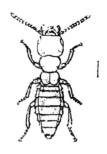

FIG. 35.

From some experiments I made with several specimens in a dish of salt water in which was a half-submerged stone, I observed that they cannot swim under water, but merely crawl on the stones, their pubescence enabling them to surround themselves with minute bubbles of air. They could not be induced to enter the water from the top of the stone. If forced to leave the stone they would swim on the surface, but seemed incapable of diving. If touched while on the side of the stone under water, however, they feigned death, and had the power of sinking readily to the bottom. Some that were left all night swimming on the surface of the water were found dead in the morning, while others which had been submerged all night were still active.

A question has been raised as to whether *L. cordicollis*, Lec. (exactly similar to the present species in form, but with the head and thorax brown), is anything more than a colour variety. I have watched both with this point in view, and speaking merely as a field observer, my belief is that they are separate species. *L. cordicollis* is the rarer of the two, but when it occurs it is in little colonies. I know, for instance, one large boulder where almost at any time I could take fifty specimens of *cordicollis*, but where I have never yet seen *brevipennis*. I have, moreover, never seen one of each *in coitu*, though pairs of one or the other are commonly met with. I may add that my view seems to receive slight confirmation from the fact that three other species of submarine beetles occur at Massett with black abdomen and limbs, but with brown head and thorax. On the other hand, however, I have noticed that the brown of *cordicollis* darkens considerably with keeping.

Tanyrhinus singularis, Mann. (Fig. 36.)

This curious insect seems to be rare in collections, for neither Mons. Fauvel nor the late Dr. Hamilton possessed a specimen till they received

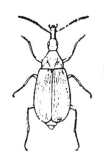

FIG. 36.

one from me. Mr. L. O. Howard, however, tells me he has a good series in the National Museum at Washington. It is by no means common at Massett, for I have only taken nine in seven years, and never more than three in one year. It has occurred always in the same spot — on the under side of a rotten spruce log on the ground. From positions I have taken it in I conclude that it feeds either on the rotten wood or on minute fungoid growths on the wood. On one occasion I obtained two specimens by pouring water into the log, which is now soft and fibrous with age, when they emerged from holes. The insect is slow and deliberate in its movements, and makes no attempt to fly when disturbed. It has occurred only in early spring ; several of my specimens were taken in the middle of February when snow was on the ground.

ON THE GENERIC POSITION OF SOME BEES HITHERTO REFERRED TO PANURGUS AND CALLIOPSIS.

BY T. D. A COCKERELL, MESILLA, N. MEX.

Having lately received from Mr. Friese, of Innsbruck, a number of European bees, I have been led to re-examine certain of our species, in order to determine their relationship to a number of old-world genera not supposed to occur in America. The result is extremely interesting, and seems to show that we have for many years been placing bees in genera to which they by no means belong. The following table may be used provisionally to separate the genera under discussion* :

A. Tongue more or less short and broad, tapering at the end. (Andreninæ).

   1. Basal nervure nearly or quite straight.

      a. Three submarginal cells................ ....*Andrena*, Fabr.

      b. Two submarginal cells................. *Parandrena*, Rob.

   2. Basal nervure strongly bent.

      a. Three submarginal cells.............. .... *Halictus*, Latr.

      b. Two submarginal cells............... .....*Hemihalictus*, Ckll.

*Mr. Friese sends me also four examples of *Nomioides pulchellus*, Schenck, taken at Pest on the second of June. This bee is a *Perdita* with the venation of an *Halictus!* It is curious to see all the ornaments, sculpture, etc., of *Perdita*, with a long tapering marginal cell and three submarginals. It is evident from this, and from the absence of *Perdita* in the American tropics, that our genus is of boreal origin, not austral, as I formerly thought.

B. Tongue narrow and more or less elongated, usually quite long. (Panurginæ).

   1. Marginal cell produced, tapering to a point, not appendiculate.

     a. Body *Colletes*-shaped, abdomen with hair-bands. *Rhophites*, Spin.

     b. Body *Halictus*-shaped, abdomen without well-formed hair-bands............ .... *Halictoides,* Nyl.

   2. Marginal cell truncate at tip, usually appendiculate.

     a. Body *Colletes*-shaped, abdomen usually with hair-bands .... *Calliopsis*, Sm.

     b. Body *Halictus*-shaped, abdomen without well-formed hair-bands............................. . *Panurginus*, Nyl.

The genera under B have but two submarginal cells ; those under A all have a marginal tapering to a point. I give the subfamilies as I find them, but it seems at least probable that the form of the tongue is an adaptive character, not to be relied upon for separating groups higher than genera. The Panurginæ, notwithstanding the tongue, appear to be certainly Andrenidæ.

Parandrena.

The type is *P. andrenoides*, a spring-flying species. The smaller stigma of the autumnal " *Panurgus*" *pectidis, rhodoceratus* and *oliviæ* is paralleled in *Andrena* by that of *A. pulchella*, also an autumnal insect. For the present I would place the three species of " *Panurgus* " named in *Parandrena*, with the reservation that they may hereafter need to be separated from it. They are much nearer to *Rhophites* than to *Panurgus*.

Hemihalictus.

The type is *H. lustrans*, described as *Panurgus*. This looks not unlike the European *Halictoides*, but differs in the tongue, which in *Halictoides* is very narrow, and by the strongly bent basal nervure and the third discoidal cell considerably narrowed above.

Rhophites.

Mr. Friese sends me *R. quinquespinosus*, Spin., and *R. canus*, Ev. These are what we should call *Panurgus*, and if there are in our fauna any " *Panurgus* " with the pointed marginal cell, of fairly robust shape, with abdominal hair-bands, these will belong to *Rophites*, provided they have the narrow elongated tongue which separates them from *Parandrena*. The stigma of *Rophites* is small, as in the autumnal species provisionally referred above to *Parandrena*.

Halictoides.

Many authors have confused this with *Rophites*, but it is fairly distinct. I have before me the following species :

H. paradoxus, Moraw.—Innsbruck, July 15th ; Sept. 13th, at *Euphrasia.* Coll. Friese.

H. dentiventris, Nyl.—Andermatt, July 9th ; " Weissnfls," Aug. 3rd; Sept. 2nd, at *Campanula,* Coll. Friese.

H. inermis, Nyl.—" Weissnfls," July 13th, at *Campanula.* Coll. Friese.

H. marginatus (Cress., as *Panurgus*).—My New Mexico insect has stood as *halictulus*, Cr., but according to Robertson that is identical with *marginatus.* It flies in August and September.

H. campanulæ, n. sp.— ♂ . Length, 9 to 10 mm. Black, shiny ; pubescence sparse ; pale cinereous, mixed with black, on head and thorax ; black, with a little cinereous, on abdomen and legs. Hair on inner side of tarsi shining orange-fulvous. Head large, very broad, a little broader than thorax, subquadrate, facial quadrangle very much broader than long, anterior edge of clypeus with a hoary fringe, clypeus and front appearing rough from very close punctures, mandibles with a well-formed inner tooth, antennæ crenulate, flagellum feebly tinged with ferruginous beneath ; mesothorax shiny, with distinct, rather close punctures ; enclosure of metathorax coarsely rugose ; tegulæ piceous, with a hyaline band ; wings smoky, nervures and stigma piceous, first recurrent nervure joining second submarginal cell considerably nearer its base than the second recurrent to its apex ; second to fourth joints of hind tarsi broadened, triangular ; abdomen shining, the surface appearing silky, hardly punctured ; no hair- or colour-bands ; sides of segments towards apex with tufts of black hair ; apex conspicuously tufted with more or less shining sooty hair ; a large tuft of sooty or black hair also arises from the sixth ventral segment, and is very conspicuous when the insect is viewed from the side. Tongue narrow.

Hab. — Four from Olympia, Washington State, June 30 ; all at flowers of *Campanula scouleri.* (T. Kincaid, coll.)

How many more of our so-called *Panurgus* will be found to belong to *Halictoides* I do not know, but it is probable that an examination of the types will show that we have at least as many *Halictoides* (six) as are known from the other side of the world.

Panurgus.

Taschenberg (" Die Gattungen der Bienen") separates *Panurgus* from *Rhophites* by its *truncate, appendiculate, marginal cell.* Three European species, now before me, all exhibit this character, which is generic. It therefore follows that none of the so-called *Panurgus* of Cresson's 1887 Catalogue belong to that genus. So far as known, we have no typical *Panurgus* in North America ; two *Panurgus*-like forms may be referred to a new group, thus :

Pseudopanurgus, n. g.

Type *Ps. æthiops* (Cr., as *Panurgus*). Includes also *Ps. fraterculus* (Ckll., as *Calliopsis*). Black, nearly naked, strongly punctured, wings fuliginous, marginal cell distinctly but obliquely truncate at tip, two sub-marginals, *first recurrent nervure joining second submarginal cell no great distance before its middle, second recurrent joining it just before its tip, basal process of labrum large, subquadrate.* In some respects this seems to resemble Provancher's *Chelynia* (which I have not seen), but it is surely not the same thing.

Panurginus.

Mr. Friese sends me *P. montanus*, Gir., collected at Airolo, Andermatt, and Innsbruck. It flies at the end of June and beginning of July ; one specimen is marked as from *Ranunculus.* The clypeus is yellow in the ♂, dark in ♀. To this genus belong *Panurginus clypeatus* (Cr.), *bidentis* (Ckll.), *margaritensis* (Fox), *compositarum* (Rob.), *albitarsis* (Cr.), *ornatipes* (Cr.), *rudbeckiæ* (Rob.), etc., all now referred in our lists to *Calliopsis.* The European *P. montanus* has the venation of our *P. clypeatus.*

Calliopsis.

This name can be retained for such species as *C. andreniformis*, *coloradensis, obscurellus*, etc. There also remain some forms which must be left in *Calliopsis* until a better place is found for them, although they seem scarcely congeneric with *andreniformis.*

Dr. Harrison G. Dyar has removed from New York to Washington, D. C., where he has accepted the position of Honorary Curator of Lepidoptera in the United States National Museum.

Mr. Arthur J. Snyder, of Evanston, Ill., has recently been appointed Principal of the North Belvidere Schools. His address is now 521 East Madison street, Belvidere, Ill.

A LIST OF MANITOBA MOTHS.

BY A. W. HANHAM, WINNIPEG, MAN.

The following list of Manitoba species, it is hoped, will prove of interest to readers of the CANADIAN ENTOMOLOGIST :

With a few exceptions, the records are from my own observations and captures. The list covers the work or collecting of three whole seasons and the latter half of a fourth ; and it is to a great extent a local one, very little collecting having been done outside of the Winnipeg district.

Last season (1896), in July, and again this year, in August, I was fortunate in being able to visit Brandon, Man.—some 130 miles west of Winnipeg—where, especially during my first visit in July, I enjoyed some very successful collecting, and I am thus enabled to add a considerable number of things to my list, many of them very desirable species.

I believe a comparison of collections made at Brandon and at Winnipeg would show some striking differences, many of the Western forms occurring at Brandon not reaching so far east as Winnipeg. This district embraces some open " rolling " prairie, a good deal of swampy land covered with willow and other bushes, plenty of thick " bush " containing no trees of any size, a little fine timber, mostly elm, along the river " bottoms," and a gravel ridge many miles in extent, more or less wooded, with some sandy tracts, commencing at Bird's Hill, some eight miles from this city.

The last described locality much resembles the general run of country around Brandon, and after Elm Park, situated in a bend of the Red River, about three miles out of Winnipeg, is much the richest collecting ground within the district. The Province of Manitoba contains numerous lakes, some of vast area, as Lakes Winnipeg and Manitoba ; none, however, come within this district, nor have any yet been visited.

The list of Sphingidæ is but a meagre one, and I think hardly representative of the district ; certainly not of Manitoba as a whole. Nearly fifty per cent. of the Bombycidæ recorded were added this year, and they were, without exception, taken "at light," at the end of June and during July. But for this my list in these too would have been equally poor.

Mr. E. F. Heath, who lives near Cartwright, in Southern Manitoba—a much better country than this, I believe, for the entomologist—could, I feel sure, supplement these records.

Most of the Bombycidæ have been submitted to Dr. H. G. Dyar, to whom I am under special obligations for his generosity in returning nearly everything sent to him (a large proportion being "uniques"). The Sesiidæ were very kindly determined for me by Mr. Beutenmüller, through Dr. Dyar. I have also received welcome assistance from both Dr. James Fletcher and the Rev. C. J. S. Bethune.

Hemaris thysbe, *Fabr.*, var. ruficaudis, *Kirby.*—In 1894 (June 17th) several were seen hovering over the blossoms of the Wild Pea, but only one was secured. Later a number were noticed (dead) in the windows of some empty shops.

Met with again this season.

Hemaris tenuis, *Grt.*—On April 19th (1897) a pupa was found in the soil under a log along the railway line at Brandon. The moth evolved on May 18th.

Deilephila gallii, *Rott.*, var. chamænerii, *Harr.*—In the collection of Mr. H. W. O. Boger, of Brandon.

Deilephila lineata, *Fabr.*—Mr. Boger reported this moth as being very abundant, on the wing, on August 25th (1896), in a market garden at Brandon, in the evening, as many as 20 or 30 being visible at the same time. It occurred here about the same date. On August 5th (1894), quite early in the afternoon, and in the bright sunshine, I noticed a Deilephila on the wing over some thistles on the prairie, but I failed to net it.

Sphinx drupiferarum, *S. & A.*—At Brandon (1897), by Mr. Boger.

Sphinx luscitiosa, *Cram.*—On July 1st (1895) I found a ♀ at rest under the loose bark of a fence post, and on June 11th (1896) a fresh ♂ was found "sitting" on a sidewalk in the heart of the city.

Sphinx chersis, *Hbn.*—July, one in a shop window, also at Brandon, in Mr. Boger's collection.

Sphinx albescens, *Tepper.*—July 1st, one at light ; another taken at Rounthwaite, Man., by Mr. L. E Marmont.

Ceratomia undulosa, *Walk.*—July 8th, one at light,

Smerinthus geminatus, *Say.*—Common at light, June 27th to July 10th. Only previous records, one at rest on a tree in Elm Park, June (1894), and July 2nd (1896) one in a spider's web on a fence near the same locality. It had, without doubt, furnished a sumptuous repast, or several.

Paonias excæcatus, *S.&.A.*—At light, June 27th and July 1st. Four specimens.

Paonias myops, *S. & A.*—An example in Mr. Boger's collection was taken at Prince Albert, N.-W.T. This species is likely to occur in Manitoba.

Cressonia juglandis, *S. & A.*—One at light, July 1st.

Albuna pyramidalis, *Barnst.*—One, July 8th (1896), Bird's Hill.

Sesia rubrofascia, *Hy. Edw.*—One, June 17th (1894).

Sesia albicornis, *Hy. Edw.*—Several, June 15th and 24th and July 13th.

Sesia sp.—July 26th, Brandon. One specimen spoilt in net, too rubbed to be determined.

All these Sesiidæ were taken when sweeping low herbage and flowers for Coleoptera, chiefly along railway tracks.

Alypia Langtonii, *Coup.*—Several at Rounthwaite, by Mr. Marmont.

Scepsis fulvicollis, *Hbn.*—One at light, middle of July.

Sarrothripa Lintneriana,*Speyer.*—My first records, Sept. 1st and 13th, etc., show it to be a late species; but as I took it this year in July, at light, it my prove to be double-brooded. One of those captured is a very handsome variety.

Argyrophyes cilicoides, *Grt.* -- According to Dr. Dyar, this is a rare species; it occurred at light from July 2nd to 20th.

Clemensia albata, *Pack.*—July 27th, etc., several at rest on trees in Elm Park, and one in the city.

Crambidia pallida, *Pack.*—A pupa found under a stone at Bird's Hill on July 21st (1895) produced the imago on Aug. 6th. Common this season at light, middle of July.

Crambidia casta, *Sanb.* (No. 988, Smith's List)—A pair evolved on Aug. 4th (1896). The larvæ were common under stones at Bird's Hill on July 7th and 8th, and a number were boxed. A day or two later during or after a journey to Brandon, most of them escaped from my jar, These larvæ were small "woolly-bears," hairs dark brown. I think they were full-grown. Dr. Dyar states that the larva of this moth has never been described, so I regret not having made some notes on its appearance. A pair, having twice the expanse and somewhat lighter secondaries, were captured on the wing at dusk, on an open hillside at Brandon, on Aug. 27th this year.

Hypoprepia fucosa, *Hbn.* (miniata, *Kirby*)—One at light on July 10th (1896) at Brandon.

Euphanessa mendica, *Walk.*—July 3rd, etc., common in Elm Park.

Crocota ferruginosa, *Walk.*—One at Brandon, July 15th (1896).

Crocota immaculata, *Reak.*—Several July 15th, 21st, etc., at Bird's Hill, and on the prairies flying during the day. Very common this season during July at light. Pupæ found under boards, etc., on June 20th and July 1st.

Crocota quinaria, *Grt.*—Several in Elm Park and dark woods, July 3rd, etc., flying during the day ; this species did not come to light.

Callimorpha clymene, *Brown.*
          "   Lecontei, *Bdv.*
          "   "   var. militaris, *Harr.*
          "   vestalis, *Pack.*—One specimen only—a beauty.

All these Callimorpha were taken in Elm Park on July 1st and 3rd (1896).

Platarctia hyperborea, *Curt.* (parthenos, *Harr.*)—A specimen of this beautiful moth was captured this season at Brandon by Mr. Boger.

Arctia virgo, *Linn.*—Common this season at light, July 3rd to middle of month. Previous records : July 15th (1895) an imago hiding at the roots of weeds in my garden ; a pupa found on July 1st, produced moth on 13th of month ; a larva taken under a log on April 22nd (1894), produced the imago on July 3rd.

Arctia Saundersii, *Grt.*—Common at light, middle to end of July. One under a stone in gravel pit at Brandon on July 31st (1896).

Arctia virguncula, *Kirby.*—One at Rounthwaite, in Mr. Marmont's collection.

An Arctian in poor condition taken this year at Brandon, by Mr. Boger, may be phalerata, *Harr.*

Pyrrharctia isabella, *S. & A.*—Larvæ seen in 1894 ; moth not taken here.

Spilosoma virginica, *Fabr.*—This moth appears to be rare here ; it did not come to light. A moth was taken July 26th (1895), and some larvæ were seen on Aug. 25th, and several pupæ were found this spring at Brandon.

Spilosoma prima, *Slosson.*—A moth evolved on May 9th (1897) from pupa found at Brandon in April.

Spilosoma antigone, *Strk.*—Several, Aug. 25th, etc.

Hyphantria cunea, *Dru.*—Several, June (1894).

Euchætes collaris, *Fitch.*—One at Brandon this season by Mr. Boger.

Halisidota maculata, *Harr.*—One, at light, July 1st. (This specimen differs considerably from my Hamilton, Ontario, examples.)

Orgyia antiqua, *Linn.*—One, Aug. 15th (1895), at rest on a window in the city.

Orgyia leucostigma, *S. & A.*—Common at light, middle to end of July, and examples taken (also at light) on Sept. 24th and 28th.

Parorgyia Clintonii, *G. & R.*—On July 23rd (1895) I found two cocoons of this species in the folds of an old newspaper in some open woods. A moth evolved from one about Aug. 1st. The other produced several handsome ichneumons.

Parorgyia plagiata, *Walk.*—Common at light, middle to end of July.

Tortricidia testacea, *Pack.*—Pairs by beating, June 10th and 14th (1894). Specimens taken at light, end of June and beginning of July, this year were all poor.

Ichthyura vau, *Fitch.*—Several, at light, middle to end of July.

Ichthyura albosigma, *Fitch.*—Common at light from July 9th to end of month.

Ichthyura Brucci, *Hy. Edw.*—One or two, at light, about 20th of July.

Datana ministra, *Dru.*—One, at light, July 2nd.

Nadata gibbosa, *S. & A.*—Several, at light, June 27th to July 6th.

Gluphisia trilineata, *Pack.*—Common at light during July.

Notodonta elegans, *Strk.* (No. 1273, Smith's List)—Four specimens at light, June 27th to July 2nd.

Lophodonta angulosa, *S. & A.*—A pair at light, beginning of July.

Macrurocampa Dorothea, *Dyar.*—One at light, beginning of July. This species was described and figured on page 176 of Vol. XXVIII. of the CANADIAN ENTOMOLOGIST.   Dr. Dyar states that my capture is only the second known specimen of this new species, and that it differs from the type in being darker and more heavily marked with yellow.

Pheosia dimidiata, *H.-S.* (rimosa, *Pack.*)—A pair at light, one on June 27th, the other on July 26th.

Edema albifrons, *S. & A.*—Several, at light, at the end of June.

Seirodonta bilineata, *Pack.*—July 8th (1894), one on a fence in the city.

Dasylophia anguina, *S. & A.*—One or two at light early in July.

Schizura ipomeæ, *Doub.*

  "   var. cinereofrons, *Pack.*

Both these forms sparingly at light, July 2nd to 25th. But one specimen (cinereofrons) taken before in the district. July 14th (1895), at rest on a fence.

Schizura eximia, *Grt.* (No. 1300, Smith's List)—Several, at light, early in July.

Schizura badia, *Pack.* (No. 1302, Smith's List) — Taken at light from the end of June until nearly the end of July, but not common.

Schizura unicornis, *S. & A.*—Three at light early in July.

Ianassa lignicolor, *Walk.*—July 19th, three at light.

Cerura occidentalis, *Lint.*—New to me this season ; took one at rest on side of house the first week in June ; examples came to light on June 27th, July 15th, 18th and 19th.

Cerura cinerea, *Walk.*—One at light, middle of July.

Dryopteris rosea, *Walk.*—Common at light from July 3rd to end of month.   On June 23rd (1894) one taken in Elm Park, at rest on a leaf. Not seen again until this season.

Dryopteris irrorata, *Pack.*—Two, at light, July 6th and 8th.

Attacus cecropia, *Linn.*—A specimen has been bred from the larva by Mr. Criddle, near Douglas, Man

Attacus columbia, *Smith.*—Recorded by Mr. E. F. Heath from Cartwright, and Mr. Marmont from Rounthwaite. Dr. Fletcher says that the food plant in the Northwest is Elœagnus argentea.

Actias Luna, *Linn.*—The Rev. W. Burman, of this city, reports the capture of a specimen in Elm Park, and last season in the same place I picked up a cocoon, most likely belonging to this species ; it contained the decayed remains of the larva.

Telea polyphemus, *Cram.*—Winnipeg and Brandon, at light in June.

Anisota virginiensis, *Dru.*—Recorded from Miami, Man., by Dr. Fletcher. The larvæ causing damage to oak trees.

Clisiocampa fragilis, *Stretch.*—July 10th (1896) and later at Brandon ; several at light and on fences. Also this season at Winnipeg, at light, in July.

Clisiocampa americana, *Harr.*—A moth evolved on July 15th (1896) from full-grown larva taken on June 20th. Several at light this season in July.

Clisiocampa disstria, *Hbn.*—One, at light, towards end of July.

Phyllodesma americana, *Harr.* (No. 1414 Smith's List)—One, at light, on July 1st.

Hepialus argenteomaculatus, *Harr.*—This moth appeared to be abundant here in 1895. I took specimens on the wing in my garden about dusk on July 11th, 15th and 17th ; they were all hovering (a most peculiar flight they have) over some high weeds. Specimens were taken at rest on July 13th and June 30th (1896). On the first mentioned occasion the moth was holding on to a tall stalk of grass within a yard or so of a railway track.

This is a very variable insect, no two of those captured agreeing in colour or markings. Mr. Marmont has one, taken at Rounthwaite, which is nearly white. The records of captures at light, where the year is not given, are all for 1897.

(To be continued.)

The readers of this magazine will deeply sympathize with PROFESSOR H. F. WICKHAM, of the State University of Iowa, who has found himself compelled, in consequence of serious trouble with his eyes, to give up the study of Entomology. He is now disposing of his splendid collection of North American Coleoptera. This is a rare opportunity for Entomologists to complete their representatives of various families of beetles. That his eyes may ere long be restored to their normal condition is the earnest wish of all his friends.

BOOK NOTICES.

THE BOOK OF BRITISH BUTTERFLIES.—A practical manual for Collectors and Naturalists : 1 vol., pp. 247. (3s. 6d.)

THE BOOK OF BRITISH HAWK-MOTHS.—A popular and practical Handbook for Lepidopterists : 1 vol., pp. 157. (3s. 6d.) By W. J. Lucas, B. A. London : L. Upcott Gill, 170 Strand, W. C.

Many excellent works on British Butterflies have been published during the last twenty-five years, and one would naturally suppose that there was little need of another book on the subject. Mr. Lucas, however, has succeeded in producing a very useful and excellent popular manual, which will be a welcome aid to those who wish to study the life-history of butterflies as well as to identify the specimens they may collect in the British Isles. As it is intended for those who have made no previous study of the subject, the author begins at the beginning, telling the reader what an insect is, what place the butterfly takes in nature, how to capture, set and care for specimens, and then describes each British species from the egg to the imago in clear and simple language, and in almost every instance gives admirable drawings of the caterpillar, chrysalis, and both surfaces of the imago. As there are no less than 266 figures in illustration of sixty-eight species, the collector should have no difficulty in determining any specimen of butterfly in any of its stages (except the egg) that he may chance to find. A book such as this should give a great impetus to the study of the preparatory stages of British butterflies, a section of entomology which is usually neglected in favour of the mere collection and arrangement of the perfect insects. A volume such as this on Canadian butterflies would be a very welcome aid to a large number of young people whose interest has been aroused by the beauty and variety of our species, but whose enthusiasm is soon dampened by the difficulty of obtaining any information about them.

"The Book of British Hawk-Moths," by the same author, deals with a somewhat less familiar group, and gives much useful information that it would otherwise be hard to find. The plan of the work is similar to that of the Butterfly book, and it is written in the same clear and simple style. As there are only seventeen species to deal with, the writer is able to go more fully into details respecting them, and to make his work all the more complete and popular. He has also provided artificial keys to the larvæ and imagines, and tables for distinguishing the species where there is more than one representative of the genus. The fifteen plates with which the volume is illustrated are very beautiful, and are admirably drawn by the author himself. Each species is represented life-size, and is shown as a caterpillar on its food-plant, chrysalis, and imago. There are also eighteen wood-cuts, for the most part illustrating details of structure. It is to be hoped that the author will continue his good work until he has completed the British Lepidoptera, or at any rate the more conspicuous and familiar families.

LIFE HISTORIES OF AMERICAN INSECTS.—By Clarence M. Weed : 1 vol.,. pp. 272. ($1.50.) New York : The Macmillan Company.

The publication of a popular book on insects is so rare an event on this side of the Atlantic that we heartily welcome an addition to the number, especially when it is so excellent and satisfactory as the volume before us. Dr. Weed has selected some five and twenty more or less familiar insects, and in a pleasant manner has given some account of their life histories. The chapters are quite independent of each other and arranged in no particular order ; the book may therefore be opened at random, and the sketch that may be hit upon read without any detriment to the continuity of the work. Some of them which deal with such creatures as the leaf-miners are naturally very brief, since so little is known about these tiny foes to vegetation, but of other species which have been subjects of particular study on the part of the author we find long and full descriptions. Among the latter may be mentioned the interesting account of the hibernation of aphides, the chapters on "harvest spiders," the "army-worm," etc. Any one, young or old, who has any desire to read about the wonderful creatures that inhabit the world, and to know something about their modes of life, cannot fail to be pleased with this book, and to be led on, we should hope, to make his own observations of their curious habits and strange doings. The volume is handsomely illustrated with 21 full-page plates and nearly 100 figures in the text.

INSECTS AND SPIDERS : their Structure, Life Histories and Habits.—By J. W. Tutt : 1 vol., pp. 116. (1 shilling.) London : George Gill & Sons, Warwick Lane, E. C.

In the Annual Report of the Entomological Society of Ontario for 1896 much attention was paid to the subject of teaching natural history, and especially entomology, in schools, and the desire was expressed that some handbook might be drawn up for the assistance of teachers in rural schools. The volume before us is the very book that is needed, if only it dealt with Canadian instead of British insects. In England "Object Lessons" are a compulsory part of the curriculum in elementary schools, and the teachers are required to give their pupils a series of simple lessons "adapted to cultivate habits of exact observation, statement, and reasoning." These lessons are to be "on objects and on the phenomena of nature and of common life," and a wide discretion is thus left in the hands of the teacher. In the country schools of Ontario no subject could be more useful than the study in this way of the commonest species of injurious and beneficial insects, and no subject is likely to compare with it in interesting the pupils. A further advantage is the ease with which specimens can be obtained and their life histories traced. Mr. Tutt's volume is admirably adapted for the use of teachers in providing lessons of this kind. After giving a general account of the external structure of insects, their internal organs and metamorphoses, he devotes the "Lessons" to typical common species of each order, giving similar particulars regarding the individuals and any general facts of interest that bear upon them. Each insect treated of is also illustrated with plates and wood-cuts. It is not, however, a text-book for pupils, but is meant for the instruction and equipment of the teachers, affording them an excellent foundation upon which to frame the instructions they are to give to those committed to their charge.

VANESSA MILBERTI.

In "The Butterflies of the Eastern Provinces of Canada," by Rev. C. J. S. Bethune (Ent. Soc. of Ont. Report, 1894), it is stated that individuals of this butterfly were seen as late as the 18th Oct. I saw two specimens on the 25th Oct., flying actively across a street near the Hotel Dieu, Montreal. This usually common butterfly is scarce within the range of my entomological field work, which is principally confined to the north-east slope of Mount Royal, and the streets of Montreal around that neighbourhood. Only one other specimen was seen by me this season, and that was also at a late date, the 19th Oct. My collection specimen was caught in 1894, and since then, I have not seen another in the same district until the above appeared.

This butterfly was common around St. Andrews East, Que., from the 1st to the 4th Aug., 1896.   CHARLES STEVENSON, Montreal.

[A specimen was seen on the wing at Port Hope on the 5th of November last.—ED. C. E.]

Mailed December 6th, 1897.

INDEX TO VOLUME XXIX.

ERRATA.

Page 100, 5th line from bottom, for
     " oratus" read " ovatus."
Page 168, 12th line from bottom, prefix
     " 7."
Page 168, 6th line from bottom, add
     " Stylurus "
Page 208, 12th line from bottom, for
     " tœta " read " læta."
Page 254, 27th line from top, insert a
     comma after "stripe."